ELECTRONICS
Principles and Applications

ELECTRONICS
Principles and Applications

CHARLES A. SCHULER
CALIFORNIA STATE COLLEGE
CALIFORNIA, PENNSYLVANIA

McGraw-Hill Book Company
Gregg Division

New York	Mexico
St. Louis	Montreal
Dallas	New Delhi
San Francisco	Panama
Auckland	Paris
Bogotá	São Paulo
Düsseldorf	Singapore
Johannesburg	Sydney
London	Tokyo
Madrid	Toronto

Library of Congress Cataloging in Publication Data

Schuler, Charles A
    Electronics, principles and applications.

    (Basic skills in electricity and electronics)
    Includes index.
    1. Electronics.   I. Title.   II. Series.
TK7816.S355   621.381   78-27296
ISBN 0-07-055572-9

Acknowledgments
Students, teachers, school administrators, and industrial trainers have contributed to the development of the *Basic Skills in Electricity and Electronics* series. Classroom testing of preliminary editions has been conducted at the following sites:

Burr D. Coe Vocational Technical High School (East Brunswick, New Jersey)
Chantilly Secondary School (Chantilly, Virginia)
Nashoba Valley Technical High School (Westford, Massachusetts)
Platt Regional Vocational Technical High School (Milford, Connecticut)
United States Steel Corporation: Edgar Thomson, Irvin Works (Dravosburg, Pennsylvania)

The publisher gratefully acknowledges the helpful comments and suggestions received from these participants and, at the same time, assumes full responsibility for any errors of omission or commission found in this series.

Electronics: Principles and Applications

4567890 VHVH 8654321

The editors for this book were Gordon Rockmaker and Mark Haas. The designer was Tracy A. Glasner. The art supervisor was George T. Resch. The production supervisor was Kathleen Morrissey. Cover photography by Martin Bough/Studios Inc. It was set in Electra by Progressive Typographers.
Printed and bound by Von Hoffman Press, Inc.

Contents

Editor's Foreword

The Gregg/McGraw-Hill *Basic Skills in Electricity and Electronics* series has been designed to provide entry-level competencies in a wide range of occupations in the electrical and electronics fields. The series consists of instructional materials geared especially for the career-oriented student. Each major subject area covered in the series is supported by a textbook, an activities manual, and a teacher's manual. All the materials focus on the theory, applications, and experiences required for those just beginning their chosen vocations.

There are two basic considerations in the preparation of educational materials for such a series: the needs of the learner and the needs of the employer. The materials in the series have been designed to meet those needs. They are based on many years of experience in the classroom and with electricity and electronics. In addition, these books reflect the needs of industry and commerce as developed through questionnaires, surveys, interviews with employers, government occupational trend reports, and various field studies.

Further refinements both in pedagogy and technical content resulted from actual classroom experience with the materials. Preliminary editions of selected texts and manuals were field tested in schools and in-plant training programs throughout the country. The knowledge gained from this testing has enhanced the effectiveness and the validity of the materials.

Teachers will find the materials in each of the subject areas well coordinated and structured around a framework of modern objectives. Students will find the concepts clearly presented with many practical references and applications. In all, every effort has been made to prepare and refine the most effective learning tools possible.

The publisher and editor welcome comments from teachers and students using this book.

Charles A. Schuler
Project Editor

BASIC SKILLS IN ELECTRICITY AND ELECTRONICS

Charles A. Schuler, Project Editor

Books in this series

Introduction to Television Servicing by Wayne C. Brandenburg
Electricity: Principles and Applications by Richard J. Fowler
Instruments and Measurements by Charles M. Gilmore
Microprocessors by Charles M. Gilmore (*in preparation*)
Motors and Generators by Russell L. Heiserman and Jack D. Burson
Small Appliance Repair by Phyllis Palmore and Nevin E. Andre
Residential Wiring by Gordon Rockmaker
Electronics: Principles and Applications by Charles A. Schuler
Your Future in Electricity and Electronics by William A. Stanton
Digital Electronics by Roger L. Tokheim

Preface

This introductory text on the principles and applications of electronics is designed for those students already having a basic understanding of Ohm's law, Kirchhoff's laws, power formulas, schematic diagrams, and electrical components such as resistors, capacitors, and inductors. A knowledge of basic algebra is the only mathematics prerequisite. This book provides an excellent foundation in electronics for those who may need or want to go into the subject in more depth.

It is thorough.

Some introductory electronics texts deal mainly with device theory. Others are devoted to applications. This book combines both theory and applications in a logical, well-paced sequence. It is important for a student's first experience in electronics to be based on such an approach. The combination of theory and applications first develops an understanding of how electronic circuits function. It then relates circuit function to how the circuits are used to solve practical problems.

I have taken a practical approach throughout this text. The devices and circuits presented are typical of those used in all phases of electronics. Theory and calculations are the same as those used by practicing technicians. Reference is made to common aids such as parts catalogs and substitution guides.

The fourteen chapters into which this book is divided progress from an introduction to the field of electronics, through basic solid-state theory, transistors and gain, amplifiers, oscillators, radio, integrated circuits, and control circuits. This use of the modern "building-block" concept insures that students will have a strong foundation on which to build in a field characterized by rapid technological advances. Where possible, future trends are discussed.

The text includes features which are designed to make the study of electronics more interesting and effective. All important facts and concepts are reviewed in a summary section at the end of each chapter. Students should read these reviews carefully. Any confusion here may indicate the need to go back and re-study a section or a portion of the section. Important terms are highlighted in the margins for quick reference. Students will also find review tests after each section. These, too, will help students check their understanding of the material and reinforce their learning.

Electronics is an exciting, rapidly changing field. These materials reflect the experience and reaction to several years of classroom testing. I welcome comments and suggestions from students and teachers alike.

Charles A. Schuler

Safety

Electric devices and circuits can be dangerous. Safe practices are necessary to prevent electric shock, fires, explosions, mechanical damage, and injuries resulting from the improper use of tools.

Perhaps the greatest hazard is electric shock. A current through the human body in excess of 10 milliamperes can paralyze the victim and make it impossible to let go of a "live" conductor. Ten milliamperes is a small amount of electrical flow: It is *ten one-thousandths* of an ampere. An ordinary flashlight uses more than 100 times that amount of current! If a shock victim is exposed to currents over 100 milliamperes, the shock is often *fatal*. This is still far less current than the flashlight uses.

A flashlight cell can deliver more than enough current to kill a human being. Yet it is safe to handle a flashlight cell because the resistance of human skin normally will be high enough to greatly limit the flow of electric current. Human skin usually has a resistance of several hundred thousand ohms. In low-voltage systems, a high resistance restricts current flow to very low values. Thus, there is little danger of an electric shock.

High voltage, on the other hand, can force enough current through the skin to produce a shock. The danger of harmful shock increases as the voltage increases. Those who work on very high-voltage circuits must use special equipment and procedures for protection.

When human skin is moist or cut, its resistance can drop to several hundred ohms. Much less voltage is then required to produce a shock. Potentials as low as 40 volts can produce a fatal shock if the skin is broken! Although most technicians and electrical workers refer to 40 volts as a *low voltage*, it does not necessarily mean *safe voltage*. You should,

therefore, be very cautious even when working with so-called low voltages.

Safety is an attitude; safety is knowledge. Safe workers are not fooled by terms such as *low voltage*. They do not assume protective devices are working. They do not assume a circuit is off even though the switch is in the OFF position. They know that the switch could be defective.

As your knowledge of electricity and electronics grows, you will learn many specific safety rules and practices. In the meantime:

1. Investigate before you act
2. Follow procedures
3. When in doubt, *do not act*: Ask your instructor

GENERAL SAFETY RULES FOR ELECTRICITY AND ELECTRONICS

Safe practices will protect you and those around you. Study the following general safety rules. Discuss them with others. Ask your instructor about any that you do not understand.

1. Do not work when you are tired or taking medicine that makes you drowsy.
2. Do not work in poor light.
3. Do not work in damp areas.
4. Use approved tools, equipment, and protective devices.
5. Do not work if you or your clothing are wet.
6. Remove all rings, bracelets, and similar metal items.
7. Never assume that a circuit is off. Check it with a device or piece of equipment that you are sure is operating properly.

8. Do not tamper with safety devices. *Never* defeat an interlocking switch. Verify that all interlocks operate properly.

9. Keep your tools and equipment in good condition. Use the correct tool for the job.

10. Verify that capacitors have discharged. Some capacitors may store a lethal charge for a long time.

11. Do not remove equipment grounds. Verify that all grounds are intact.

12. Do not use adaptors that defeat ground connections.

13. Use only an approved fire extinguisher. Water can conduct electric current and increase the hazards and damage. Carbon dioxide (CO_2) and certain halogenated extinguishers are preferred for most electrical fires. Foam types may also be used in some cases.

14. Follow directions when using solvents and other chemicals. They may explode, ignite, or damage electric circuits.

15. Certain electronic components affect the safe performance of the equipment. Always use the correct replacement parts.

16. Use protective clothing and safety glasses when handling high-vacuum devices such as television picture tubes.

17. Do not attempt to work on complex equipment or circuits before you are ready. There may be many hidden dangers.

18. Some of the best safety information for electric and electronic equipment is in the literature prepared by the manufacturer. Find it and use it!

Any of the above rules could be expanded. As your study progresses, you will learn many of the details concerning proper procedure. Learn them well, because they are the most important information available.

Remember, always practice safety; your life depends on it.

Introduction

■ The purpose of this chapter is to introduce you to electronics. Electronics is such a huge field that beginners may be confused as to what they are studying. They may also wonder how it relates to what they have already studied.

In this book, *electronics* refers to the study of *active* devices and their uses. An active device is a diode, a transistor, or an integrated circuit. *Passive* devices are resistors, conductors, capacitors, and inductors. Generally, courses that deal mainly with active devices are called electronics courses.

Electronics is really a brand new field. Its entire history is limited to this century. Yet, it has grown so rapidly and become so important that several branches have developed. One of the branches is digital electronics. This chapter will define this term and compare it to *analog* electronics. The rapid growth of electronics in recent years will not slow down in the near future. Thus, this chapter will also review some major trends in the field.

1-1 A BRIEF HISTORY

Electronics is very young. It is hard to place an exact date on its beginning. Two important developments at the beginning of this century made people interested in electronics. The first was in 1901 when Marconi sent a message across the Atlantic Ocean using *wireless* telegraphy. Today, we call wireless communication *radio*. The second development came in 1906 when De Forest invented the *audion* vacuum tube. The term "audion" related to its first use to make sounds ("audio") louder. It was not long before the wireless inventors used the vacuum tube to improve their equipment.

There was another development in 1906 worth mentioning. Pickard used the first crystal radio detector. This was a great improvement. It helped to make radio and electronics more popular. It also suggested the use of *semiconductors* (crystals) as materials with great promise for the future of the new field of radio and electronics.

Commercial radio was born in Pittsburgh, Pennsylvania, at station KDKA in 1920. This marked the beginning of a new era with electronic devices appearing in the average home. Commercial television began around 1946. Television receivers were very complex compared to most radios. Other complicated electronic devices were now in use, too. The vacuum tube, which had served so well, was now making engineers and technicians wish for something better.

Vacuum tube equipment was not as reliable as people wanted it to be. Complicated equipment used many vacuum tubes. The tubes made a lot of heat. The equipment was large and expensive. Luckily, something better was just around the corner.

Scientists knew that many of the jobs done by vacuum tubes could be done more efficiently by semi-conducting crystals. They knew this for quite some time but they could not make crystals that were pure enough to do the job. They kept working at it. The breakthrough came in 1947. Three scientists working for Bell Laboratories made the first working transistor. This was such a major contribution to science and technology that the three men—Bardeen, Brittain, and Shockley—were awarded the Nobel Prize.

Improvements in the transistor came rap-

1

From page 1:
Audion

Semiconductor

Vacuum tube

Transistor

On this page:
Integrated circuit

Digital circuit

Analog circuit

idly, and now transistors have all but completely replaced the vacuum tube. "Solid state" has become a household word. Many people say that the transistor is one of the greatest developments of all time.

Solid-state circuits were small, efficient, and more reliable. But the scientists still were not satisfied. Nor were the engineers satisfied. So it was with Jack Kilby of Texas Instruments. His work led to the development of the integrated circuit in 1958. Integrated circuits are complex combinations of several kinds of devices on a common base, called a *substrate*, or in a tiny piece of silicon. They offer low cost for the performance, as well as good efficiency, small size, and good reliability. The complexity of some integrated circuits allows a single chip of silicon only 0.64 centimeter (cm) [0.25 inch (in)] square to replace huge pieces of equipment. The chip can hold thousands of transistors. You would not think any room would be left, but you might also find some diodes, resistors, and capacitors too!

The integrated circuit is producing an electronics explosion. Now electronics is being applied in more ways than ever before. At one time, radio was just about the only application. Today, electronics makes a major contribution to every part of our society and in every field of human endeavor. It affects us in many ways that we may not even be aware of. We are living in the electronic age.

Review Questions

Determine whether each statement is true or false.

1. The entire history of electronics is limited to the twentieth century.

2. The early histories of radio and of electronics are the same.

3. Transistors were invented before vacuum tubes.

4. A modern integrated circuit can contain thousands of transistors.

1-2 DIGITAL OR ANALOG

Today, electronics is such a huge field that it is often necessary to divide it into smaller subfields. You will hear terms such as "medical electronics," "instrumentation electronics," "automotive electronics," "avionics," "consumer electronics," and others. One way that electronics can be divided is into *digital* and *analog*.

A digital electronic device or circuit will recognize or produce an output of only several limited states. For example, most digital circuits now in use will respond to only two conditions: high or low voltages. Digital circuits can be built that will operate with more than two states, but most are limited to only two.

An analog circuit can respond to or produce an output of an infinite number of states. An analog voltage might vary between 0 and 10. The actual value might be 1.5, 2.8, or even 7.653 volts (V). In theory at least, an infinite number of voltages are possible.

For a long time, almost all electronic devices and circuits operated in the analog fashion. This seemed to be the most obvious way to do a particular job. After all, most of the things that we measure are analog in nature. Your height, your weight, and the speed at which you travel in an automobile are all analog quantities. Your voice is analog. It contains an infinite number of levels and frequencies. Therefore, if you wanted a circuit to amplify your voice, that is, make it sound louder, your first thought would probably be to use an analog circuit.

Telephone switching and computer circuits forced engineers to explore digital electronics. They needed circuits and devices that could make logical decisions based on certain input conditions. They needed highly reliable circuits that would operate in the same way every time. By limiting the number of conditions or states in which the circuits must operate, the circuits could be made more reliable. An infinite number of states—the analog circuit—was not what they needed.

Digital circuits can do most of the things we once thought of as strictly analog. We can amplify and otherwise process sound with digital circuits. Which way something will actually be done is simply a question of efficiency. Circuit designers choose the best way to do a job while keeping the cost as low as possible.

Analog electronics involves techniques and concepts different from those of digital electronics. In this book, you will be studying analog electronics. However, what you learn about analog electronics will serve as a useful

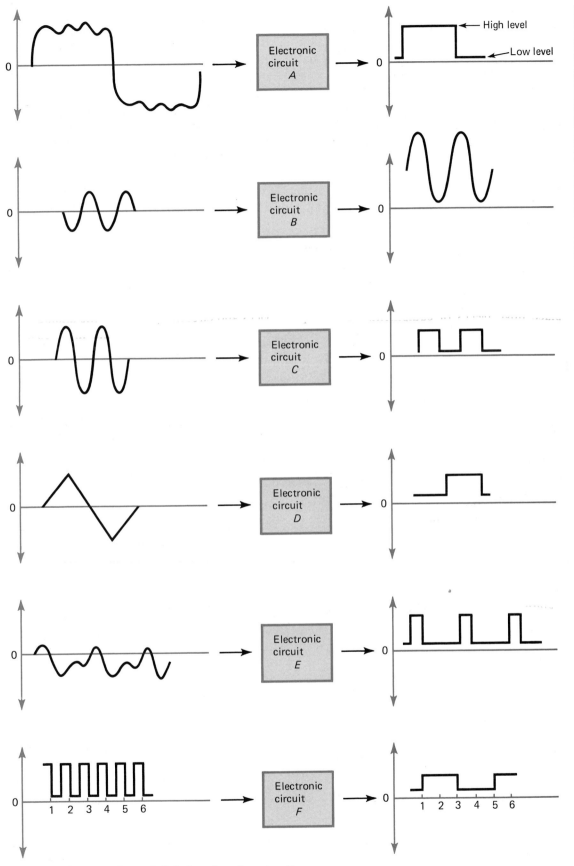

High level

Low level

Fig. 1-1 A comparison of digital and analog circuits.

Linear circuit

background for later studies of digital electronics.

Figure 1-1 gives some examples of circuit behavior that will help you to identify digital or analog operation. The signal going into the circuit is on the left, and the signal coming out of the circuit is shown on the right. The circuit marked A is an example of a digital device. The output signal is a square wave while the input signal is not exactly a square wave. Square waves show only *two* voltage levels and are very common in digital devices.

Circuit B in Fig. 1-1 is an analog device. The input is a sine wave, and the output is a sine wave. The output is larger than the input, and it has been shifted above the zero axis. The most important feature is that the output signal is a combination of an infinite number of voltages. In a *linear circuit*, the output is an exact replica of the input. Though circuit B is linear, *not all analog circuits are linear*.

Circuits C through F are all digital. Note that the outputs are all square waves (two levels of voltage). Circuit F deserves special attention. Its input is a square wave. This could be an analog circuit responding to only two voltage levels except that something has happened to the signal which did not occur in any of the other examples. The output *frequency* is different from the input frequency. Digital circuits that accomplish this are called *counters*, or *dividers*.

Review Questions

Determine whether each statement is true or false.

5. Electronic circuits can be divided into two categories, digital and analog.

6. A digital circuit can respond to an infinite number of conditions.

7. Digital circuits became popular before analog circuits.

8. This book is limited to the study of analog circuits.

9. A linear circuit can be put in the digital category.

1-3 TRENDS IN ELECTRONICS

Electronics will change as rapidly as any field in the future. Those who will work in elec-

tronics must be willing to accept this rapid change. They will have to want to improve themselves and always learn more about electronics. It will be challenging but very exciting for those involved.

The integrated circuit (IC) will be the key to most of the changes coming in electronics. Some new circuits will be expensive, but they will almost always replace equivalent circuits that cost even more. They will operate more quickly and be able to do very complex things. Some integrated circuits, called *microprocessors*, permit automation in very small devices and in applications unheard of only a few years ago.

Mechanical control and timing devices are being replaced in many cases by digital circuits. People will be able to set up a machine or appliance to do a series of steps, make adjustments, and have it all happen at the right time. Such machines and devices are said to be *programmable*.

Modern automobiles use a wide range of electronic circuits. In the future the engine will be exactly controlled to save fuel and reduce pollution. Braking systems will be controlled to reduce skids. Automobiles will use radar to warn the driver of excess speed and other hazards.

Business and industry will use more computers and more computer-controlled equipment. Numerical control devices will increase production and reduce costs. Most printed material will be prepared by electronic means. Many of the skilled people in business and industry will have to learn new skills as new equipment is installed.

Homes will be equipped with many complex devices used not only to control and operate ordinary household functions, but also to entertain, educate, and exercise those who live in the home. People will play programmable electronic games and have more electronics-related hobbies. The television receiver will provide entertainment as well as be a window to a world of information. The telephone or some other device may serve as the *data terminal* for the average home. People will be able to get information, buy things, and transact business—all from their own terminals.

Medical electronic devices will help doctors diagnose health problems. Other devices may help the sick live a normal life. Electronic monitoring stations will watch our water and air and help keep them clean. The public health and welfare will be very dependent on electronics.

Electronics has always been a part of space exploration and communication. In the future, electronics will bring us information from the farthest reaches of space.

Finally, electronics itself is changing. Equipment will monitor itself. If a malfunction occurs, electronic equipment will be able to make repairs automatically. Instruments will make calculations and be at least partially automatic. Technicians working on electronic circuits will have to know more, but there will be better tools to help them.

Determine whether each statement is true or false.

10. Integrated circuits will be used less in the future.

11. Devices that can be set up by the operator to do a series of tasks automatically are said to be programmable.

12. In the future many mechanical timing and control devices will be replaced by digital circuits.

Summary

1. Electronics is a young field. Its entire history is contained in the twentieth century.
2. The key developments in electronics have been

    1901   radio
    1906   vacuum tube
    1947   transistor
    1958   integrated circuit

3. Electronic circuits can be classified as digital or analog.

4. The number of states or voltage levels is limited in a digital circuit (usually to two).
5. An analog circuit has an infinite number of voltage levels.
6. In a linear circuit, the output signal is a replica of the input.
7. All linear circuits are analog, but not all analog circuits are linear.
8. Digital circuits will find many new applications and in areas now thought of as analog.

Chapter Review Questions

Determine whether each statement is true or false.

1-1. Radio was invented in 1854.

1-2. Integrated circuits contain several vacuum tubes.

1-3. "Solid state" means that the equipment uses vacuum tubes.

1-4. Most digital circuits can output only two states, high and low.

1-5. Digital circuit outputs are usually sine waves.

1-6. The output of a linear circuit is an exact replica of the input.

1-7. Linear circuits are classified as analog.

1-8. All analog circuits are linear.

1-9. Because of electronics, many future devices will become automatic.

1-10. Future applications of electronics will be limited to just a few areas in engineering.

Answers to Review Questions

1. T	3. F	5. T	7. F	9. F	11. T
2. T	4. T	6. F	8. T	10. F	12. T

Semiconductors

- After a brief review of conductors and insulators you are going to study materials with properties that fall between conductors and insulators. These materials are called *semiconductors*.
    Semiconductors have a unique way of supporting the flow of electric current. Some of the current carriers are *not* electrons. High temperatures can create new carriers in semiconductors. These are important differences between semiconductors and conductors.
    The transistor has been called one of the most important developments of all time. It is a semiconductor device. To make your study of electronics more effective, you will learn some of the properties of semiconductors.

2-1 CONDUCTORS

Conductors are the fundamental component of electronic circuits and devices. They easily and efficiently allow the flow of electric charges. Figure 2-1 shows how a copper wire supports the flow of electrons. A copper atom contains a positively charged nucleus and negatively charged electrons that orbit around the nucleus. Figure 2-1 is simplified to show only the outermost orbiting electron, the *valence electron*. The valence electron is very important since it acts as the *current carrier* for the atom.

Even a very small wire will contain billions of atoms, each with one valence electron. These electrons are only *weakly* attracted to the nucleus of the atom. They are very easy to move. If an electromotive force (a voltage) is applied across the wire, the valence electrons will respond to this force and begin drifting toward the positive end of the source voltage. Since there are so many valence electrons and since they are so easy to move, we can expect tremendous numbers of electrons to be set in motion by even a small voltage. Thus, we say copper is an excellent electric conductor. We can also say that it has *low resistance*.

Heating the wire will change its resistance. As the wire becomes warmer, the valence electrons become more active. They move farther away from the nucleus, and they move more rapidly. This increases the chance for collisions as the electrons drift toward the pos-

Valence electrons

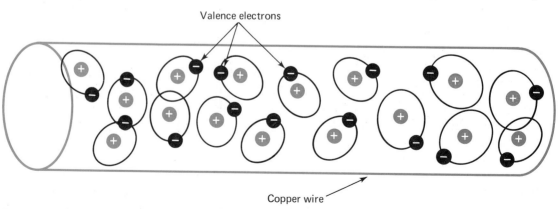

Copper wire

Fig. 2-1 The structure of a copper conductor.

6

itive end of the wire. These collisions resist the current. Therefore the resistance of the wire is a little greater than it was before.

All conductors show this effect. As they become hotter, they conduct less efficiently and their resistance increases. Such materials are said to have a *positive temperature coefficient*. This simply means that the relationship between temperature and resistance is positive—that is, they increase together.

Copper is the most important conductor in electronics. Almost all the wire used in electronics is made from copper. Printed circuits use copper foil to act as circuit conductors. Copper is a good conductor, and it is easy to solder. This makes it very popular.

Silver is the best conductor because it has the lowest resistance. It is also easy to solder. The cost of silver makes it less popular than copper. However, silver-plated conductors are sometimes used in critical electronic circuits to minimize resistance.

Gold is a good conductor. It is very stable and does not corrode as badly as copper and silver. Sometimes sliding and moving electronic contacts will be gold-plated. This makes the contacts very reliable.

Review Questions

Determine whether each statement is true or false.

1. Valence electrons are located in the nucleus of the atom. F

2. In conductors, the valence electrons are strongly attracted to the nucleus. weakly F

3. The current carriers in conductors are valence electrons. T

4. Cooling a conductor will decrease its resistance. T

5. Gold cannot be used in electronic circuits because of its high resistance. F

2-2 SEMICONDUCTORS

Semiconductors do not allow current to flow as easily as conductors do. Under some conditions semiconductors can conduct so poorly that they behave as insulators.

Silicon is a very popular semiconductor. It is used to make diodes, transistors, and integrated circuits. These and other components make modern electronics possible. It is very

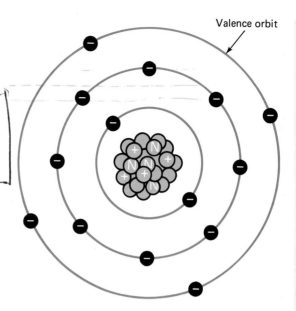

Valence orbit

Fig. 2-2   The structure of a silicon atom.

important to understand some of the details about silicon.

Figure 2-2 shows atomic silicon. The compact bundle of particles in the center of the atom contains protons and neutrons. This bundle is called the *nucleus* of the atom. The protons show a positive (+) electric charge, and the neutrons show no electric charge (N). Negative electrons travel around the nucleus in *orbits*. The first orbit has two electrons. The second orbit has eight electrons. The last, or outermost, orbit has four electrons. The outermost orbit is the most important atomic feature when you are studying the electrical behavior of materials. It is called the *valence orbit*.

Because we are interested mainly in the valence orbit, it is possible to simplify the drawing of the silicon atom. Figure 2-3 shows only the nucleus and the valence orbit of a sili-

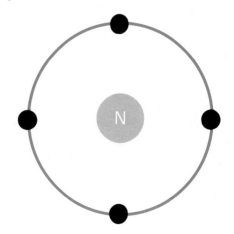

Fig. 2-3   A simplified silicon atom.

From page 6:
Conductor

Valence electron

On this page:
Positive temperature coefficient

Semiconductor

Silicon

Nucleus

Valence orbit

7

Ionic bond

Covalent bond

Negative
temperature
coefficient

Germanium

con atom. Remember that there are four electrons in the valence orbit.

Materials with four valence electrons are not stable. They tend to combine chemically with other materials. They can be called *active* materials. This activity can lead them to a more stable state. A law of nature makes certain materials tend to form combinations that will make *eight* electrons available in the valence orbit. Eight is an important number because it gives stability.

One possible combination for silicon is with oxygen. A single silicon atom can join, or link, with two oxygen atoms to form silicon dioxide. This linkage is called an *ionic bond*. The new structure is much more stable than either silicon or oxygen. It is interesting to note that chemical, mechanical, and electrical properties often run parallel. Silicon dioxide is stable chemically. It does not react easily with other materials. It is stable mechanically. It is a hard, glasslike material. It is stable electrically. It does not conduct; in fact, it is used as an insulator in integrated circuits and other solid-state devices.

Sometimes oxygen or any other material is not available for silicon to combine with. The silicon still wants the stability given by eight valence electrons. If the conditions are right, silicon atoms will arrange to *share* valence electrons. This process of sharing is called *covalent bonding*. The structure that results is called a *crystal*. Figure 2-4 is a sym-

bolic diagram of a crystal of pure silicon. The dots represent valence electrons.

Count the valence electrons around the nucleus of one of the atoms shown in Fig. 2-4. Select one of the internal atoms as represented by the circled N. You will count *eight* electrons. Thus, the silicon crystal is very stable. At room temperature, *pure silicon is a very poor conductor*. If a moderate voltage is applied across the crystal, very little current will flow. The valence electrons that normally would support current flow are all tightly locked up in covalent bonds.

Pure silicon crystals behave as insulators. Yet they are classified as semiconductors. They can be made to semi-conduct. One way to do this is to heat them. Heat adds energy to the crystal. An electron can absorb some of this energy and move to a higher orbit level. The high-energy electron will break its covalent bond. Figure 2-5 shows a high-energy electron in a silicon crystal. This electron is free to move and will serve as a current carrier. Now, if a voltage is placed across the crystal, current will flow.

Silicon has a *negative temperature coefficient*. As temperature increases, resistance decreases in silicon. It is difficult to predict exactly how much the resistance will change in a given case. One rule of thumb is that the resistance will be cut in half for every 6°C rise is temperature.

The semiconductor *germanium* is used to

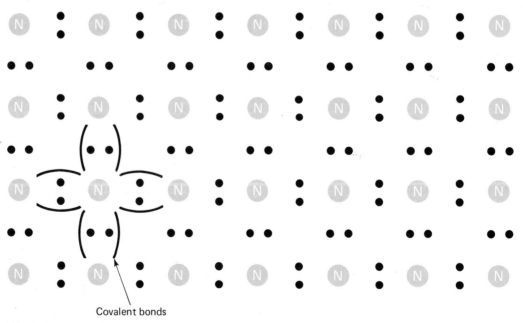

Covalent bonds

Fig. 2-4 A crystal of pure silicon.

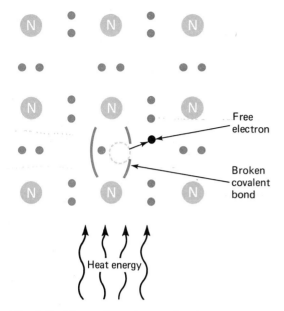

Fig. 2-5 Thermal carrier production.

make transistors and diodes, too. Germanium has four valence electrons and can form the same type of crystalline structure as silicon. It is interesting to observe that the first transistors were all made of germanium. The first silicon transistor was not developed until 1954. Now, silicon has almost replaced germanium in most solid-state applications. One of the major reasons for this shift from germanium to silicon is the temperature response. Germanium also has a negative temperature coefficient. The rule of thumb for germanium is that the resistance will be cut in half for every 10°C rise in temperature. This would seem to make germanium more stable with temperature change.

The big difference between germanium and silicon is the amount of heat energy needed to move one of the valence electrons to a higher orbit level, breaking its covalent bond. This is far easier to do in a germanium crystal. A comparison between two crystals, one germanium and one silicon, of the same size and at room temperature will show about a 1000:1 ratio in resistance. The silicon crystal will actually have 1000 times the resistance of the germanium crystal. So even though the resistance of silicon changes more rapidly with increasing temperature than that of germanium, silicon is still going to show greater resistance than germanium.

It is easy to see why circuit designers prefer silicon devices for most uses. The thermal, or heat, effects are usually a source of trouble.

Temperature is not easy to control, and we do not want circuits to be influenced by it. However, all circuits are changed by temperature. Good designs minimize that change.

Sometimes heat-sensitive devices are necessary. A probe for measuring temperature might take advantage of the temperature coefficient of semiconductor germanium. So germanium's temperature coefficient is not always a disadvantage.

Review Questions

Determine whether each statement is true or false.

6. Silicon is a conductor. F
7. Silicon has four valence electrons. T
8. Silicon dioxide is a good conductor. F
9. A silicon crystal is formed by covalent bonding. T
10. A pure silicon crystal is a very poor conductor at room temperature. T
11. Heating semiconductor silicon will decrease its resistance. T *neg.*
12. Semiconductor germanium has less resistance than semiconductor silicon. T
13. Silicon transistors and diodes are not used as often as germanium devices. F

2-3 N-TYPE SEMICONDUCTORS

Thus far, we have seen that pure semiconductor crystals are very poor conductors. High temperatures can make them semiconduct because carriers are produced. There has to be another way to make them semi-conduct.

Doping is a process of adding other materials to the silicon crystal to change its electrical characteristics. One such doping material is arsenic. Figure 2-6 shows a simplified arsenic atom. Arsenic is different from silicon in several ways, but the important difference here is in the valence orbit. Arsenic has five valence electrons.

When an arsenic atom enters a silicon crystal, a free electron will result. Figure 2-7 shows what happens. The covalent bonds will capture four of the arsenic atom's valence electrons, just as if it were another silicon atom. This tightly locks the arsenic atom into

Hole

P-type
semiconductor

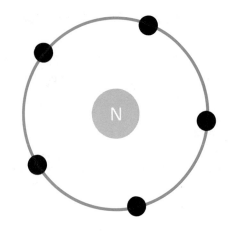

Fig. 2-6   A simplified arsenic atom.

the crystal.   The fifth valence electron cannot form a bond.   It is a free electron as far as the crystal is concerned.   This makes the electron very easy to move.   It can serve as a current carrier.   Silicon with some arsenic atoms will semi-conduct even at room temperature.

Doping lowers the resistance of the silicon crystal.   When materials with five valence electrons are added, free electrons are produced.   Since electrons have a negative charge, we say that an *N-type semiconductor material* results.

Review Questions

Supply the missing word in each statement.

14. Arsenic has ____?__5__ valence electrons.

15. When silicon is doped with arsenic, each arsenic atom will give the crystal one free ____?____. *free electron*

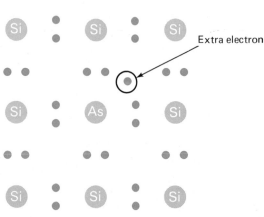

Fig. 2-7   N-type silicon.

16. Free electrons in a silicon crystal will serve as current ____?____. *carrier*

17. When silicon is doped, its resistance ____?____. *decreases*

2-4 P-TYPE SEMICONDUCTORS

Doping can involve the use of other materials. Figure 2-8 shows a simplified boron atom. Note that boron has only *three* valence electrons.   If a boron atom enters the silicon crystal, another type of current carrier will result.   Figure 2-9 shows that one of the covalent bonds will not be formed.   This produces a *hole*, or missing electron.   The hole is assigned a positive charge since it is capable of attracting, or being filled by, an electron.

Holes are current carriers just as electrons are.   In a conductor or N-type semiconductor, the carriers are electrons.   The free electrons are set into motion by an applied voltage, and they drift toward the positive terminal.   But in a P-type semiconductor, the holes move toward the negative terminal of the voltage source.   Hole current is *equal to electron current but opposite in direction*. Figure 2-10 illustrates the difference between N-type and P-type semiconductor materials. In Fig. 2-10(*a*) the carriers are *electrons*, and they drift toward the *positive* end of the voltage source.   In Fig. 2-10(*b*) the carriers are *holes*, and they drift toward the *negative* end of the voltage source.

Figure 2-11 shows a simple analogy for hole current.   Assume that a line of cars is stopped for a red light, but there is space for the first car to move up one position.   The driver of that car takes the opportunity to do so, and

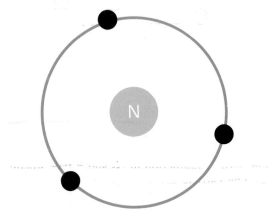

Fig. 2-8   A simplified boron atom.

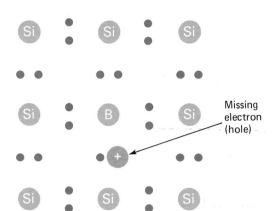

Fig. 2-9   P-type silicon.

this moves the space for a car back one position. The driver of the second car also moves up one position. This continues with the third car, the fourth car, and so on down the line. The cars are moving from left to right. Note that the space is moving from right to left. A hole may be considered as *a space for an electron*. This is why hole current is opposite in direction to electron current.

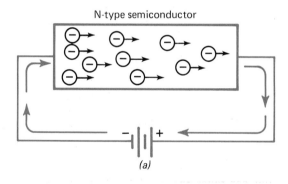

(a)

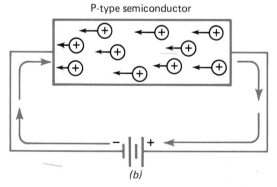

(b)

Fig. 2-10   Conduction in N- and P-type silicon. (*a*) The carriers are electrons and drift toward the positive end of the voltage source. (*b*) The carriers are holes and drift toward the negative end of the voltage source.

Review Questions

Supply the missing word in each statement.

18. Boron has ___?___ valence electrons.

19. Electrons are assigned a negative charge, and holes are assigned a ___?___ charge.

20. Doping a semiconductor crystal with boron will produce current carriers called ___?___.

21. Electrons will drift toward the positive end of the energy source, and holes will drift toward the ___?___ end.

2-5 MAJORITY AND MINORITY CARRIERS

When N- and P-type semiconductor materials are made, the doping levels can be as small as one part per million or one part per billion. Only a tiny trace of materials having five or three valence electrons enters the crystal. It is not possible to make the silicon crystal absolutely pure. Thus, it is easy to imagine that an occasional atom with *three* valence electrons might be present in an N-type semiconductor. An unwanted hole will exist in the semiconductor. This hole will be called a *minority carrier*. The free electrons are the *majority carriers*.

In a P-type semiconductor, one expects holes to be the carriers. They are in the *majority*. A few free electrons might also be present. They will be the *minority* carriers in this case.

The majority carriers will be electrons for N-type material and holes for P-type material. Minority carriers will be holes for N-type material and electrons for P-type material.

Extremely pure silicon can be prepared. Although this keeps the minority carriers to a very small number, at high temperatures they can increase. This can be quite a problem in electronic circuits. To understand how heat produces minority carriers, refer to Fig. 2-5. As heat energy enters the crystal, more and more electrons will gain energy and break their bonds. The broken bond represents a hole. *Heat produces carriers in pairs.* Each broken bond gives a free electron and a free hole. If the crystal was originally an N-type material, then the hole becomes a minority carrier and the electron joins the other majority carriers. If the crystal was a P-type mate-

Minority carrier

Majority carrier

11

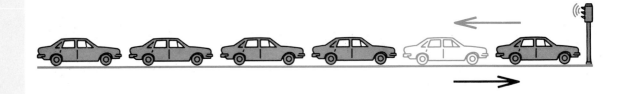

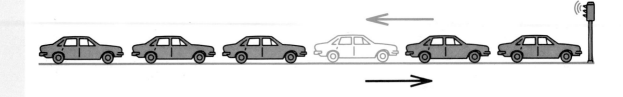

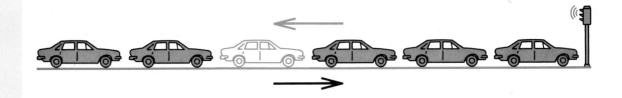

Fig. 2-11   Hole-current analogy.   The white car is a space.

rial, then the hole joins the majority carriers and the electron becomes a minority carrier.

Carrier production by heat decreases the crystal's resistance.   The heat also produces minority carriers.   These minority carriers can have a very bad effect on the way semiconductor devices work.

Review Questions

Determine whether each statement is true or false.

22. In making N-type semiconductor material, a typical doping level is about 90 percent silicon and 10 percent arsenic. *F*

23. A free electron in a P-type crystal is called a majority carrier. *F*

24. A free hole in an N-type crystal is called a minority carrier. *T*

25. As P-type semiconductor material is heated, one can expect the number of minority carriers to increase. *T*

26. As P-type semiconductor material is heated, the number of majority carriers decreases. *F   increases*

27. Heat decreases resistance in semiconductors. *T   neg coeff.*

99-1

Summary

1. Good conductors, such as copper, contain a large number of current carriers.

2. In a conductor, the valence electrons are weakly attracted to the nucleus of the atom.

3. Heating a conductor will increase its resistance. This effect is called a positive temperature coefficient.

4. Silicon atoms have four valence electrons. They will form ~~covalent bonds~~ which result in a stable crystal structure.

5. Heat energy can break covalent bonds, making free electrons available to conduct current. This gives silicon a negative temperature coefficient.

6. Germanium crystals are more sensitive to heat than silicon crystals are. This makes germanium transistors and diodes less useful than silicon devices for most applications.

7. The process of adding impurities to a semiconductor crystal is called doping.

8. Doping a semiconductor crystal changes its electrical characteristics.

9. If the impurity has five valence electrons, a free electron appears in the crystal.

10. The free electron will serve as a current carrier. This forms N-type semiconductor material.

11. Doping with a material that has only three valence electrons produces a hole in the semiconductor crystal.

12. Holes in semiconductor materials serve as current carriers.

13. Hole current is opposite in direction to electron current.

14. Semiconductors with free holes are classified as P-type materials.

15. Materials with five valence electrons produce N-type semiconductors.

16. Materials with three valence electrons produce P-type semiconductors.

17. Holes drift toward the negative end of a voltage source.

18. Majority carriers are electrons for N-type material. Holes are majority carriers for P-type material.

19. Minority carriers are holes for N-type material. Electrons are minority carriers for P-type material.

20. The number of minority carriers increases with temperature.

Chapter Review Questions

Determine whether each statement is true or false.

2-1. The current carriers in conductors such as copper are valence electrons.

2-2. It is easy to move the valence electrons in conductors.

2-3. A positive temperature coefficient means the resistance goes up as temperature goes down. *up*

2-4. Conductors have a positive temperature coefficient.

2-5. Semiconductors do not semi-conduct unless they are doped or heated.

2-6. Silicon has *four* five valence electrons.

2-7. A silicon crystal is built by ionic bonding.

2-8. Materials with eight valence electrons tend to be unstable. *stable*

2-9. Semiconductors have a negative temperature coefficient.

2-10. Silicon is usually preferred to germanium because it is less affected by temperature.

2-11. When a semiconductor is doped with arsenic, free electrons can be found in the crystal.

13

2-12. An N-type material has free electrons available to support the flow of current.

2-13. Doping a crystal increases its resistance.

2-14. Doping with boron produces many free electrons in the crystal.

2-15. Hole current is opposite in direction to electron current.

2-16. Holes are current carriers and are assigned a positive charge.

2-17. If a P-type semiconductor shows a few free electrons, the electrons are called minority carriers.

2-18. If an N-type semiconductor shows a few free holes, the holes are called minority carriers.

Answers to Review Questions

1. F	10. T	19. Positive
2. F	11. T	20. Holes
3. T	12. T	21. Negative
4. T	13. F	22. F
5. F	14. Five	23. F
6. F	15. Electron or carrier	24. T
7. T	16. Carriers	25. T
8. F	17. Decreases	26. F
9. T	18. Three	27. T

Junction Diodes

■ This chapter introduces you to the most basic semiconductor device, the *junction diode*. Diodes are very important in electronic circuits. They perform many functions. Everyone working in electronics must be familiar with diodes.

Your study of diodes will enable you to predict when they will be on and when they will be off. You will be able to read their characteristic curves. Also you will be able to identify their symbols and their terminals.

This chapter will also introduce some of the ways that diodes are used. You will find them very popular electronic components.

3-1 THE PN JUNCTION

A basic use for P- and N-type semiconductor materials is in *diodes*. Diodes are one of the most important devices in modern electronics since they can be used to do many jobs. They are generally two-terminal devices.

Figure 3-1 shows a representation of a PN-junction diode. Notice that it contains a P-type region with free holes and an N-type region with free electrons. The diode structure is continuous from one end to the other. It is one complete crystal of silicon or germanium.

The junction shown in Fig. 3-1 is the boundary, or dividing line, that marks the end of one section and the beginning of the other. It does not represent a mechanical joint. In other words, the *junction* of a diode is that region of the crystal where the P-type material ends and the N-type material begins.

Because the diode does have a continuous crystalline structure, the electrons can move across the junction. After the diode is formed, some of the electrons cross the junction to fill some of the holes. Figure 3-2 shows this effect. The result is that a *depletion region* is formed. With the electrons gone and the holes filled, no free carriers are left. This region around the junction has become *depleted*.

The depletion region will not continue to grow for very long. An electric potential, or force, forms along with the depletion region and prevents all the electrons from crossing over and filling all the holes in the P-type material.

Figure 3-3 shows why this potential is formed. Any time an atom loses an electron it becomes unbalanced. It now has more protons in its nucleus than it has electrons in orbit. This gives it an overall positive charge.

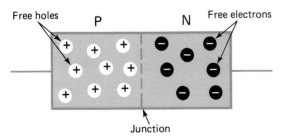

Fig. 3-1 The structure of a junction diode.

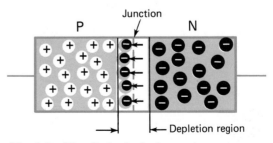

Fig. 3-2 The diode depletion region.

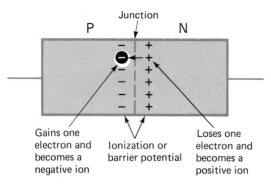

Fig. 3-3 Formation of the barrier potential.

From page 15:
Diode

PN junction

Depletion region

On this page:
Positive ion

Negative ion

Barrier potential

Forward bias

Reverse bias

It is called a *positive ion*. In the same way, if an atom gains an extra electron, it shows an overall negative charge and is a *negative ion*. When one of the free electrons in the N-type material leaves its parent atom, that atom becomes a positive ion. When the electron joins another atom on the P-type side, that atom becomes a negative ion. The ions form a charge that prevents any more electrons from crossing the junction.

So, when a diode is first made, some of the electrons cross the junction to fill some of the holes. Soon the action stops because a negative charge forms on the P-type side to repel any other electrons that might try to cross over. This negative charge is called the *ionization*, or *barrier potential*. "Barrier" is a good name since it is a barrier against more electrons that might cross the junction.

Now that we know what happens when a PN junction is formed, we can try to guess how it will behave electrically. Figure 3-4 shows a summary of the situation. There are two regions with free carriers. Since there are carriers, we can expect these regions to *semi-conduct*. But right in the middle there is a region with no carriers. With no carriers

it would be expected to conduct very poorly. It would *insulate*.

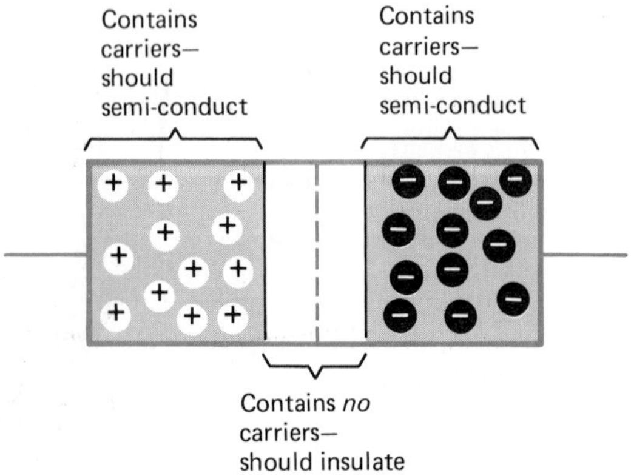

Any device with an insulator in the middle should not conduct much electric current. So we can guess that PN-junction diodes are very poor conductors. However, a depletion region is not exactly the same as other insulators. It was formed in the first place by electrons moving and filling holes. An externally applied voltage can remove the effect of the depletion region.

In Fig. 3-5 a PN-junction diode is connected to an external battery in such a way that the depletion region is collapsed. The positive terminal of the battery will repel the holes on the P-type side and push them toward the junction. The negative terminal of the battery will repel the electrons and push them toward the junction. This actually collapses the depletion region.

With the depletion region gone, the diode can semi-conduct. Figure 3-5 shows electron current leaving the negative side of the battery, flowing through the diode, through the current limiter (a resistor), and returning to the positive side of the battery. The current-limiting resistor is needed in some circuits to keep the current flow at a safe level. Diodes can be destroyed by excess current.

The condition of Fig. 3-5 is called *forward bias*. In electronics, a *bias* is a voltage or a current applied to a device. Forward bias indicates that the voltage or current is so applied *to turn on the device*. The diode in Fig. 3-5 has been turned on by the battery, so it is an example of forward bias.

Reverse bias is another possibility. With zero bias connected to the diode, the deple-

Contains
carriers—
should
semi-conduct

Contains
carriers—
should
semi-conduct

Contains *no*
carriers—
should insulate

Fig. 3-4 Depletion region as an insulator.

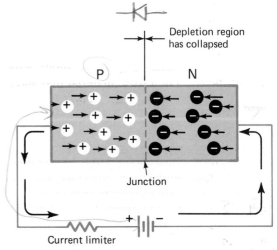

Fig. 3-5 Forward bias.

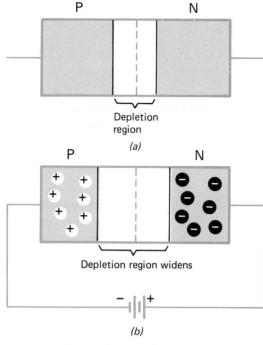

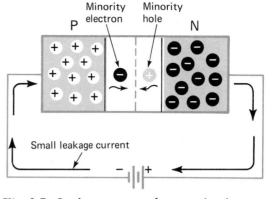

Fig. 3-7 Leakage current due to minority carriers.

Leakage current

Fig. 3-6 Effect of bias on depletion region. (*a*) Zero bias. (*b*) Reverse bias increases the depletion region.

tion region is as shown in Fig. 3-6(*a*). When reverse bias is applied to a junction diode, the depletion region does not collapse. In fact, it becomes wider than it was. Figure 3-6(*b*) shows a diode with reverse bias applied. The positive side of the battery is applied to the N-type material. This attracts the free electrons away from the junction. The negative side of the battery attracts the holes in the P-type material away from the junction. This makes the depletion region wider than it was when no voltage was applied.

Because reverse bias widens the depletion region, it can be expected that no current flow will result. The depletion region is an insulator, and it will block the flow of current. Actually, some current flow will occur because of minority carriers. Figure 3-7 shows why this happens. The P-type material may have a few minority electrons. These will be pushed to the junction by the repulsion of the negative side of the battery. The N-type material may have a few minority holes. These will also be pushed toward the junction. Reverse bias forces the minority carriers together, and a small *leakage current* will result. Diodes are not perfect, but modern silicon diodes may show a leakage current so small that it cannot even be measured with ordinary meters. At reasonable temperatures

there are few minority carriers in silicon, and often the reverse leakage can be ignored.

Germanium diodes show more leakage. The reason is that germanium has many more thermal carriers than silicon at a given temperature. Silicon diodes cost less, show very low leakage current, and are more desirable for most applications. Germanium diodes do have certain advantages, and they are still used in some applications.

In summary, the PN-junction diode will conduct well in one direction and very little in the other. The direction of good conduction is from the N-type material to the P-type material. If a voltage is applied across the diode to move the current in this direction, it is called forward bias. If the voltage is reversed, it is called reverse bias. The diode is very useful because it can steer current in a given direction. It can also be used as a switch and as a means of changing alternating current (ac) to direct current (dc). Other diodes can perform many special jobs in electric and electronic circuits. The diode is a very important part of modern circuitry.

Review Questions

Determine whether each statement is true or false.

1. A junction diode is made up of N-type semiconductor material only.

2. The depletion region is formed by electrons crossing over from the P-type side of the junction to fill holes on the N-type side of the junction.

3. The barrier potential prevents all the electrons from crossing the junction and filling all the holes.

17

Electronics:
Principles and
Applications
CHAPTER 3

Volt-ampere
characteristic
curve

Linear device

4. The depletion region is a good conductor.

5. Once the depletion region forms, it cannot be removed.

6. Forward bias expands the depletion region.

7. Reverse bias collapses the depletion region and turns on the diode.

8. A reverse-biased diode may show a little current because of minority carrier action.

9. High temperatures will increase diode leakage current.

3-2 CHARACTERISTIC CURVES OF DIODES

Diodes conduct well in one direction but not in the other. This is probably the most important property of diodes. They have other characteristics, too, and some of these must be understood in order to have a complete, working knowledge of electronic circuits.

Characteristics of electronic devices can be shown in several ways. One way is to list the amount of current flow for each of several values of voltage. These values could be presented in a table. A better way to do it is to show the values on a graph. Graphs are often easier to use than tables of data. Trends and patterns show up well. For example, a graph of the average temperature for each month of the year would show the reader at a glance how the pattern changes from season to season.

One of the most frequently used graphs in electronics is the *volt-ampere characteristic curve*. Units of voltage make up the horizontal axis, and units of current make up the vertical axis. Figure 3-8 shows a volt-ampere characteristic curve for a 100-ohm (Ω) resistor. The *origin* is the point where the two axes cross. This point usually indicates zero voltage and zero current. Note that the resistor curve passes exactly through the origin. This simply means that with zero voltage across a resistor we can expect zero current through it. Ohm's law will verify this:

$$I = \frac{V}{R} = \frac{0}{100} = 0 \text{ A}$$

At 5 V on the horizontal axis, the curve passes through a point exactly opposite 50 milliamperes (mA) on the vertical axis. By looking at the curve, we can quickly and easily find

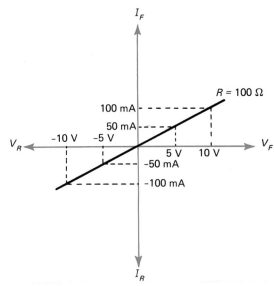

Fig. 3-8   A volt-ampere characteristic curve for a resistor.

the current for any value of voltage. At 10 V the current is 100 mA. We can check this using Ohm's law:

$$I = \frac{V}{R} = \frac{10}{100} = 0.1 \text{ A} = 100 \text{ mA}$$

Moving to the left of the origin in Fig. 3-8, we can obtain current levels for values of reverse voltage. Reverse voltage is indicated by V_R, and V_F indicates the forward voltage. At -5 V the current through the resistor will be -50 mA. This simply means that when the voltage across a resistor is reversed in polarity, the resistor current will reverse (change direction). Forward current is indicated by I_F, and I_R indicates reverse current.

The characteristic curve for a resistor is a straight line. For this reason, it is said to be a *linear device*. Resistor curves are not really necessary in electronics since they are linear devices. With Ohm's law to help us, we can easily obtain any data point without a graph.

Diodes are more complicated than resistors. Their volt-ampere characteristic curves give information that cannot be obtained with a simple linear equation. Figure 3-9 shows the volt-ampere characteristic curve for a typical PN-junction diode. With 0 V across the diode, the diode will not conduct. It will continue not to conduct even as V_F is increased from 0 V. In fact, the diode will not conduct until a few tenths of a volt are applied across it. This may seem strange, but if you re-

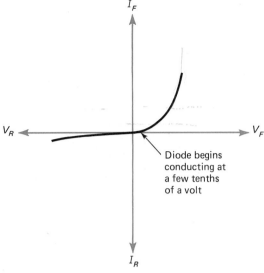

Fig. 3-9 A volt-ampere characteristic curve of a diode.

Diode begins conducting at a few tenths of a volt

Nonlinearity

Avalanche voltage

member the depletion region and the barrier potential, it is easy to understand. It requires about 0.2 V to *turn on* a germanium diode, that is, to cause it to conduct, and about 0.6 V to turn on a silicon diode. This voltage is needed to overcome the barrier and collapse the depletion region.

Figure 3-9 also shows what happens when reverse bias is applied to a diode. At increasing levels of V_R, the curve shows some reverse current I_R. This leakage current is due to minority carriers. It may be very small, so that sometimes the I_R scale is marked in microamperes (μA). Leakage current often does not become significant until there is a large reverse bias across the diode. For that reason, the V_R scale is sometimes marked in tens or hundreds of volts.

The diode characteristic curve shows that the diode is not a linear device. This nonlinearity can be an advantage or a disadvantage depending on the exact circuit application. In later chapters we will examine diodes used in radio receivers for detection. Their nonlinearity makes them very useful in this application.

A comparison of the characteristic curves for a silicon diode and a germanium diode is shown in Fig. 3-10. It is clear that the germanium diode requires much less forward bias to conduct. This can be an advantage in low-voltage circuits. Also, note that the germanium diode will show a lower voltage drop for any given level of current than the silicon

diode will. Germanium diodes have less resistance for forward current. Germanium is a better conductor than silicon. However, the silicon diode is still superior for most applications because of its low cost and lower leakage current at high temperatures.

Figure 3-10 also shows how silicon and germanium diodes compare under conditions of reverse bias. At reasonable levels of V_R, the leakage current of the silicon diode is very low. The germanium diode shows much more leakage. However, if a certain critical value of V_R is reached, the silicon diode will show a rapid increase in reverse current. This is shown as the *reverse breakdown point*. It is also referred to as the *avalanche voltage*. Avalanche breakdown occurs when minority carriers gain enough energy to collide with valence electrons and knock them loose. This causes an "avalanche" of carriers, and the current flow increases tremendously.

Avalanche in a silicon diode will occur when the reverse bias reaches from 50 to over 1000 V, depending on how the diode was manufactured. If the reverse current at avalanche is not limited, the diode will be destroyed. Avalanche is usually avoided by choosing the right diode or by controlling the amount of reverse bias across the diode. Some diodes have a controlled avalanche characteristic and can safely withstand brief periods of reverse breakdown.

Figure 3-11 shows how volt-ampere characteristic curves can be used to indicate the ef-

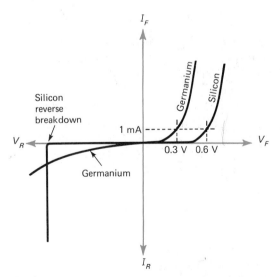

Fig. 3-10 Comparison of silicon and germanium diodes.

19

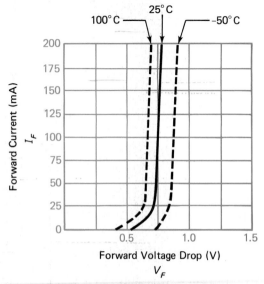

Fig. 3-11 Characteristic curves showing the effect of temperature on a typical silicon diode.

fects of temperature on diodes. The temperatures are in degrees Celsius. Electronic circuits may have to work over a range of temperatures from -50 to $+100°C$. At the low end mercury will freeze; at the high end water will boil. The range for military-grade electronic circuitry is -65 to $+125°C$. To be able to operate in such a wide temperature range, extreme care must be taken in the selection of materials, the manufacturing processes used, and the handling and testing of the finished

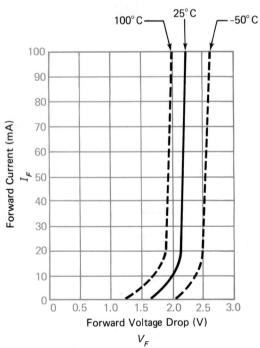

Fig. 3-12 High-voltage diode curves.

product. This is why military-grade devices tend to be much more expensive than industrial- and consumer-grade devices.

By examining the curves in Fig. 3-11, you can conclude that silicon conducts better at elevated temperatures. Since the forward voltage drop V_F decreases as temperature goes up, then its resistance must be going down. This agrees with silicon's negative temperature coefficient.

Figure 3-12 shows the characteristic curves for special silicon diodes made to operate safely at high voltage levels. Notice that the voltage drops are much higher than those in Fig. 3-11. High-voltage diodes have a much greater resistance than ordinary diodes. This is important to remember in testing diodes with an ohmmeter.

Review Questions

Supply the missing word in each statement.

10. A volt-ampere characteristic curve for a resistor always passes through the _____?_____. *origin*

11. A volt-ampere characteristic curve for a resistor is shaped as a _____?_____ *straight line*

12. A volt-ampere characteristic curve for a 5000-Ω resistor will, at 100 V on the horizontal axis, pass through a point opposite _____?_____ on the vertical axis. *20 mA*

13. The volt-ampere characteristic curve for an open circuit will be a straight line on the _____?_____ axis. *Hor.*

14. The volt-ampere characteristic curve for a short circuit (0 Ω) will be a straight line on the _____?_____ axis. *Ver*

15. Resistors are linear devices. Diodes are _____?_____ devices. *nonlinear*

16. A silicon diode does not begin conducting until _____?_____ V of forward bias is applied. *.6*

17. Diode avalanche, or reverse breakdown, is caused by excess reverse _____?_____ *Bias Voltage*

3-3 DIODE LEAD IDENTIFICATION

Diodes behave very differently when forward-biased as compared to reverse-biased. Obviously, it is critical that they be wired into the circuit properly. Connecting a diode backwards can destroy it and may also destroy many other parts of the circuit. A technician must always be absolutely sure that the diode is connected properly.

Technicians often refer to schematic diagrams when checking diode polarity. Figure 3-13 shows the schematic symbol for a diode. The P-type material makes up the _anode_ of the diode. The word "anode" is used in electric and electronic devices to identify the terminal that attracts electrons. The N-type material makes up the _cathode_ of the diode. The word "cathode" refers to the terminal that gives off, or emits, electrons. Note that the forward current moves _from the cathode to the anode._

Diodes are available in many physical forms or _packages._ Some typical examples are shown in Fig. 3-14. Manufacturers may use plastic, glass, metal, ceramic, or a combination of these to package the PN junction. They use many sizes and shapes. They employ several techniques for indicating which lead is the cathode and which is the anode. If a band or a bevel can be found at one end of the package, then it is fairly certain that this marks the cathode lead. If a flange or a plus (+) sign can be found, then this marks the cathode end of the device. Notice that the stud mount diode may have the diode symbol printed on its side. In this figure, the stud, or screw, is the cathode. However, be careful when replacing this type of diode because the stud can be the cathode lead or the anode lead. They are made both ways. Check the marking on the diode itself to be sure.

Since there are a few case styles where confusion can occur, the technician often must check the diode and identify the leads, using a volt-ohm-milliammeter (VOM) or a vacuum-tube voltmeter (VTVM). This check uses the ohmmeter function of the meter. The ohmmeter is connected across the diode, and the resistance is noted as in Fig. 3-15. The R × 100 range usually is used for these readings. It is necessary to note only whether the resistance is high or low. Exact readings are not needed. Then the ohmmeter leads are reversed as in Fig. 3-15(b). The resistance should change drastically. If it does not, the diode is probably defective. In Fig. 3-15, we can conclude that the diode is good and that the cathode lead is at the left. When the positive lead of the ohmmeter was on the right lead, the diode was turned on. Forward cur-

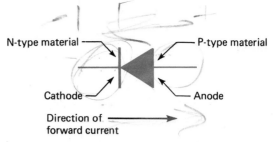

Fig. 3-13 Diode schematic symbol.

rent is from cathode to anode. Making the anode positive is necessary if the anode is going to attract electrons. Remember, in order to turn on the diode, the anode must be _positive with respect to the cathode._

Using ohmmeters to check diodes and identify the leads is very effective. However, there are two traps that the technician must know about in order to avoid mistakes. First, some ohmmeters have reversed polarity. This may occur on some ranges or on all ranges. _The only way to be sure is to check the ohmmeter with a separate dc voltmeter. Know your meter._ When using an unfamiliar meter, do not assume the common, or black, lead is negative on the ohms function. The second trap is that the ohmmeter supply voltage might not be high enough to turn on the

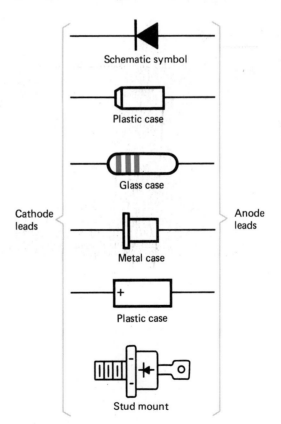

Schematic symbol

Plastic case

Glass case

Cathode leads   Anode leads

Metal case

Plastic case

Stud mount

Fig. 3-14 Diode package styles.

21

Diode polarity

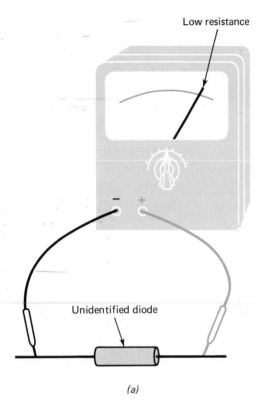

(a)

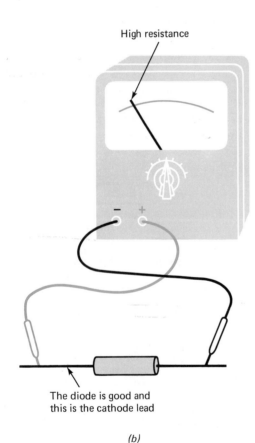

(b)

Fig. 3-15 Testing a diode with an ohmmeter.

diode. This would indicate that the diode is open (shows a high resistance in both directions). In fact, some modern ohmmeters are purposely designed to use a very low voltage to avoid turning on PN junctions. Remember, it takes about 0.2 V to turn on a germanium diode and about 0.6 V to turn on a silicon diode. Some high-voltage rectifiers may even require much more than this. High-voltage rectifiers should not be tested with ordinary ohmmeters if exact results are needed.

One last comment about diode checking is in order. A few ohmmeters may supply enough voltage or current to damage certain diodes. Some diodes used in high-frequency detection circuits can be very delicate. Usually, it is not a good idea to check these diodes with an ohmmeter. It pays to find out how much voltage is used on the various ranges of your ohmmeter and how much current is supplied to the device being checked.

Beginners are almost always confused by diode polarity. There is a good reason, too. Look at Fig. 3-14 again. Note that one way that manufacturers mark the cathode end is to use a plus (+) symbol. Yet, we have said that the diode is turned on when its anode lead is made positive. This seems a little confusing. The reason the plus sign is used to indicate the cathode lead is related to *how the diode behaves in a rectifier circuit.* Rectifier circuits are covered in detail in Chap. 4.

Review Questions

Supply the missing word in each statement.

18. Assume that a diode is forward-biased. The diode terminal that attracts the electrons is called the _____?_____.

19. The diode lead near the band or bevel on the package is the _____?_____ lead.

20. A plus (+) sign on a diode indicates the _____?_____ lead.

21. An ohmmeter is connected across a diode. A low resistance is shown. The leads are reversed. A low resistance is still shown. The diode is _____?_____.

22. When the positive lead from an ohmmeter is applied to the anode lead of a diode, the diode is turned _____?_____.

23. When the positive lead of an ohmmeter is applied to the cathode lead of a diode, the diode is turned _____?_____.

22

3-4 DIODE TYPES AND APPLICATIONS

There are many diode types and applications in electronic circuits. Some of the important ones will be discussed here briefly.

Rectifier diodes are very important. A rectifier is a device that changes alternating current (ac) to direct current (dc). Since a diode will conduct easily in just one direction, only half of the ac cycle will pass through the diode. A diode can be used to supply dc in a simple battery charger (Fig. 3-16). A secondary battery can be charged by passing a direct current through it that is opposite in direction to its discharge current. If an ac source is available, the diode will permit only that direction of current that will restore the battery.

Notice that in Fig. 3-16 the diode is connected so that the current flow during charging is opposite to the current flow during discharging. The cathode of the diode must be connected to the positive terminal of the battery. A mistake in this connection would discharge the battery or damage the diode. It is very important to connect diodes correctly.

Diodes also are used to keep voltages constant. This application is called *voltage stabilization, or voltage regulation*. A type of diode called a *Zener* diode is used as a voltage regulator. The characteristic curve for a typical Zener diode is shown in Fig. 3-17. It is very similar to the silicon characteristic curve of Fig. 3-10. The symbol for the Zener diode is similar to that of the regular diode except that the cathode is drawn as a bent line representing the letter Z. The Zener breakdown process is a bit different from avalanche in a rectifier diode. Zener breakdown is normally expected to occur at much lower voltages. However, there are Zener diodes available for use up to about 200 V.

The important difference between Zener diodes and rectifier diodes is in how they are used in electronic circuits. Zeners are available in many voltage ratings. As long as they are operated over their normal range, the voltage across the diode will equal the rated voltage plus or minus a small error voltage. They are operating backward compared to a rectifier diode. In a rectifier, the major current is from cathode to anode. Zeners are operated with reverse bias. They conduct during breakdown from anode to cathode.

A large change in diode current will corre-

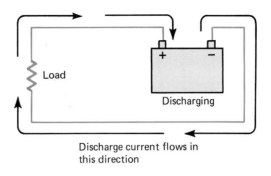

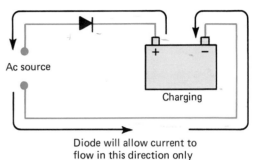

Fig. 3-16 Battery charging with a diode.

Rectifier diode

Voltage regulation

Zener diode

spond to only a very small change in diode voltage. This can be seen clearly in Fig. 3-18(a). Over the normal operating range, the Zener voltage changes only slightly.

Figure 3-18(b) shows how a Zener diode can be used to stabilize a voltage. A current-limiting resistor must be included to prevent the Zener diode from conducting too much and overheating. The stabilized output is available across the diode itself. Conduction is from anode to cathode.

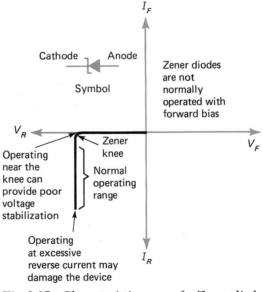

Fig. 3-17 Characteristic curve of a Zener diode.

23

Light-emitting
diode (LED)

Seven-segment
display

Varicap

Varactor diode

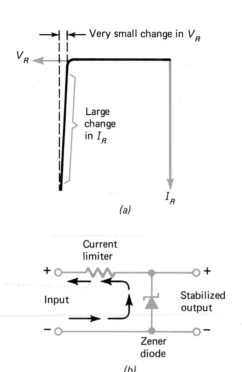

(a)

Current
limiter

Input

Stabilized
output

Zener
diode

(b)

Fig. 3-18 A Zener diode used as a voltage regulator. (*a*) Volt-ampere characteristic curve. (*b*) Voltage regulator circuit.

Another important diode type is the *light-emitting diode*, or LED. Its schematic symbol is as shown in Fig. 3-19(*a*). As the electrons of the LED cross the junction, they combine with holes. This changes their status from one energy level to a lower energy

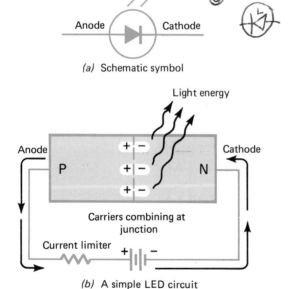

(a) Schematic symbol

(b) A simple LED circuit

Fig. 3-19 Light-emitting diode. (*a*) Schematic symbol. (*b*) A simple LED circuit.

level. The extra energy they had as free electrons must be released. Some of the excess energy is given off as heat. Some is given off as light as in Fig. 3-19(*b*). Special materials such as gallium arsenide and gallium phosphide are used to make LEDs.

Light-emitting diodes are rugged, small, and have a very long life. They can be switched rapidly since there is no thermal lag caused by gradual cooling or heating in a filament. They lend themselves to certain photochemical fabrication methods and can be made in various shapes and patterns. They are much more flexible than incandescent lamps and are used in many different ways in electronics. LEDs have wide application as the numeric displays used to indicate numerals 0 through 9. A typical seven-segment display is shown in Fig. 3-20. By selecting the correct segments, the desired number is displayed. These displays are used in digital meters, clocks, and many forms of measuring instruments.

The last type of diode to be covered in this chapter is the *varicap*, or *varactor*, diode. This device is a solid-state replacement for the variable capacitor. Much of the tuning and adjusting of electronic circuits involves changing capacitance. Variable capacitors are often large, delicate, and expensive parts. If the capacitor must be adjusted from the front panel of the equipment, a metal shaft or a complicated mechanical connection must be used. This causes some design problems. The varicap diode can be controlled by voltage. No control shaft or mechanical linkage is needed. The varicap diodes are small, rugged, and inexpensive. They are used to replace ordinary variable capacitors and some variable inductors in modern electronic equipment.

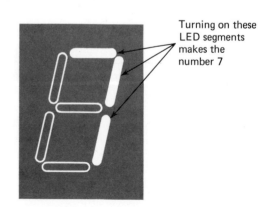

Turning on these
LED segments
makes the
number 7

Fig. 3-20 LED numeric display.

The capacitor effect of a PN junction is shown in Fig. 3-21. A capacitor consists of two conducting plates separated by a dielectric material or insulator. Its capacitance depends on the area of the plates as well as on their separation. A diode with reverse bias has a similar arrangement. The P-type material semi-conducts and forms one plate. The N-type material also semi-conducts and forms the other plate. The depletion region is an insulator and forms the dielectric. By adjusting the reverse bias, the width of the depletion region, that is, the dielectric, is changed; and this changes the capacity of the diode. With a high reverse bias, the capacity will be low because the depletion region widens. This is the same effect as moving apart the plates of a variable capacitor. With little reverse bias, the depletion region is narrow. This makes the diode capacitance increase.

Figure 3-22 shows the capacitance in picofarads versus reverse bias for a varicap tuning diode. Capacitance decreases as reverse bias increases. The varicap diode can be used in a simple LC tuning circuit, as shown in Fig. 3-23. The tuned circuit is formed by an inductor (L) and two capacitors. The top capacitor C_2 is usually much higher in value than the bottom varicap diode capacitor C_1. This makes the resonant frequency of the tuned circuit mainly dependent on the inductor and the varicap capacitor. The formula for two capacitors in series is

$$\frac{1}{C_S} = \frac{1}{C_1} + \frac{1}{C_2}$$

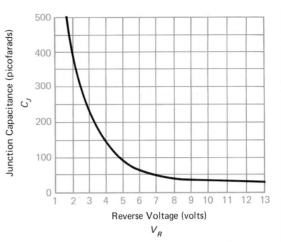

Tuning diode

Fig. 3-22 Junction capacitance versus reverse voltage characteristic curve of a varicap diode.

Solving for C_S, this becomes

$$C_S = \frac{C_1 \times C_2}{C_1 + C_2}$$

If the varicap changes from 400 to 100 picofarads (pF), the net effect across the inductor can be calculated for a fixed capacitor C_2 of 0.005 μF (or 5000 pF):

For $C_1 = 400$ pF

$$C_S = \frac{400 \times 5000}{400 + 5000} = 370.37 \text{ pF}$$

For $C_1 = 100$ pF

$$C_S = \frac{100 \times 5000}{100 + 5000} = 98.04 \text{ pF}$$

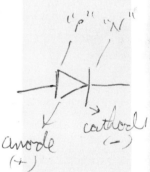

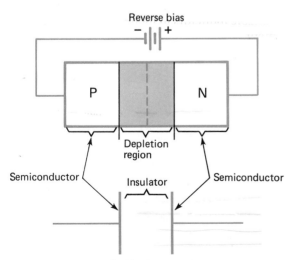

Fig. 3-21 Diode capacitance effect.

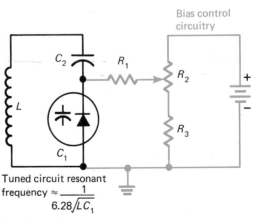

Fig. 3-23 Tuning with a varicap diode.

25

These two actual values are close to the value of the varicap capacitor alone. The resonant frequency can therefore be determined by the equation

$$f_r = \frac{1}{6.28\sqrt{LC_1}}$$

where f_r is the resonant frequency of a circuit with inductance L and capacitance C_1.

Resistor R_1 of Fig. 3-23 is a high value of resistance and isolates the tuned circuit from the bias control circuit. This prevents the Q of the tuned circuit, that is, the sharpness of the resonance, from being lowered by resistive loading. High resistance gives light loading and high Q. Resistors R_2 and R_3 form the variable bias divider. As the wiper arm on the resistor is moved up, the reverse bias across the diode will increase. This will decrease the capacitance of the varicap diode and raise the resonant frequency of the tuned circuit. You should inspect the resonant frequency formula and verify this trend. Without R_3, the diode bias could be reduced to zero. In a varicap tuning diode, zero bias is not usually acceptable. An ac signal in the tuned circuit could switch the diode into forward conduction. This would cause undesired effects. A circuit such as the one shown in Fig. 3-23

could be used for many tuning purposes in electronics.

Review Questions

Determine whether each statement is true or false.

24. A rectifier is a device used to change alternating current to direct current.

25. A normally operating Zener diode will conduct from its anode to its cathode.

26. A normally operating rectifier diode will conduct from its anode to its cathode.

27. Zener diodes are often used as voltage regulators.

28. The schematic symbol for an LED is the same as that for a Zener diode.

29. Varactor diodes show large inductance change with changing bias.

30. The depletion region serves as the dielectric in a varicap diode capacitor.

31. Increasing the bias (reverse) across a varicap diode will increase its capacitance.

32. Heavily loading a tuned circuit will raise its Q.

33. Decreasing the capacitance in a tuned circuit will raise its resonant frequency.

Summary

1. One of the most basic and useful electronic components is the PN-junction diode.
2. When the diode is formed, a depletion region appears that acts as an insulator.
3. Forward bias forces the majority carriers to the junction and collapses the depletion region. The diode conducts. (Technically speaking, it *semi-conducts*.)
4. Reverse bias widens the depletion region. The diode does not conduct.
5. Reverse bias forces the minority carriers to the junction. This causes a small leakage current to flow. It can usually be ignored.
6. Volt-ampere characteristic curves are used very often to describe the behavior of electronic devices.
7. The volt-ampere characteristic curve of a resistor is linear (a straight line).
8. The volt-ampere characteristic curve of a diode is nonlinear.

9. It takes about 0.2 V of forward bias to turn on a germanium diode and about 0.6 V of forward bias to turn on a silicon diode.
10. A silicon diode will avalanche at some high value of reverse voltage.
11. High-voltage diodes have higher voltage drops than ordinary diodes.
12. Diode leads are identified as the cathode lead and the anode lead.
13. The anode must be made positive with respect to the cathode to make a diode conduct.
14. Manufacturers mark the cathode lead with a band, bevel, flange, or plus (+) sign.
15. If there is doubt, the ohmmeter test can identify the cathode lead. It will be connected to the negative terminal. A low resistance reading indicates that the negative terminal of the ohmmeter is connected to the cathode.

16. Caution should be used when applying the ohmmeter test. Some ohmmeters have reversed polarity. The voltage of some ohmmeters is too low to turn on a PN-junction diode. Some ohmmeters' voltages are too high and may damage delicate PN junctions.

17. A diode used to change ac to dc is called a rectifier diode.

18. A diode used to stabilize or regulate voltage is the Zener diode.

19. Zener diodes conduct from anode to cathode when they are working as regulators. This is just the opposite from the way rectifier diodes conduct.

20. Varicap diodes are solid-state, variable capacitors. They are operated under conditions of reverse bias.

21. Varicap diodes show minimum capacitance at maximum bias. They show maximum capacitance at minimum bias.

Chapter Review Questions

Determine whether each statement is true or false.

3-1. A PN-junction diode is made by mechanically joining a P-type crystal to an N-type crystal.

3-2. The depletion region forms only on the P-type side of the PN junction in a solid-state diode.

3-3. The barrier potential prevents all the N-type side electrons from crossing the junction to fill all the holes in the P-type side.

3-4. The depletion region acts as an insulator.

3-5. Forward bias collapses the depletion region.

3-6. Reverse bias drives the majority carriers toward the junction.

3-7. It takes a few tenths of a volt to collapse the depletion region and turn on a solid-state diode.

3-8. A diode has a linear volt-ampere characteristic curve.

3-9. Excessive reverse bias across a silicon diode may cause avalanche.

3-10. Silicon is a better conductor than germanium.

3-11. Less voltage is required to turn on a germanium diode than to turn on a silicon diode.

3-12. The behavior of electronic devices such as diodes changes with temperature.

3-13. The Celsius temperature scale is used in electronics.

3-14. Diodes have two leads, the cathode lead and the anode lead.

3-15. Forward current in a diode is from the cathode to the anode.

3-16. Diode manufacturers usually mark the package in some way so as to identify the cathode lead.

3-17. Making the diode anode negative with respect to the cathode will turn on the diode.

3-18. It is possible to test diodes with an ohmmeter and identify the cathode lead.

3-19. Rectifier diodes are used in the same way as Zener diodes.

3-20. Zener diodes are normally operated with the cathode positive with respect to the anode.

3-21. LEDs emit light by heating a tiny filament red hot.

27

3-22. The capacitance of a varicap diode is determined by the reverse bias across it.

3-23. Germanium diodes cost less and are therefore more popular than silicon diodes in modern circuitry.

Answers to Review Questions

1. F	12. 20 mA [0.02 A]	23. Off
2. F	13. Horizontal	24. T
3. T	14. Vertical	25. T
4. F	15. Nonlinear	26. F
5. F	16. 0.6	27. T
6. F	17. Bias (voltage)	28. F
7. F	18. Anode	29. F
8. T	19. Cathode	30. T
9. T	20. Cathode	31. F
10. Origin	21. Shorted	32. F
11. Straight line	22. On	33. T

Power Supplies

■ Diodes have many uses in electronics. Perhaps their most obvious use is in the area of power supplies. Power supplies use diodes as rectifiers, to convert alternating current to direct current. They also use diodes as regulators, to stabilize voltage. This chapter covers the circuits which use diodes in these ways.

Power supplies also rely on other components that you have already studied. Most electronic circuits use a great variety of devices. It is very important to understand how they can perform together as a system.

As you begin to understand how systems such as power supplies work in electronics, you can begin to understand the principles of *troubleshooting*. Troubleshooting is an analytical process by which you can determine the cause of circuit problems. It is much easier to repair circuits when you know how and why they work.

4-1 THE POWER-SUPPLY SYSTEM

This chapter marks your beginning study of electronic circuits. The power supply is the most fundamental circuit of any electronic system. The power supply changes the available electric energy (usually ac) to the form required by the various circuits within the system (usually dc). Without electric energy, circuits cannot function. *One of the first steps in troubleshooting any electronic device is to check the supply voltages at various stages in the circuitry.*

Power supplies can be very simple or very complicated depending on the requirements of the system. A simple power supply may be required to furnish 12 V dc. A more complicated supply may provide several voltages, some positive and some negative with respect to the chassis ground. Some power supplies may have a wide tolerance regarding voltage. The actual voltage output may vary ± 20 percent. Another supply may have to keep its output voltage within ± 0.01 percent. Obviously, this strict tolerance complicates the design of the supply.

Figure 4-1 shows a power supply as part of

an electronic system. Drawings of this type are called *block diagrams*. Block diagrams are very useful for showing the various major circuits within electronic systems. The power supply in Fig. 4-1 occupies one of the blocks within the system. It is the most critical since it energizes the rest of the system. For example, if a problem develops in the power supply, the fuse might "blow" (open). In that case, none of the voltages could be supplied to

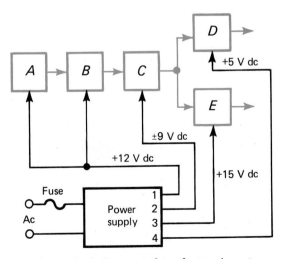

Fig. 4-1 Block diagram of an electronic system.

From page 29:
Block diagram

On this page:
Rectification

the other circuits. Another type of problem might involve the loss of only one of the outputs of the power supply. Suppose the +12-V dc output drops to zero because of a component failure in the power supply. Circuit A and circuit B would no longer work.

The second output of the power supply shown in Fig. 4-1 develops both positive and negative dc voltages with respect to the common point (usually the metal chassis). This output could fail, too. It is also possible that only the negative output could fail. In either case, circuit C would not work normally under such conditions.

Troubleshooting electronic systems can be made much easier with block diagrams. If the symptoms indicate the failure of one of the blocks, then the technician can devote most attention to that part of the circuit. Since the power supply energizes most or all the other blocks, *it is one of the first things to check when troubleshooting.*

Review Questions

Supply the missing word in each statement.

1. Power supplies usually change ac to _____?_____ . DC

2. Power-supply voltages are usually specified by using the chassis _____?_____ as a reference. grm

3. Some power-supply voltages may be positive with respect to the chassis while others may be _____?_____ . neg

4. On a block diagram, the circuit that energizes most or all the other blocks is called the _____?_____ . power Supply

4-2 RECTIFICATION

Most electronic circuits need dc. Ac is supplied by the power companies. The purpose of the power supply is to change ac to dc by *rectification*. Alternating current flows in both directions, and direct current flows in only one direction. Since diodes will conduct in only one direction, they are good rectifiers.

Typical ac voltages supplied by the power companies to residential and small commercial customers are 115 V, 120 V, 208 V, and 240 V. Electronic circuits often require much lower voltages. Transformers are

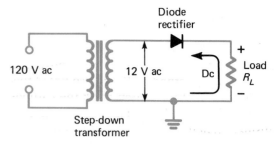

Fig. 4-2 A simple dc power supply.

used to step down the voltage to the level needed. Figure 4-2 shows a simple power supply using a step-down transformer and a diode rectifier.

The load for the power supply of Fig. 4-2 could be an electronic circuit, a battery being charged, or some other device. In this chapter, the power-supply circuits will be shown with load resistors (R_L).

The transformer in Fig. 4-2 has a voltage ratio of 10:1. With 120 V across the primary, 12 V ac is developed across the secondary. If it were not for the diode, there would be 12 V ac across the load resistor. The diode will allow the current to flow through the load only in the direction of the arrow. Since current is flowing in only one direction, it can be called direct current. *When direct current flows through a load, a dc voltage appears across the load.*

Note the polarity across the load in Fig. 4-2. Electrons always move from negative to positive through a load. The positive end of the load is connected to the cathode end of the rectifier. In any rectifier circuit, *the positive end of the load will be that end which contacts the cathode of the rectifier.* It can also be stated that the *negative* end of the load will be in contact with the *anode* of the rectifier. Figure 4-3 shows the negative side of a load re-

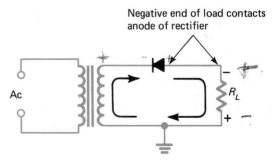

Fig. 4-3 Establishing polarity in a rectifier circuit.

sistor connected to the anode end of the rectifier.

In Chap. 3 it was stated that to forward-bias a diode, the anode must be made positive with respect to the cathode. It was also noted that diode manufacturers often mark the cathode with a plus (+) sign. This may seem confusing. When the diode acts as a rectifier, the function of the plus sign becomes clear. _The plus sign is placed on the cathode end to show the technician which end of the load will be positive._ Look at Fig. 4-2 to see if this is so.

Figure 4-4(a) shows the input waveform to the rectifier circuits of Figs. 4-2 and 4-3. Two complete cycles are shown. In Fig. 4-4(b), the waveform that appears across the load resistor of Fig. 4-2 is shown. The negative half

Ac input to rectifier

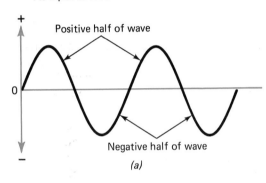

(a)

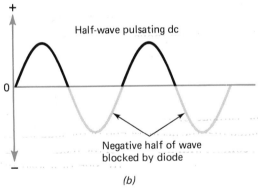

(b)

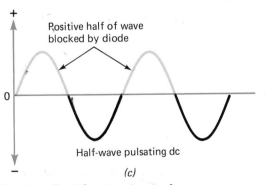

(c)

Fig. 4-4 Rectifier circuit waveforms.

of the cycle is missing since the diode blocks it. This waveform is called _half-wave, pulsating dc._ It represents only the positive half of the ac input to the rectifier.

In Fig. 4-3 the diode has been reversed. This causes the positive half of the cycle to be blocked [Fig. 4-4(c)]. The waveform is also half-wave, pulsating dc. Both circuits, Figs. 4-2 and 4-3, are classified as _half-wave rectifiers._

The ground reference point will determine which way the waveform will be shown for a rectifier circuit. For example, in Fig. 4-3 the positive end of the load is grounded. If an oscilloscope is connected across the load, the ground lead of the oscilloscope will be positive and the probe tip will be negative. Most oscilloscopes show positive as "up" and negative as "down." The actual waveform seen on the screen will then appear as that shown in Fig. 4-4(c). Waveforms can appear up or down depending on circuit polarity, instrument polarity, and the connection between instrument and circuit.

Half-wave rectifiers are usually limited to low-power applications. They take useful output from the ac source for only half the input cycle. They are actually not supplying any load current half the time. This limits the amount of electric energy they can deliver over a given period of time. High power means delivering large amounts of energy in a given time. A half-wave rectifier is a poor choice in high-power applications.

Review Questions

Determine whether each statement is true or false.

5. Current that flows in both directions is called alternating current. T

6. Diodes make good rectifiers because they will conduct in only one direction. T

7. A rectifier can be used in a power supply to step up voltage. F

8. In a rectifier circuit, the positive end of the load will be connected to the cathode of the rectifier. T

9. A single-diode rectifier provides half-wave rectification. T

10. Half-wave rectifiers are usually used in high-power applications. F
Low

**Full-wave
rectifier**

Center tap

**Full-wave,
pulsating dc**

4-3 FULL-WAVE RECTIFICATION

A full-wave rectifier is shown in Fig. 4-5(*a*). It uses a center-tapped transformer secondary and two diodes. The transformer center tap is located at the electrical center of the secondary winding. If, for example, the entire secondary winding has 100 turns, then the center tap will be located at the 50th turn. The waveform across the load in Fig. 4-5(*a*) is full-wave, pulsating dc. Both alternations of the ac input are used to energize the load. Thus, a full-wave rectifier can deliver twice the power of a half-wave rectifier.

The ac input cycle is divided into two parts: a positive alternation and a negative alternation. The positive alternation is shown in Fig. 4-5(*b*). The induced polarity at the secondary is such that D_1 is turned on. Electrons leave the center tap, flow through the load, through D_1, and back into the top of the secondary. Note that the positive end of the load resistor is in contact with the cathode of D_1.

On the negative alternation, the polarity

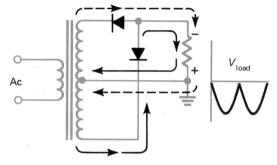

Fig. 4-6 Reversing the rectifier diodes.

across the secondary is reversed. This is shown in Fig. 4-5(*c*). Electrons leave the center tap, flow through the load, through D_2, and back into the bottom of the secondary. The load current is the same for both alternations: it flows up through the resistor. Since the direction never changes, the load current is dc.

Figure 4-6 shows a full-wave rectifier with the diodes reversed. This reverses the polarity across the load resistor. Note that the output waveform shows both alternations

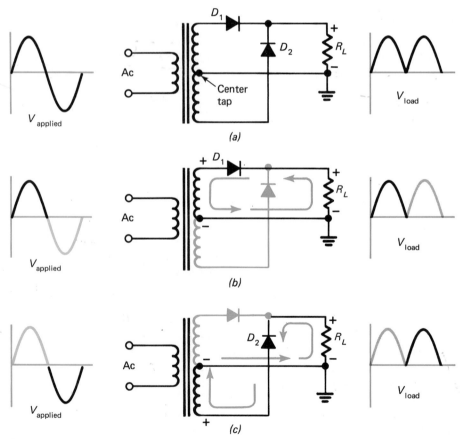

Fig. 4-5 A full-wave rectifier circuit.

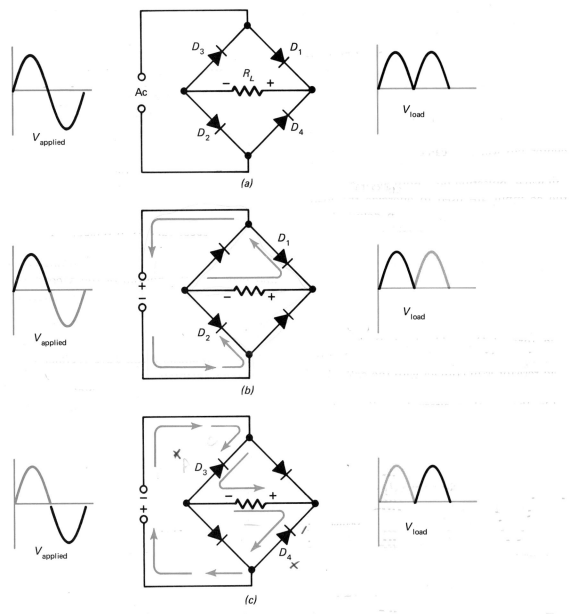

Fig. 4-7 A bridge rectifier circuit.

going in a negative direction. This is what would be seen on an oscilloscope since the output is negative with respect to ground. The diode rule regarding polarity holds true in Fig. 4-6. The negative end of the load is in contact with the anodes of the rectifiers.

Full-wave rectifiers have one slight disadvantage. The transformer must be center-tapped. This may not always be possible. In fact, there are occasions when the use of any transformer is not desirable because of size, weight, or cost restrictions. Figure 4-7(a) shows a rectifier circuit that gives full-wave performance without the transformer. It is

called a *bridge* rectifier. It uses four diodes to give full-wave rectification.

Figure 4-7(b) traces the circuit action for the positive alternation of the ac input. The current moves through D_2, through the load, through D_1, and back to the source. The negative alternation is shown in Fig. 4-7(c). The current is always moving from left to right through the load. Again, the positive end of the load is in contact with the rectifier cathodes. This circuit could be arranged for either ground polarity simply by choosing the left or the right end of the load as the common point.

Electronics:
Principles and
Applications
CHAPTER 4

Damping

Average value

Rms value

Review Questions

Supply the missing word in each statement.

11. A transformer secondary is center-tapped. If 80 V is developed across the secondary, the voltage from one end to the center tap should be ____?____ 40 V

12. Each cycle of the ac input has two ____?____. alter

13. In a rectifier circuit, the load current never changes ____?____ direction

14. A bridge rectifier can eliminate the need for a ____?____. transformer

4-4 CONVERSION OF RMS VALUES TO AVERAGE VALUES

There is quite a difference between pure dc and rectified ac. Meter readings taken in rectifier circuits can be confusing if the differences are not understood. Figure 4-8 compares a pure dc waveform with a pulsating dc waveform. A meter used to make measurements in a pure dc circuit can respond to the steady value of the dc. In the case of pulsating dc, the meter has a problem. At one instant in time, the meter tries to read zero. At another instant in time, the meter tries to read the peak value. Meter movements cannot react to the rapid changes because of the ~~damping~~ in their mechanism. Damping in a meter limits the speed with which the pointer can change position. Eventually, the meter will settle on the *average* value of the waveform.

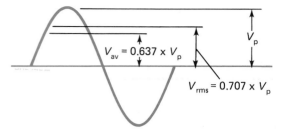

Fig. 4-9 Measuring sinusoidal ac.

Ac supply voltages are specified by their rms values. Thus it would be convenient to have a way of converting rms values to average values when working with rectifier circuits. Figure 4-9 relates three ways of measuring sinusoidal ac. An rms-to-average conversion factor can be obtained by a little manipulation of the formula from Fig. 4-9. Since $V_{av} = 0.637 \times V_p$ and $V_{rms} = 0.707 \times V_p$, we can solve for V_{av} in terms of V_{rms}:

$$V_{av} = \frac{0.637 \times V_{rms}}{0.707} = \boxed{0.9 \times V_{rms}}$$

Thus, the average value of a sine wave is 0.9, or 90 percent, of the rms value. This means that a dc voltmeter connected across the output of a rectifier should indicate 90 percent of the rms voltage input to the rectifier. Figure 4-10 shows one problem, however. Half-wave dc has a much lower average content than full-wave dc. Half of the original ac waveform is gone. So, for a half-wave rectifier, the average value of the waveform is 0.45, or 45 percent, of the rms value.

Figure 4-11 is the schematic for a half-wave

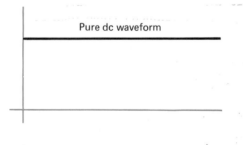

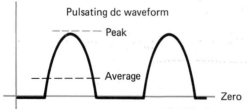

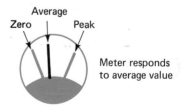

Fig. 4-8 Comparing dc waveforms.

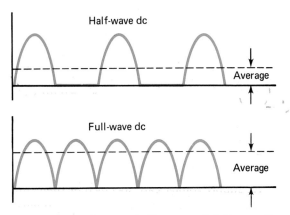

Fig. 4-10   Comparing half-wave and full-wave dc.

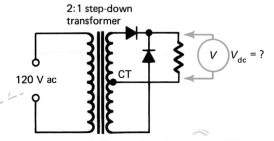

Fig. 4-12   Calculating dc voltage in a full-wave rectifier circuit.

rectifier. We wish to calculate the dc voltmeter reading. The first step is to take into account the transformer action. It is specified as a 10:1 step-down type. The voltage across the secondary should be

$$V_{secondary} = \frac{120}{10} = 12 \text{ V}$$

Now we can calculate the average dc voltage across the load. For half-wave rectifiers

$$V_{av} = V_{rms} \times 0.45 = 12 \times 0.45 = 5.4 \text{ V}$$

Thus, the meter in Fig. 4-11 can be expected to read 5.4 V.

If the circuit in Fig. 4-11 were constructed, how close could we expect the actual reading to be? The actual reading will be influenced by several factors: (1) the actual line voltage, (2) transformer winding tolerance, (3) meter accuracy, (4) rectifier loss, (5) transformer *IR* losses and core losses, and (6) line *IR* losses. The actual line voltage and the actual transformer secondary voltage can be accounted for by accurate measurements. The meter accuracy can be high with a quality meter that

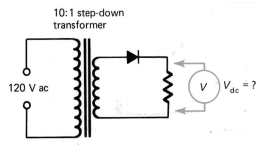

Fig. 4-11   Calculating dc voltage in a half-wave rectifier circuit.

has been checked against a standard. The rectifier loss is caused by the 0.6-V forward drop needed for conduction in a silicon diode. At high current levels, the drop will be greater. Line and transformer losses also increase at high current levels. Thus, *readings can be expected to be a little on the low side, especially at high load current levels*.

Figure 4-12 presents a full-wave rectifier circuit. Again, we wish to calculate the dc voltmeter reading. The step-down transformer has a 2:1 ratio. Thus we can expect 60 V across the secondary. *But what about the center tap?* A quick review of Fig. 4-5 will prove that only *half* the secondary is conducting at any given time. This means that the effective value of ac input to the rectifiers is *half* of 60, or 30 V. Now, we convert to average:

$$V_{av} = V_{rms} \times 0.90 = 30 \times 0.90 = 27 \text{ V}$$

The meter should read 27 V. If the load takes a high current, then the actual voltage might be less. What would happen if one of the diodes should "burn out" (open)? This would change the circuit from full-wave to half-wave. The dc voltmeter could then be expected to read

$$V_{av} = V_{rms} \times 0.45 = 30 \times 0.45 = 13.5 \text{ V}$$

Figure 4-13 shows a bridge rectifier circuit. The transformer will step down the line voltage to 30 V. The voltmeter should read

$$V_{av} = V_{rms} \times 0.90 = 27 \text{ V}$$

The diode loss in a bridge rectifier is twice that of the other circuits. A review of Fig. 4-7 will show that *two* diodes are always conducting in *series*. The 0.6-V drop will be doubled to 1.2 V. *In low-voltage rectifier circuits, this can be significant.*

35

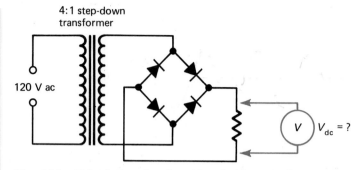

4:1 step-down
transformer

120 V ac

V $V_{dc} = ?$

Fig. 4-13   Calculating dc voltage in a bridge rectifier circuit.

Ac component

Ripple

Filter

Capacitive-input
filter

Review Questions

Supply the missing word in each statement.

15. A transformer has 10 times as many primary turns as secondary turns. If 120 V ac is across the primary, the secondary voltage should be ___?___ *12.v*

16. The ac input to a half-wave rectifier is 80 V. A dc voltmeter connected across the load should read ___?___ *36 V*

17. The ac input to a bridge rectifier is 200 V. A dc voltmeter connected across the load should read ___?___ *180 v*

18. In rectifier circuits, one can expect the output voltage to drop as load current ___?___ *increases*

19. The _Diode_ loss or drop can be significant in low-voltage rectifier circuits.

4-5 FILTERS

Half-wave or full-wave, pulsating dc is not directly usable in most electronic circuits. A purer and smoother dc is required. Batteries produce pure dc. But battery operation is usually limited to low-power and portable types of equipment. Figure 4-14 shows a battery connected to a load resistor. The voltage waveform across the load resistor is a straight line. There are no pulsations. Power supplies can also produce pure dc.

Pulsating dc is not pure because it contains an ac component. Figure 4-15 shows how both can appear across the same load. An ac generator and a battery are series-connected across a load. The voltage waveform across the load shows both ac and dc content. This situation is similar to the output of a rectifier. It is dc because of the rectification, but it still contains an ac component (the pulsations).

The ac component in a dc power supply is called *ripple*. Much of the ripple must be removed for most applications. The circuit used to remove the ripple is called a *filter*. Filters can produce a very smooth waveform that will approach the waveform produced by a battery.

Power-supply filters can be divided into two categories: (1) capacitive input and (2) choke, or inductive, input. Figure 4-16 shows a simple capacitive-input filter that has been added to a full-wave rectifier circuit. The voltage waveform across the load resistor shows that the ripple has been greatly reduced by the addition of the capacitor.

Capacitors are energy storage devices. They can take a charge and then later deliver that charge to a load. In Fig. 4-17(a) the rectifiers are producing peak output, load current is flowing, and the capacitor is charging. Later, when the rectifier output drops off, the capacitor takes over and furnishes the load current [Fig. 4-17(b)]. Since the current through the load has been maintained, the voltage across the load will be maintained

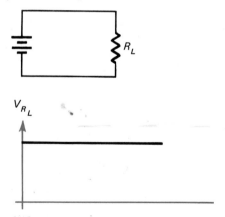

R_L

V_{R_L}

Fig. 4-14   Pure dc.

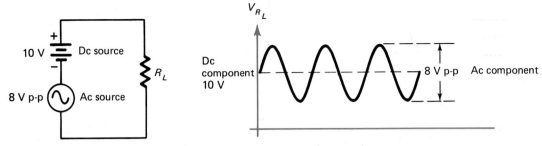

Fig. 4-15 An ac source in series with a battery. The voltage that results is ac and dc combined.

also. This is why the voltage waveform shows less ripple.

The effectiveness of a capacitive filter is determined by three factors:

1. The size of the capacitor
2. The value of the load
3. The time between pulsations

These three factors are related by the formula

$$T = R \times C$$

where T = time in seconds(s)
R = resistance in ohms(Ω)
C = capacitance in farads(F)

The product RC is called the *time constant* of the circuit. A charged capacitor will lose 63.2 percent of its voltage in T seconds. It takes approximately $5 \times T$ seconds to completely discharge the capacitor.

Hopefully, the capacitor will be only slightly discharged between peaks. This will mean a small voltage change across the load and thus little ripple. The time constant will have to be long when compared to the time between peaks. This makes it interesting to compare half-wave and full-wave filtering. The time between peaks for full-wave and half-wave rectifiers is shown in Fig. 4-18. Obviously, in a half-wave circuit the capacitor has more time to discharge, and the ripple will be greater. Full-wave rectifiers are desirable when most of the ripple must be removed. This is because it is easier to filter a wave whose peaks are closer together.

Figure 4-18 is based on the 60-hertz (Hz) power line frequency. By using a much higher frequency, the job of the filter could be made even easier. For example, if the fre-

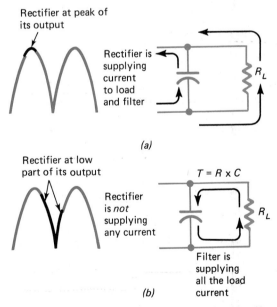

Fig. 4-17 Filter capacitor action. (*a*) The filter capacitor is being charged through the rectifier. (*b*) The filter capacitor is discharged through the load.

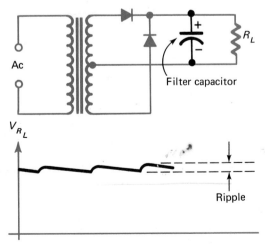

Fig. 4-16 A full-wave rectifier with a capacitive filter.

37

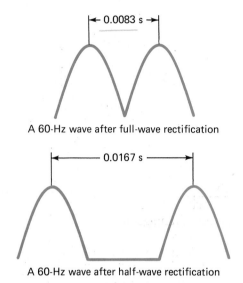

A 60-Hz wave after full-wave rectification

A 60-Hz wave after half-wave rectification

Fig. 4-18 A rectified 60-Hz wave.

quency were 1 kHz, the time between peaks in a full-wave rectifier output would be only 0.0005 s. In this short period of time, the filter capacitor would be only very slightly discharged. Another interesting point about high frequencies is that transformers can be made much smaller. Some equipment will convert the power line frequency to a much higher frequency to gain these advantages. Power supplies of this type are sometimes called *switching-mode supplies*.

Another way to get good filtering is to use a very high value of capacitance. This means that it will take longer for the capacitor to discharge. If the load resistance is low, the capacitance will have to be very high to give good filtering. Inspect the time constant formula, and you will see that if R is made lower, then C must be made higher if T is to remain the same. So, with heavy current demand (a low value of R), the capacitor value must be quite high.

Electrolytic capacitors are available with very high values of capacitance. However, a very high value in a capacitive-input filter can cause problems. Figure 4-19 shows waveforms that might be found in a capacitive-input power supply. The unfiltered waveform is shown in Fig. 4-19(a). In Fig. 4-19(b) the capacitor supplies energy between peaks. Note that the rectifiers do not conduct until their peak output exceeds the capacitive voltage. The rectifier turns off when the peak output passes. *The rectifiers are conducting for only a short time*. Figure 4-19(d) shows

the rectifier *current* waveform. Notice the high peak-to-average ratio.

In some power supplies, the peak-to-average current ratio in the rectifiers may exceed 100:1. This will cause the rms rectifier current to be greater than 8 times the current delivered to the load. The rms current determines the actual heating effect in the rectifiers. This is why diodes may have to be rated at 10 A when the power supply is designed to deliver only 2 A.

The dc output voltage of a filtered power supply may be higher than the output of a nonfiltered supply. Figure 4-20 shows a bridge rectifier circuit with a switchable filter capacitor. Before the switch is closed, the meter will read the average value of the waveform (ignoring diode loss):

$$V_{av} = 0.9 \times V_{rms}$$
$$= 0.9 \times 10$$
$$= 9 \text{ V}$$

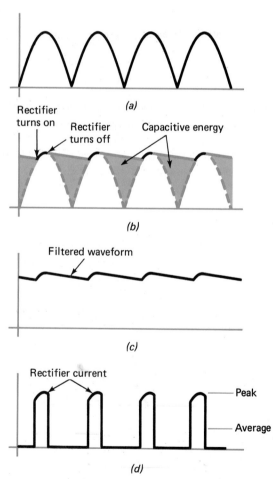

Fig. 4-19 Capacitor filter circuit waveforms.

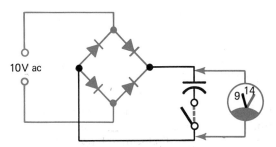

Fig. 4-20   Calculating dc with a capacitive filter.

After the switch is closed, the capacitor charges to the *peak* value of the waveform:

$$V_p = 1.414 \times V_{rms}$$
$$= 1.414 \times 10$$
$$= 14.14 \text{ V}$$

This represents a significant change in output voltage. However, as the supply is loaded, the capacitor may not be able to maintain the peak voltage, and the output tends to drop. The more heavily it is loaded (the more current there is), the lower the output voltage will be.

Choke-input power-supply filters are also common. They have some advantages over capacitive-input filters. Unfortunately, chokes tend to be heavy, large, and expensive. These drawbacks eliminate them from many designs. A choke-input filter is shown in Fig. 4-21. The filter capacitor is still included, but it is connected *after* the choke. Chokes are seldom used alone in power-supply filters.

The filter in Fig. 4-21 has several advantages over the simple capacitive filter. The choke and the capacitor work together to produce less ripple. Inductance is that circuit property opposing any change in current. Since the inductance (choke) is in series with the load, the load current is smoothed (pulsations are removed). Capacitance is that circuit property opposing any change in voltage. The capacitor is in parallel with the load and holds the load voltage constant. The combined ac-

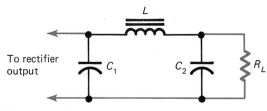

Fig. 4-22   Pi filter.

tion of the capacitor and inductor gives a smoother and purer dc than can be obtained with a single capacitor.

Another advantage of the choke-input filter is less heating in the rectifiers and transformer. With capacitive-input filters, the rectifier current takes the form of brief pulses that cause the rms currents to be high. With choke-input filters, the rectifiers conduct for a much longer time. This means that the diode ratings and perhaps the transformer ratings can be lower.

The third advantage of the choke-input filter is less change in output voltage with increased load current. In a simple capacitive filter, the output voltage will reach the peak value of the waveform with a light load. As the supply is loaded, the voltage tends to drop. With the choke-input filter, the capacitor cannot charge to the peak value of the waveform even when small load currents are present. This makes the output voltage lower, but *more stable*.

The power-supply filter shown in Fig. 4-22 is sometimes called a *pi filter*. This name comes from its schematic layout which looks like the Greek letter pi (π). The pi filter has the characteristics of capacitive-input filters: (1) a high output voltage that drops rapidly with increasing load and (2) a high rms rectifier current. The pi filter does an excellent job of reducing ac ripple. Its use is limited to high-power circuits.

The disadvantages of the pi filter can be eliminated by adding a second inductor, as in the circuit of Fig. 4-23. This filter will show a

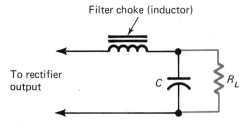

Fig. 4-21   Choke-input filter.

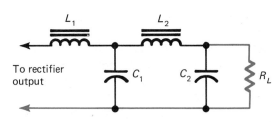

Fig. 4-23   A pi filter with two inductors.

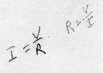

Choke-input filter

Pi filter

Voltage
multiplier

Full-wave
voltage doubler

lower but ~~more stable output voltage.~~ The
rectifier rms current will be much less because
it is a choke-input filter. Pi filters with two inductors are very heavy, large, and expensive.
Their use is also limited to high-power circuits.

Review Questions

Supply the missing word in each statement.

20. Pure dc contains no ____?____ *ripple*

21. Rectifiers provide *pulsating* dc.

22. A circuit added after the rectifier to pro-
    vide pure dc is the ___*filter*___.

23. Capacitors and inductors are useful in
    filter circuits because they store electric
    ___*energy*___.

24. In a power supply with a capacitor filter,
    the effectiveness of the filter is deter-
    mined by the size of the capacitor, the ac
    frequency, and the ___*load current*___

25. Half-wave rectifiers are more difficult to
    filter because the filter has more time to
    ___*discharge*___

26. Heating effect is related to the ___*R M S*___
    value of the current.

27. In a filtered power supply, the dc output
    voltage can be as high as ___*1.414*___ times
    the ac input voltage.

28. A choke-input filter will give ___*less*___
    output voltage than a capacitive-input
    filter.

29. A pi filter uses an inductor and two
    ____?____. *capacitors*

4-6 VOLTAGE MULTIPLIERS

The typical, general-purpose line voltage in
this country is about 115 to 120 V ac.
Usually, solid-state circuits require lower voltage for operation. Sometimes, higher voltages are required. One way to obtain a higher
voltage is to use a step-up transformer.
Unfortunately, transformers are expensive devices. They are also relatively large and
heavy. For these reasons, designers may not
want to use them to obtain high voltages.

Voltage multipliers can be used to produce
higher voltages without the need for a transformer. Figure 4-24(a) shows the diagram for
a full-wave voltage doubler. This circuit can
produce an output voltage as high as 2.8 times
the rms input voltage. The output will be a dc
voltage with some ripple.

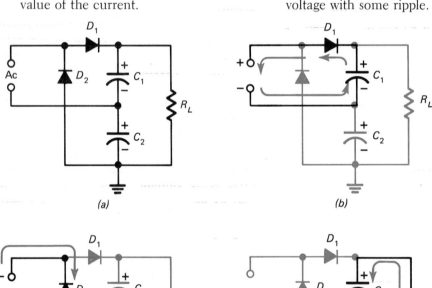

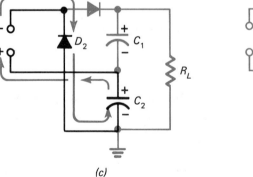

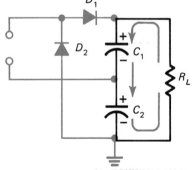

Fig. 4-24 Full-wave voltage doubler.

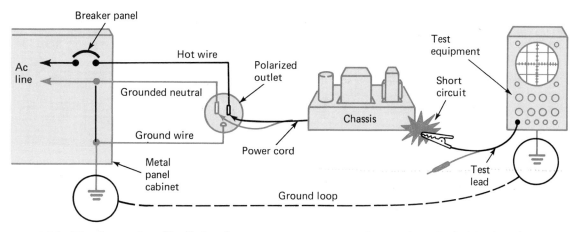

Fig. 4-25   The danger in a "hot" chassis.

Figure 4-24(*b*) shows the operation of the full-wave doubler. It shows that C_1 charges through D_1 when the ac line is on its positive alternation. Capacitor C_1 can be expected to charge to the peak value of the ac line. Assuming the input voltage is 120 V, we have

$$V_p = 1.414 \times V_{rms}$$
$$= 1.414 \times 120$$
$$= 169.68 \text{ V}$$

On the negative alternation of the ac line voltage [Fig. 4-24(*c*)], C_2 charges through D_2 to the peak value of 169.68 V. Now both C_1 and C_2 are charged. In Fig. 4-24(*d*) it can be seen that C_1 and C_2 are in series. Their polarities are series-aiding and they will produce double the peak line voltage across the load:

$$V_{R_L} = V_{C_1} + V_{C_2}$$
$$= 169.68 + 169.68$$
$$= 339.68 \text{ V}$$

Voltage doublers can come close to producing 3 times the line voltage. As they are loaded, their output voltage tends to drop rapidly. Thus, a voltage doubler energized by a 120-V ac line might produce a voltage near 240 V dc when it is delivering current to a load. A voltage multiplier is a poor choice when stable output voltages are required.

Lack of line isolation is the greatest problem with transformerless supplies. Most electronic equipment is fabricated on a metal framework or chassis. Often, this chassis is the common conductor for the various circuits. If the chassis is not isolated from the ac line, it can present an extreme shock hazard. The chassis is usually inside a nonconducting cabinet. The control knobs and shafts are made of nonconducting materials such as plastic. This gives some protection. However, a technician working on the equipment may be exposed to a shock hazard. Figure 4-25 shows a situation that has surprised more than one technician. Most test equipment is wired with a three-conductor power cord that automatically grounds its frame or case. If the test lead from the side of the test equipment touches a live, or "hot," chassis, then a ground loop results. The power cords, the chassis, and the test leads form a short circuit across the ac line. Connecting the test leads into parts of the circuitry may cause circuit damage. The technician may wind up in the loop and receive a very serious shock. Working on equipment that is not isolated from the ac line is dangerous.

Figure 4-26 shows a polarized power plug that keeps the chassis connected to the grounded neutral side of the ac line. However, some equipment and some buildings may be improperly wired so that the chassis

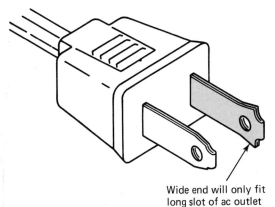

Wide end will only fit
long slot of ac outlet

Fig. 4-26   A polarized plug.

Isolation
transformer

Half-wave
voltage doubler

would still be hot. The only sure solution to the problem is to use an isolation transformer when working on electronic equipment. The transformer isolates all the circuitry, including the chassis, from the ac power line. With a good isolation transformer, there will be no chance for ground loops to occur.

The half-wave voltage doubler shown in Fig. 4-27(a) offers some improvement in safety over the full-wave voltage doubler. Compare Figs. 4-24(a) and 4-27(a). The chassis is always hot in the full-wave doubler. In the half-wave doubler, the chassis will be hot only if the connection to the ac outlet is wrong.

The half-wave doubler works a little differently from the full-wave doubler. On the negative alternation, C_1 will be charged [Fig. 4-27(b)]. Then, in Fig. 4-27(c), C_1 adds in series with the ac line's positive alternation, and C_2 will be charged to twice the peak line voltage. Load resistance R_L is in parallel with C_2 and will see a peak voltage of about 340 V with a line voltage of 120 V ac. The key differences are the capacitor voltage ratings and

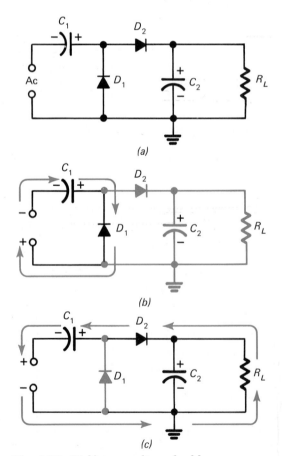

Fig. 4-27   Half-wave voltage doubler.

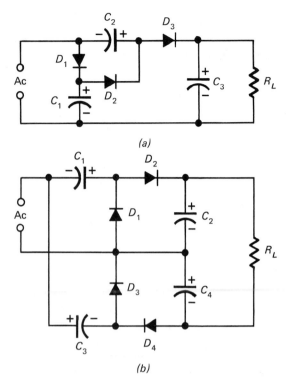

(a)

(b)

Fig. 4-28   Voltage multipliers.

the ripple frequency across the load. Full-wave doublers use two identical capacitors. Each would have to be rated at least equal to the peak line voltage. Half-wave doublers require the load capacitor to be rated at least equal to *twice* the peak line voltage. The ripple frequency in a full-wave doubler will be twice the line frequency. Half-wave doublers will show a ripple frequency equal to the line frequency.

It is possible to build voltage multipliers that triple, quadruple, and multiply even more. Figure 4-28(a) shows a voltage tripler. On the first negative alternation, C_1 is charged through D_1. On the next alternation, C_2 is charged to twice the peak line voltage through D_2 and C_1. Finally, C_3 is charged to 3 times the peak line voltage through D_3 and C_2. With 120-V ac input, the load would see a peak voltage of 509 V. A voltage quadrupler is shown in Fig. 4-28(b). This circuit is actually two half-wave doublers connected back to back and sharing a common input. The voltages across C_2 and C_4 will add to produce 4 times the peak ac line voltage. Assuming a line input of 120 V, R_L would see a peak voltage of 679 V.

All voltage multipliers tend to produce a high output voltage that drops quite a bit when the load is increased. This drop can be offset

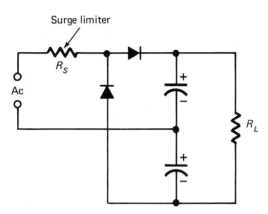

Fig. 4-29 A voltage multiplier with a surge-limiting resistor.

4-7 RIPPLE AND REGULATION

A power-supply filter reduces ripple to a low level. The actual effectiveness of the filter can be checked with a measurement and then a simple calculation. The formula for calculating the percentage of ripple is

$$\% \text{ Ripple} = \frac{ac}{dc} \times 100$$

For example, assume the ac ripple remaining after filtering is measured and found to be 1 V in a 20-V dc power supply. The percentage of ripple is

$$\% \text{ Ripple} = \frac{ac}{dc} \times 100$$
$$= \frac{1}{20} \times 100$$
$$= 5\%$$

Ripple should be measured *only when the supply is delivering its full rated output*. At zero load current, even a poor filter will reduce the ripple to almost zero. Ripple can be measured with an oscilloscope or a voltmeter. The oscilloscope will easily give the peak-to-peak value of the ac ripple. A meter will indicate the *rough*, or *approximate*, value of the rms ripple content. It will not be exact since the ripple waveform is nonsinusoidal. In a simple capacitive filter, the ripple is similar to a sawtooth waveform. This causes error with most meters since they are calibrated in rms values for *only pure sine waves*. There are meters that will read the true rms value of nonsinusoidal ac, but they are generally too expensive for most applications.

To measure the ac ripple riding on a dc waveform, the meter may have to be switched to a special function, or one of the test leads may have to be moved to a special jack. The special function or jack may be labeled *output*. The output jack is connected to the meter circuitry through a *coupling capacitor*. This capacitor is selected to have a low reactance at 60 Hz. Thus, 60- or 120-Hz ripple will reach the meter circuits with little loss. Capacitors have infinite reactance for dc (0 Hz). This means the dc content of the waveform will be blocked and will not interfere with the measurement. If an unusually high ripple content is measured, *the meter circuit should be checked to be certain the dc component is not upsetting the reading.*

somewhat by using large values of filter capacitors. Using large capacitors may cause the peak rectifier current to be very high. The surge is usually worst when the supply is first turned on. If the supply should happen to be switched just as the ac line is at its peak, the surge may damage the diodes. *Surge limiting* must be added to some multiplier circuits. A surge limiter is usually a low-value resistor connected into the circuit so that it can limit the surge current to a value safe for the rectifiers. Figure 4-29 shows a surge limiter in a full-wave doubler circuit.

Review Questions

Supply the missing word in each statement.

30. The typical ac line voltage in this country is ____?____ V.

31. Voltage doublers may be used to obtain higher voltages and eliminate the need for a(n) ____?____ .

32. A lightly loaded voltage doubler will give a dc output voltage that is ____?____ times the rms input.

33. Voltage doublers have poor voltage ____?____ .

34. To reduce shock hazard and equipment damage, a technician should use a(n) ____?____ transformer.

35. The ripple frequency in a 60-Hz half-wave doubler supply will be ____?____ Hz.

36. The ripple frequency in a 60-Hz full-wave doubler supply will be ____?____ Hz.

37. Voltage multipliers may use surge-limiting resistors to protect the ____?____ .

Percentage of
voltage
regulation

Bleeder resistor

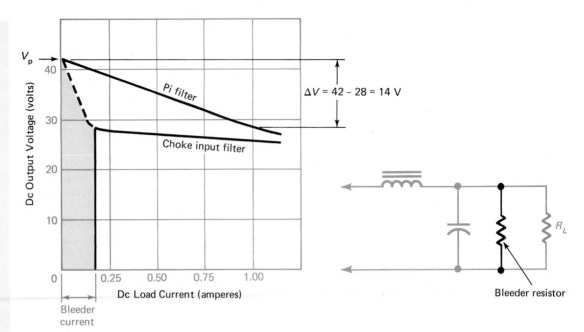

Fig. 4-30   Voltage regulation curves.

The regulation of a power supply means the ability to hold the output steady under conditions of changing input or changing load. As power supplies are loaded, the output voltage tends to drop to a lower value. The quality of the voltage regulation can be checked with two measurements and then a simple calculation. The formula for calculating the percentage of voltage regulation is

$$\% \text{ Regulation} = \frac{\Delta V}{V_{FL}} \times 100$$

where ΔV = *voltage change* from no load
                    to full load
      V_{FL} = output voltage at full load

For example, a power supply is checked with a dc voltmeter and shows an output of 14 V when no load current is supplied. When the supply is loaded to its rated maximum current, the meter reading drops to 12 V. The percentage of voltage regulation is

$$\begin{aligned}
\% \text{ Regulation} &= \frac{\Delta V}{V_{FL}} \times 100 \\
&= \frac{2}{12} \times 100 \\
&= 16.7\%
\end{aligned}$$

Power supplies with choke-input filters can show improved voltage regulation by keeping the supply loaded with a fixed resistor, called a *bleeder resistor,* across the output. The

bleeder resistor draws some current from the supply even if all external loads are removed. Thus, the supply is always providing some minimum current to the bleeder. Figure 4-30 shows voltage regulation curves for two power supplies. One of the supplies has a pi-type filter, and the other has a choke-input filter. The supply with the pi filter produces an output equal to the peak value of the ac input when no load current is flowing. As the load is increased to the maximum, the output drops to about 28 V. The voltage change ΔV for this supply is 14 V, and the percent voltage regulation is

$$\% \text{ Regulation} = \frac{14}{28} \times 100 = 50\%$$

The choke-input supply would also produce an output equal to the peak value of the ac input if its load current were permitted to drop to zero. By keeping the minimum load current at about 0.2 A, the output voltage will not go above about 28 V. The change will then be from 28 to about 25 V, and the percent voltage regulation will be

$$\% \text{ Regulation} = \frac{3}{25} \times 100 = 12\%$$

This is a much smaller percentage change. It is fair to say that the choke-input supply shows better voltage regulation than the pi-filter supply. Without the bleeder resistor, the choke-input supply would show very poor voltage regulation.

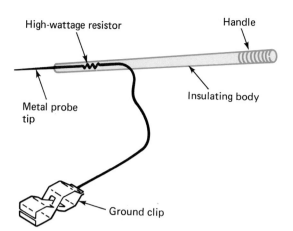

Fig. 4-31   A shorting rod.

Bleeder resistors perform another important function. They drain the filter capacitors after the power is turned off. Some filter capacitors can store a charge for months. They would present a shock hazard if they remained charged. *It is not safe to assume the capacitors have been drained even if there is a bleeder resistor across them.* Technicians who work on high-voltage supplies use a shorting rod or a shorting stick to be certain that all the filters are drained before working on the equipment. High-energy capacitors can discharge violently, so the shorting rod should contain a high-wattage resistor of around 50 Ω to keep the discharge current reasonable. Figure 4-31 shows such a device.

Review Questions

Supply the missing word in each statement.

38. As the load current increases, the ac ripple tends to ___?___ *increase*

39. As the load current increases, the dc output voltage tends to ___?___ *decrease*

40. A power supply develops 90 V dc with 5-V ac ripple. The percentage of ripple is ___?___. *5.56*

41. A power supply develops 100 V under no-load conditions and drops to 90 V when loaded. The percent regulation is ___?___. *11.11*

4-8 ZENER REGULATORS

Power-supply output voltage tends to change as the load on the supply changes. The output also tends to change as the ac input voltage changes. This can cause some electronic circuits to operate improperly. When a very stable voltage is required, the power supply must be *regulated*. The block diagram of a power supply (Fig. 4-32) shows where the regulator would be located in the system.

Regulators can be elaborate circuits using integrated circuits and transistors. For some applications, however, a simple Zener shunt regulator does the job (Fig. 4-33). The regulator is a Zener diode, and it is connected in shunt (parallel) with the load. If the voltage across the diode is constant, then the load voltage must also be constant.

The design of a shunt regulator using a Zener diode is based on a few simple calculations. For example, suppose a power supply develops 16 V and 12 V regulated is required for the load. A simple calculation tells us that we have to drop 4 V (16 − 12 = 4). This voltage will be dropped across R_z in Fig. 4-33. Assume the load current is 100 mA. Also assume that we want the Zener current to be 50 mA. Now we can calculate a value for R_z using Ohm's law:

$$R_z = \frac{V}{I}$$
$$= \frac{4\text{ V}}{0.100\text{ A} + 0.050\text{ A}}$$
$$= 26.67\ \Omega$$

The nearest standard value of resistor is 27 Ω, which is very close to the calculated value. The power dissipation in the resistor can be calculated:

$$P = V \times I$$
$$= 4 \times 0.150$$
$$= 0.6\text{ watt (W)}$$

We can use a 1-W resistor though a 2-W resistor may be required for greatest reliability.

Regulator

Shunt regulator

Zener diode

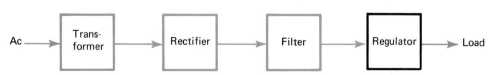

Ac — Transformer → Rectifier → Filter → Regulator → Load

Fig. 4-32   Location of a regulator in a power supply.

Zener diod

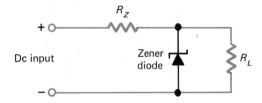

Fig. 4-33   Zener-diode shunt regulator.

Next, the power dissipation in the diode is

$$P = V \times I$$
$$= 12 \times 0.050 = 0.6 \text{ W}$$

A 1-W Zener diode would seem do the job. If the load were disconnected, however, the Zener would have to dissipate quite a bit more power. All the current (150 mA) would flow through the diode. The dissipation would be

$$P = V \times I$$
$$= 12 \times 0.150 = 1.8 \text{ W}$$

Obviously, a shunt regulator must be capable of handling more power if there is a possibility of the load being disconnected.

Another possibility is that the load might demand more current. Suppose that the load increases to 200 mA. Resistor R_Z would now drop more than 4 V:

$$V = I \times R$$
$$= 0.200 \times 27$$
$$= 5.4 \text{ V}$$

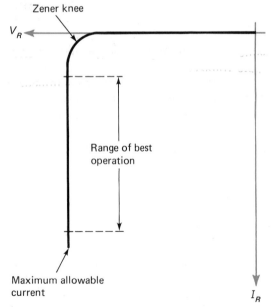

Fig. 4-34   Characteristic curve of a Zener-diode.

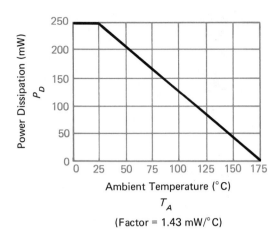

Fig. 4-35   Derating curve of a Zener diode.

This would cause a decrease in voltage across the load:

$$V_{\text{load}} = V_{\text{supply}} - V_{R_Z}$$
$$= 16 - 5.4$$
$$= 10.6 \text{ V}$$

The regulator is no longer working. Shunt regulators only work *up to the point where the device stops conducting.* Usually, the Zener current should not be allowed to become very low. As shown in Fig. 4-34, the region of the characteristic curve near the Zener knee shows poor regulation.

Operation at high temperatures creates another problem. The power rating of Zener diodes and other solid-state devices *must be decreased as the temperature goes up.* The temperature inside the cabinet of an electronic device might increase from 25 to 50°C after hours of continuous operation. This increase in temperature decreases the safe dissipation levels of the devices in the cabinet. Figure 4-35 shows a typical power derating curve for a Zener diode. Note that at a temperature of 175°C the allowable power dissipation is 0. Obviously, the device is not usable at this temperature.

Review Questions

Supply the missing word in each statement.

42. A Zener diode shunt regulator uses the Zener connected in ~~parallel~~ with the load.

43. A power supply develops 9 V. Regulated 5 V is required at a load current of 1 A. A Zener diode shunt regulator will be used.

The diode current must be at least 100 mA. The value of R_Z should be ____?____.

44. The minimum wattage rating for R_Z in question 43 should be ____?____.

45. If the load current were interrupted in question 43, the Zener would dissipate ____?____ W.

4-9 TROUBLESHOOTING

One of the major skills of an electronic technician is troubleshooting. The process involves the following steps:

1. Carefully observing the symptoms
2. Analyzing the possible causes
3. Limiting the possibilities by tests and measurements

Good troubleshooting is an orderly process. To help keep things in order, remember the word "GOAL." GOAL stands for Good, Observe, Analyze, and Limit.

Electronic equipment that is broken usually shows very definite symptoms. These symptoms are extremely important. Good technicians try to note all the symptoms before proceeding. This demands a knowledge of the equipment. You must know what the normal performance of a piece of equipment is in order to be able to identify what is abnormal. It is often necessary to make some adjustments or run some checks to be sure that the symptoms are clearly identified. For example, if a radio receiver has a hum or a whistle on one station, several other stations should be tuned in to determine if the symptom persists. Another example might involve checking the sound from a tape recorder with a tape recorded on another machine known to be working properly. These kinds of adjustments and checks will help the technician to properly observe the symptoms.

Analyzing possible causes comes after the symptoms are identified. This part of the troubleshooting process involves a general knowledge of the block diagram of the equipment. Certain symptoms are closely tied to certain blocks on the diagram. Experienced technicians "think" the block diagram. They do not have to have one in front of them. Their experience tells them how the major sections of the circuit work and what happens when one section is not working prop-

erly. For example, suppose a technician is troubleshooting a radio receiver. There is only one major symptom. There is no sound of any kind coming from the speaker. Experience and knowledge of the block diagram will tell the technician that two major parts of the circuit can cause this symptom: the power supply and/or the audio output section.

After the possibilities are established, it is time to limit them by tests and measurements. A few voltmeter checks generally will tell the technician if the power-supply voltages are correct. If they are not correct, then the technician must further limit the possibilities by making more checks. Circuit breakdown is usually limited to *one component*. Of course, one component failure may damage several others because of the way they are connected. A resistor that has *burned black* is almost always a sure sign that another part has shorted.

Power-supply troubleshooting follows the same general process. The symptoms that can be observed are

1. No output voltage
2. Low output voltage
3. Excessive ripple voltage
4. High output voltage

— *filter*

Note that the symptoms are all limited to voltages. This is the way technicians work. Voltages are easy to measure. Current analysis is rarely used because it is necessary to break into the circuit and insert the ammeter. It is also worth mentioning that two of the power-supply symptoms might appear at the same time: low output voltage and excessive ripple voltage.

Once the symptoms are clearly identified, it is time to analyze possible causes. For no output voltage, the possibilities include

1. Open fuse or circuit breaker
2. Defective switch, line cord, or outlet
3. Defective transformer
4. Open surge-limiting resistor
5. Open diode or diodes (rare)
6. Open filter choke or doubler capacitor

The last step is to limit the list of possibilities to one or two defects. This is accomplished by making some measurements. Figure 4-36 is the schematic diagram for a half-wave

Troubleshooting

GOAL

Symptoms

Power-supply troubleshooting

47

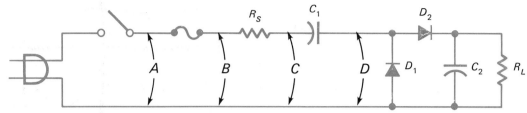

Fig. 4-36   A half-wave doubler schematic.

Electrolytic
capacitor

doubler power supply.   The technician can make ac voltage measurements as shown at A, B, C, and D to find the cause of no output voltage.   For example, suppose the measurement at A is 120 V ac but 0 V at B.   This indicates a blown fuse.   Suppose A and B show line voltage and C shows zero.   This would indicate an open surge-limiting resistor.   If measurements A, B, and C are 120 V ac and if measurement D is zero, then the capacitor C_1 is open.

Some defects show the need for more checking.   Again referring to Fig. 4-36, if the surge-limiting resistor R_S is open, it may be because another component is defective.   Simply replacing R_S may only cause it to burn out again.   It is a good idea to check the diodes and the capacitors when a surge limiter opens or a fuse blows.   One of the capacitors or diodes could be shorted.

Solid-state rectifier diodes usually do not *open* (show a very high resistance in both directions).   There are exceptions, of course.   They normally will show a shorted condition or excessive leakage.   Ohmmeter tests are normally good for checking diodes.   But this requires disconnecting at least one side of the diode.   Sometimes, it is possible to obtain a rough check with the diode still in the circuit.   *Always remove power before making ohmmeter tests and make sure the filter capacitors are discharged.*   Figure 4-37 is the schematic for a full-wave power supply.   An ohmmeter test across the diodes will show a low resistance when the diode is forward-biased and a higher resistance roughly equal to the load resistance

when the diode is reverse-biased.   This will prove that the diode is not shorted but could have excessive leakage.   The *sure* method is to remove one end of the diode from the circuit.   Bridge rectifier diodes can also be checked in circuit with similar results and limitations.

Many of the filter capacitors used in modern power supplies are of the electrolytic type.   These capacitors can short, develop leakage, and may lose much of or all their capacity.   They can be tested on a capacitor tester, or a rough check can be made with an ohmmeter.   *Observe polarity* when testing them.   A good electrolytic capacitor will show a momentary low resistance as it draws a charging current from the ohmmeter.   The larger the capacitor, the longer the low resistance will be shown.   After some time, the ohmmeter should show a high resistance.   It may not be infinite.   All electrolytic capacitors have some leakage, and it is more pronounced in the very high values.   A large capacitor may show a leakage resistance of 100,000 Ω.   Usually this is not significant in a power supply.   This same leakage in a smaller capacitor used elsewhere in an electronic circuit could cause trouble.

The symptom of low output voltage in a power supply could be caused by

1. Excessive load current
2. Low input (line) voltage
3. Defective surge-limiting resistor
4. Defective filter capacitors
5. Defective rectifiers

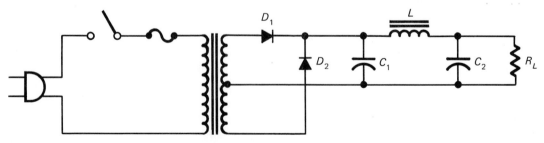

Fig. 4-37   A full-wave power-supply schematic.

Power supplies are part of an electronic system. Some part of the system can change and demand excess current from the power supply. This will cause the power-supply output voltage to drop. There may not be anything wrong with the supply itself. It is a good idea to first make sure that the current demand is normal when the power-supply output is low. This is one case where a current measurement may be required.

If the load is normal, then the supply itself must be checked. Some of the defects that might cause the half-wave doubler of Fig. 4-36 to produce low output voltage are

1. Resistor R_S increased in value.
2. Capacitor C_1 lost much of its capacity.
3. Capacitor C_2 lost much of its capacity.
4. Defective rectifiers.

Low output voltage may be accompanied by excessive ripple. For example, in Fig. 4-37 suppose C_1 is open. This changes the filter from a capacitive-input type to a choke-input type. This will cause a drop in the output voltage. It will also cause the ripple voltage to increase. Excessive loading on the supply will also increase the ripple. Again, a current measurement may be required.

Excessive ripple usually is caused by defective filter capacitors. If we assume the output voltage is normal, too much ripple usually means that one or more of the capacitors has lost its capacity. Some technicians use clip leads to connect a test capacitor in parallel with the one they suspect. This will restore the circuit to normal operation. Be *very careful* when making this kind of test. Remember, the supply can store quite a charge. Be sure to observe the correct polarity with the test capacitor. If the test proves the capacitor is defective, it should be removed from the circuit. It is poor practice to leave the original capacitor in the circuit with a new one soldered across it.

The last power-supply symptom is high output voltage. Usually this is caused by low load current. The trouble is not in the power supply but somewhere else in the circuit. It may be that a bleeder resistor is open. This decreases the load on the supply, and the output voltage goes up. High output in a regulated supply would indicate a defect in the regulator.

Review Questions

Determine whether each statement is true or false.

46. A skilled troubleshooter will use a trial-and-error technique to find circuit faults.

47. In troubleshooting, usually it is not possible to limit the problem to one area of the block diagram by observing the symptoms.

48. A burned-out resistor may indicate that another component in the circuit has failed.

49. High output current will usually produce the symptom of excessive output voltage in a power supply.

50. Refer to Fig. 4-29. Resistor R_S burns out (opens). The symptom will be zero output voltage.

51. Refer to Fig. 4-33. The Zener diode burns out. The symptom will be excessive output voltage.

52. Refer to Fig. 4-36. The fuse blows repeatedly. The rectifier D_1 is probably open.

53. Refer to Fig. 4-36. The output voltage is low. Capacitor C_1 could be defective.

54. Refer to Fig. 4-37. The output voltage is low, and there is too much ripple. Capacitor C_1 could be open.

55. Refer to Fig. 4-37. The output voltage is low, and there is too much ripple. The choke coil could be open.

4-10 REPLACEMENT PARTS

After the defective parts are located, it is time to choose replacement parts. Exact replacements are the safest choice. If exact replacements are not available, it may be possible to make substitutions. Parts that are substituted for the originals should have ratings *at least equal* to those of the originals. It would never do to replace a 2-W resistor with a 1-W resistor. The replacement resistor would probably fail in a short time. It may not be a good idea to replace a resistor with one having a higher power rating. In some circuits, the resistor may protect another more expensive part by increasing in value under overload conditions. Also, a fire hazard can result in some circuits if a carbon-composition resistor

Low output voltage

Excessive ripple

Exact replacement

Physical
characteristics

Substitution
guides

Registered EIA
number

House number

is substituted for a film resistor. It is easy to see why *exact* replacements are the safest.

Rectifier diodes have several important ratings. They are rated for *average current* and their capacity for *current peaks*. The current peaks can be much higher than the average current with capacitive-input filters. The maximum reverse-bias voltage that the diode can withstand is also an important rating. In a half-wave supply with a capacitive-input filter or in a full-wave supply with a center-tapped transformer, the diodes are subjected to a reverse voltage equal to 2 *times the peak value of the ac input*. In these supplies, the rectifier peak reverse voltage (PRV) rating will have to be greater.

Electrolytic capacitors are rated to a dc working voltage (Vdcw or dcwv or WVdc). This voltage must not be exceeded. In a capacitive-input filter, the capacitor can charge to the peak value of the rectified wave. Such a capacitor's dcWV rating must be at least equal to the peak voltage value.

The capacity of the electrolytic filters is also very important. Substituting a lower value may result in low output voltage and excessive ripple. Substituting a much higher value may cause excess peak rectifier current. A value close to the original is desirable.

Transformers and filter chokes may also have to be replaced. The replacements should have the same voltage ratings, the same current ratings, and the same taps.

Sometimes, the physical characteristics of the parts are just as important as the electrical characteristics. A replacement transformer may be too large to fit in the same place on the chassis, or the mounting bolt pattern may be different. A replacement filter capacitor may not fit in the space taken by the old one. The stud on a power rectifier may be too large for the hole in the heat sink. It pays to check into the mechanical details when choosing replacement parts.

Technicians use substitution guides to help them choose replacement parts. These are especially helpful for finding replacements for solid-state devices. The guides list many device numbers and the numbers for the replacement parts. The guides often include some of the ratings and physical characteristics for the replacement parts. Even though the guides are generally very good, at times the recommended part will not work properly. Some circuits are critical, and the recom-

mended replacement part may be just different enough to cause trouble. There may also be some physical differences between the original and the replacement recommended by the guide.

Solid-state devices have two types of part numbers. The first type is the registered EIA number. The letters EIA stand for the Electronic Industries Association. It is an association of manufacturers. When a solid-state device maufacturer uses a registered EIA part number, that device must conform to registered specifications. This means that a diode or a transistor could be purchased from any of several manufacturers and its registered number will guarantee its similarity to the original part. The EIA part numbers for solid-state devices have the prefix 1N, 2N, or 3N. Examples of EIA registered numbers for solid-state devices are 1N4002, 2N2712, and 3N128.

The second type of part number is the nonregistered or so-called "house number." These part numbers are "invented" by the manufacturers. They do not conform to any agreed-upon standards. They do not even indicate who the manufacturer was. Nonregistered device numbers can cause problems for technicians. Luckily, substitution guides include nonregistered as well as registered part numbers. *A good assortment of up-to-date substitution guides is a very valuable part of the technician's library.* Some examples of nonregistered part numbers for solid-state devices are MR1816, MCB5405F, CA200, and 2000287-28. There is no design or pattern to these numbers, and only through experience and searching can they be traced to the original manufacturer.

Review Questions

Determine whether each statement is true or false.

56. It is always safe to replace a 1-W resistor with a 2-W resistor.

57. It is always safe to replace a film resistor with a carbon-composition resistor.

58. It is always safe to replace a 1000-μF filter capacitor with a 2000-μF capacitor.

59. A transistor is marked 2N2712. This is a house number.

60. The safest replacement part is the exact replacement having the same ratings and physical characteristics.

61. Substitution guides have no value for electronics technicians.

$V = I \times l$

Summary

1. The power supply provides the various voltages for the circuits in an electronic system.

2. Power supplies may develop both polarities with respect to the chassis ground.

3. Diagrams that show the major sections of electronic systems and how they are related are called block diagrams.

4. Power supplies usually change voltage levels and rectify ac to dc.

5. In a diode rectifier circuit, the positive end of the load will be in contact with the cathode of the rectifier. The negative end of the load will be in contact with the anode of the rectifier.

6. A single diode forms a half-wave rectifier.

7. Half-wave rectification is generally limited to low-power applications.

8. A full-wave rectifier utilizes both alternations of the ac input.

9. One way to achieve full-wave rectification is to use a center-tapped transformer secondary and two diodes.

10. It is possible to achieve full-wave rectification without a transformer by using four diodes in a bridge circuit.

11. A dc voltmeter or a dc ammeter will read the average value of a pulsating waveform.

12. The average value of half-wave, pulsating dc is 45 percent of the rms value.

13. The average value of full-wave, pulsating dc is 90 percent of the rms value.

14. Pulsating dc contains an ac component called ripple.

15. Ripple can be reduced in a power supply by adding filter circuits after the rectifiers.

16. Filters can be classified as capacitor input or choke input.

17. Capacitor-input filters give a high output voltage which drops as the load is increased.

18. Capacitor-input filters produce more heating in the rectifiers.

19. Choke-input filters reduce rectifier heating and give a more stable output voltage.

20. Full-wave rectifiers are easier to filter than half-wave rectifiers.

21. Line-operated equipment should always be operated with an isolation transformer to protect the technician and the equipment being serviced.

22. A surge-limiting resistor may be included in power supplies to protect the rectifiers from damaging current peaks.

23. Ripple should be measured when the power supply is delivering its rated full-load current.

24. Ripple may be nonsinusoidal.

25. The percent regulation is a comparison of the no-load voltage and the full-load voltage.

26. Bleeder resistors can improve voltage regulation and drain the filter capacitors when the supply is off.

27. A voltage regulator can be added to a power supply to keep the output voltage constant.

28. Zener diodes are useful as shunt regulators.

29. Limiting the possible causes to one or two defects usually involves making tests with meters and other equipment. The schematic diagram is very helpful in this phase of the troubleshooting process.

30. Defects may be in groups. One part shorting out could damage several others.

31. In troubleshooting power supplies, no output voltage usually is caused by *open* components.

32. Open components can be isolated by voltage measurements or resistance checks *with the circuit turned off*.

33. Electrolytic capacitors can short, develop excess leakage, or open (lose much of their capacity).

34. Power-supply voltages are affected by load current.

35. Excessive ripple is usually caused by defective filter capacitors.

36. Maximum ratings of parts must never be exceeded. A substitute part should be at least equal to the original.

37. Substitution guides are very helpful in choosing replacement parts.

Determine whether each statement is true or false.

F 4-1. A schematic shows the major sections of an electronic system in block form.

T 4-2. In troubleshooting, one of the first checks that should be made is power-supply voltages.

T 4-3. Rectification is changing ac to dc.

T 4-4. Diodes make good rectifiers.

T 4-5. A transformer has 120 V ac across its primary and 40 V ac across its secondary. It is a step-down transformer.

F 4-6. The positive end of the load will be in contact with the anode of the rectifier.

F 4-7 A single diode can give full-wave rectification.

T 4-8. Half-wave rectifiers are limited to low-power applications.

T 4-9. A full-wave rectifier uses two diodes and a center-tapped transformer.

T 4-10. A bridge rectifier can eliminate the need for the center-tapped transformer.

T 4-11. A bridge rectifier uses four diodes.

F 4-12. The average value of a sine wave is 0.637 times its rms value.

F 4-13. With pulsating dc, a dc voltmeter will read the rms value of the waveform.

F 4-14. The ac input to a half-wave rectifier is 100 V. A dc voltmeter connected across the load should read 90 V.

T 4-15. Increasing the load current taken from a power supply will tend to make the output voltage drop.

F 4-16. Diode losses can always be ignored when they are used as rectifiers.

T 4-17. The input to a half-wave power supply is 50 V. A filter capacitor is connected across the output. The dc voltage can be as high as 70 V if the load current is small.

T 4-18. A filter capacitor loses much of its capacity. The symptoms could be excess ripple and low output voltage.

T 4-19. Capacitive-input filters increase the heating effect in the rectifiers.

F 4-20. A pi filter uses two chokes and a capacitor.

T 4-21. Choke-input filters give better voltage regulation than capacitor-input filters.

T 4-22. Pure dc means that no ac ripple is present.

T 4-23. A lightly loaded voltage doubler may give a dc output voltage near 2.8 times the ac input voltage.

F 4-24. An isolation transformer eliminates all shock hazards for an electronics technician.

T 4-25. The ripple frequency for a full-wave doubler will be twice the ac line frequency.

F 4-26. A 12-V dc power supply shows 2 V of ac ripple. The ripple percentage is 9.2.

T 4-27. From no load to full load, the output of a supply drops from 40 to 38 V. The regulation is 5.26 percent.

F 4-28. Ac ripple can be measured with a dc voltmeter.

F 4-29. It is not necessary to load a power supply to measure its ripple and regulation.

F 4-30. The main function of a bleeder resistor is to protect the rectifiers from surges of current.

F 4-31. A Zener diode shunt regulator is generally used to filter out ac ripple.

F 4-32. The dissipation in a shunt regulator goes down as the load current goes down.

F 4-33 A power supply blows fuses. The trouble could be an open filter capacitor.

T 4-34. A power supply develops too much output voltage. The problem might be low load current.

T 4-35. A burned-out surge resistor is found in a voltage doubler circuit. It might be a good idea to check the diodes and filter capacitors before replacing the resistor.

T 4-36. A shorted capacitor can be found with an ohmmeter check.

T 4-37. A shorted diode can be found with an ohmmeter check.

F 4-38. There is no way to locate data on parts using house numbers.

Answers to Review Questions

1. Dc
2. Ground (common)
3. Negative
4. Power supply
5. T
6. T
7. F
8. T
9. T
10. F
11. 40
12. Alternations
13. Direction
14. Transformer
15. 12 V ac
16. 36 V
17. 180 V
18. Increases
19. Diode
20. Ripple (ac)
21. Pulsating
22. Filter
23. Energy
24. Load current
25. Discharge
26. Rms
27. 1.414
28. Less
29. Capacitors
30. 120
31. Transformer
32. 2.8
33. Regulation
34. Isolation
35. 60
36. 120
37. Diodes
38. Increase
39. Decrease
40. 5.56
41. 11.11
42. Parallel
43. 3.64 Ω
44. 10 W (*must at least double 4.4 for safety*)
45. 5.5 W
46. F
47. F
48. T
49. F
50. T
51. T
52. F
53. T
54. T
55. F
56. F
57. F
58. F
59. F
60. T
61. F

Junction Transistors

■ This chapter introduces the transistor. Transistors are solid-state devices similar in some ways to the diodes you have studied. Transistors are more complex and can be used in many more ways. They are very important and can be found in almost all modern electronic equipment. The most important feature of transistors is their ability to amplify signals. Amplification can make a weak signal strong enough to be useful in some electronic applications. For example, an audio amplifier can be used to supply a strong signal to a loudspeaker.

5-1 AMPLIFICATION

Amplification is one of the most basic ideas in electronics. Amplifiers make sounds louder and signal levels greater and, in general, provide a function called *gain*. Figure 5-1 shows the general function of an amplifier. Note that the amplifier must be provided with two inputs: a dc power supply and the input signal. The signal is the electrical quantity that is too small in its present form to be usable. With gain, it can become usable. As shown in Fig. 5-1, the output signal is greater because of the gain of the amplifier.

Gain can be measured in several ways. If an oscilloscope is used to measure the amplifier input voltage and output voltage (not the power supply), then the *voltage gain* can be observed. A certain amplifier may provide an output voltage that is 10 times greater than the input voltage. The voltage gain of the ampli-

fier is 10. If an ammeter is used to measure amplifier input and output currents, the *current gain* can be obtained. With a 0.1-A input signal, an amplifier might produce a 0.5-A output signal for a current gain of 5. If the voltage gain and the current gain are known, the *power gain* can be established. An amplifier that produces a voltage gain of 10 and a current gain of 5 will give the following power gain:

$$P = V \times I$$

or

$$
\begin{aligned}
P_{\text{gain}} &= V_{\text{gain}} \times I_{\text{gain}} \\
&= 10 \times 5 \\
&= 50
\end{aligned}
$$

Only amplifiers provide a power gain. Other devices might give a voltage gain or a current gain. A step-up transformer provides voltage gain. Why can it *not* be considered an amplifier? The answer is that it does not provide any power gain. If the transformer steps *up* the voltage 10 times, then it steps *down* the current 10 times. The power gain, ignoring loss in the transformer, will be

$$
\begin{aligned}
P_{\text{gain}} &= V_{\text{gain}} \times I_{\text{gain}} \\
&= 10 \times 0.1 \\
&= 1
\end{aligned}
$$

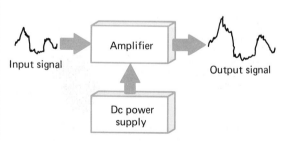

Input signal | Amplifier | Output signal

Dc power supply

Fig. 5-1 Amplifiers provide gain.

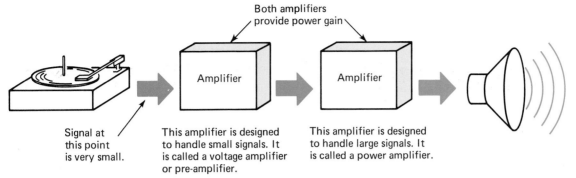

Fig. 5-2 Small-signal and large-signal amplifiers.

A step-down transformer provides a current gain. It *cannot* be considered an amplifier. The current gain is offset by a voltage loss, and, thus there is no power gain.

Even though power gain seems to be the important idea, some amplifiers are classified as voltage amplifiers. In some circuits, only the voltage gain is mentioned. This is especially true in amplifiers designed to handle very small electric signals. You will run across many voltage amplifiers or small-signal amplifiers in electronic devices. You should remember that they provide power gain, too. Amplifiers designed to handle *large signals* are usually called *power amplifiers*. In the electronic system of Fig. 5-2, the speaker requires several watts for good volume. The signal from the pick-up arm is a fraction of a milliwatt (mW). A total power gain of thousands is needed. However, only the final large-signal amplifier is called a power amplifier.

Review Questions

Determine whether each statement is true or false.

1. An amplifier needs an input signal and a power supply to develop an output signal.

2. An amplifier has a voltage gain of 50. If the input signal is 1 mV, the output signal should be 50 mV.

3. The input signal to an amplifier is 1 mA. The output signal is 10 mA. The amplifier has a current gain of 10 W.

4. A step-up transformer is a true voltage amplifier.

5. A step-down transformer is a true current amplifier.

5-2 TRANSISTORS

Transistors make good amplifiers. They can provide the power gain that is needed. There are several important types of transistors. This chapter will be mainly concerned with the *bipolar junction transistor.*

Bipolar junction transistors are similar to junction diodes, but one more junction is included. Figure 5-3 shows one way to make a transistor. A P-type semiconductor region is located between two N-type regions. The polarity of these regions is controlled by the valence of the materials used in the doping process. If you have forgotten this process and how it works, review the information in Chap. 2.

The transistor regions shown in Fig. 5-3 are named emitter, base, and collector. The emitter is very rich in current carriers. Its job is to send its carriers into the base region and then on to the collector. The collector *collects* the carriers. The emitter *emits* the carriers. The base acts as the control region. As will be seen later, the base can allow few or many carriers to flow from the emitter to the collector.

The transistor of Fig. 5-3 is *bipolar* because both holes and electrons will take part in the current flow through the device. The N-type

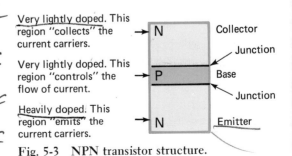

Very lightly doped. This region "collects" the current carriers.

Very lightly doped. This region "controls" the flow of current.

Heavily doped. This region "emits" the current carriers.

Fig. 5-3 NPN transistor structure.

NPN transistor

PNP transistor

Base-emitter
junction

Collector-base
junction

regions contain free electrons which are negative carriers. The P-type region contains free holes which are positive carriers. Two (*bi*) polarities of carriers are present. Note that there are also two PN junctions in the transistor. It is a bipolar *junction* transistor.

The transistor shown in Fig. 5-3 would be classified as an NPN transistor. Another way to make a bipolar junction transistor is to make the emitter and collector of P-type material and the base of N-type material. This type would be classified as a PNP transistor. Figure 5-4 shows both possibilities and the schematic symbols for each. You should memorize the symbols. Remember that the emitter lead is always the one with the arrow. Also remember that if the arrow is Not Pointing iN, the transistor is an NPN type.

The two transistor junctions must be biased properly. This is why you cannot replace an NPN transistor with a PNP transistor. The polarities would be wrong. Transistor bias is shown in Fig. 5-5. The collector-base junction must be reverse-biased for proper operation. In an NPN transistor, the collector will have to be *positive* with respect to the base. In a PNP transistor, the collector will have to be *negative* with respect to the base. PNP and NPN transistors are *not* interchangeable.

The base-emitter junction must be forward-biased, as shown in Fig. 5-5. This makes the resistance of the base-emitter junction very low as compared with the resistance

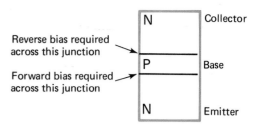

Fig. 5-5 Biasing the transistor junctions.

of the collector-base junction. A forward-biased semiconductor junction has low resistance. A reverse-biased junction has high resistance. Figure 5-6 compares the two junction resistances.

The large difference in junction resistance makes the transistor capable of power gain. Assume that a current is flowing through the two resistances shown in Fig. 5-6. Power can be calculated by

$$P = I^2 \times R$$

The power gain from R_{BE} to R_{CB} could be established by calculating the power in each and dividing:

$$P_{gain} = \frac{I^2 \times R_{CB}}{I^2 \times R_{BE}}$$

If the current through R_{CB} were equal to the current through R_{BE}, I^2 would be canceled out and the power gain would be

$$P_{gain} = \frac{R_{CB}}{R_{BE}}$$

Actually, the currents are not equal in transistors, but *they are very close*. Thus, we are making only a small error. A typical resistance value for R_{CB} might be 10,000 Ω. It is high since the collector-base junction is reverse-biased. A typical value for R_{BE} might be 100 Ω. It is low because the base-emitter

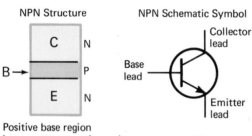

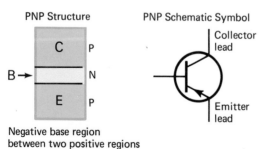

Fig. 5-4 Transistor structures and symbols.

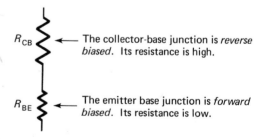

Fig. 5-6 Comparing junction resistances.

junction is forward-biased. The power gain for this typical transistor would be

$$P_{gain} = \frac{R_{CB}}{R_{BE}} = \frac{10,000}{100} = 100$$

Perhaps the biggest puzzle is why the current through the reverse-biased junction is as high as the current through the forward-biased junction. Diode theory tells us to expect *almost no current* through a reverse-biased junction. This is true in a diode but not true in the collector-base junction of a transistor.

Figure 5-7 shows why the collector-base current is high. The collector-base voltage V_{CB} produces a reverse bias across the collector-base junction. The base-emitter voltage V_{BE} produces a forward bias across the base-emitter junction. If the transistor were simply two diode junctions, the results would be

- I_B and I_E would be high
- I_C would be zero

The base region of the transistor is very narrow (about 0.0025 cm, or 0.001 in). The base region is lightly doped. It has only a few free holes. *It is not likely that an electron coming from the emitter will find a hole in the base with which to combine.* With so few electron-hole combinations in the base region, *the base current is very low.* The collector is an N-type region but is charged positively by V_{CB}. Since the base is such a narrow region, the positive field of the collector is quite strong and *the great majority of the electrons coming from the emitter are attracted and collected by the collector.* Thus

- I_E and I_C are high
- I_B is low

Low in resistance

The emitter current of Fig. 5-7 is the highest current in the circuit. The collector current is just a bit less. Typically, about 99 percent of the emitter carriers go on to the collector. About 1 percent of the emitter carriers combine with carriers in the base and

Base current

Collector current

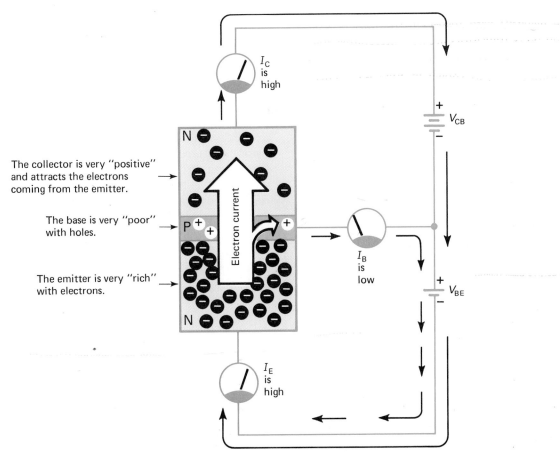

Fig. 5-7 NPN transistor currents.

The collector is very "positive" and attracts the electrons coming from the emitter.

The base is very "poor" with holes.

The emitter is very "rich" with electrons.

β (beta)

h_{fe}

Hole current

become base current. The current equation for Fig. 5-7 is

$$I_E = I_C + I_B$$

By using typical percentages it can be stated as

$$100\% = 99\% + 1\%$$

The base current is quite small but very important. Suppose, for example, that the base lead of the transistor in Fig. 5-7 is opened. With the lead open, there can be no base current. The two voltages V_{CB} and V_{BE} would add in series to make the collector positive with respect to the emitter. You might guess that current would continue to flow from the emitter to the collector, *but it does not*. With no base current, there will be no emitter current and no collector current. The base-emitter junction must be forward-biased for the emitter to emit. Opening the base lead removes this forward bias. If the emitter is not emitting, there is nothing for the collector to collect. Even though the base current is very low, it must be present for the transistor to conduct from emitter to collector.

The fact that a low base current controls much higher currents in the emitter and collector is very important. This shows how the transistor is capable of good *current gain*. Quite often, the current gain from the base terminal to the collector terminal will be specified. This is one of the most important transistor characteristics. This characteristic is called β (beta), or h_{fe}:

$$\beta = \frac{I_C}{I_B} \quad \text{or} \quad h_{fe} = \frac{I_C}{I_B}$$

What is the β of a typical transistor? If the base current is 1 percent and the collector current is 99 percent, then

$$\beta = \frac{99\%}{1\%} = 99$$

Note that the percent symbol cancels since it appears in both the numerator and the denominator. This is also the case if actual current readings are used. The unit of current will cancel, leaving β as a pure number. It does not have a dimension of percentage or current.

The β of actual transistors varies greatly.

Certain power transistors can have a β as low as 20. Small-signal transistors can have a β as high as 400. A good average value for β might be considered 100.

The value of β varies among transistors with the same part number. A 2N2712 is a registered transistor. One manufacturer of this particular device lists a typical β range of 80 to 300. Thus, if three seemingly identical 2N2712 transistors are checked for β, values of 98, 137, and 267 might be obtained.

The value of β is important but unpredictable. Luckily, there are ways to use transistors that make the actual value of β less important than other, more predictable circuit characteristics. This will become clear in a later chapter. For now, concentrate on the idea that the current gain from the base terminal to the collector terminal tends to be high. Also, remember that the base current *controls* the collector current.

Figure 5-8 shows what happens in a PNP transistor. Again, the base-emitter junction must be forward-biased. Note that V_{BE} is reversed in polarity when compared to Fig. 5-7. The collector-base junction of the PNP transistor must be reverse-biased. Note also that V_{CB} has been reversed in polarity. This is why PNP and NPN transistors are *not* interchangeable. If one were substituted for the other, both the collector-base and the base-emitter junctions would be biased incorrectly.

Figure 5-8 shows the current from emitter to collector as *hole current*. In an NPN transistor, it is electron current. The two transistor structures operate about the same in most ways. The emitter is very rich with carriers. The base is quite narrow and has only a few carriers. The collector is charged by the external bias source and attracts the carriers coming from the emitter. The major difference between PNP and NPN transistors is *polarity*.

The NPN transistor has become more popular than the PNP transistor. Electrons have better *mobility* than holes; that is, they can move more quickly through the crystal structure. This gives NPN transistors an advantage in high-frequency circuits where things have to happen quickly. Transistor manufacturers have more NPN types in their line. This makes it easier for circuit designers to choose the exact characteristics they need from the NPN group. Finally, it is often

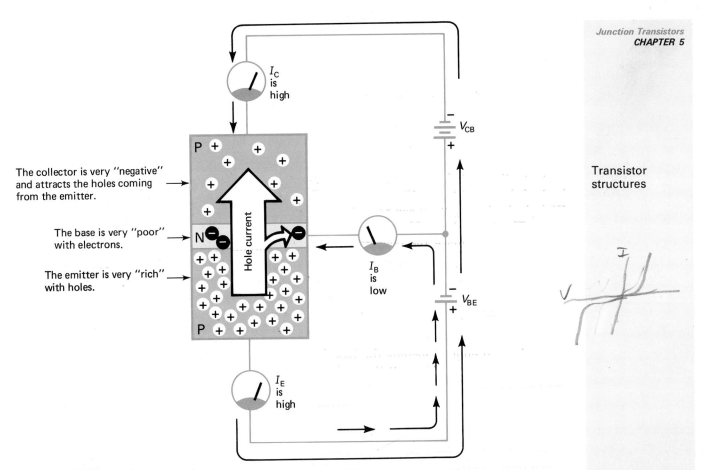

Fig. 5-8   PNP transistor currents.

The collector is very "negative" and attracts the holes coming from the emitter.

The base is very "poor" with electrons.

The emitter is very "rich" with holes.

Transistor structures

more convenient to use NPN devices in negative ground systems. Negative ground systems are very popular.

You will find both types of transistor in use. Many devices use both PNP and NPN transistors in the same circuit. It is very convenient to have both polarities available. This adds flexibility to circuit design.

Junction transistors are manufactured in many ways. Figure 5-9 shows the structure of three different transistors. These three ways are by no means the only processes used. They merely show that the physical details of transistors vary quite a bit from one manufacturing process to another.

Review Questions

Determine whether each statement is true or false.

6. The emitter region of a junction transistor has many current carriers.

7. The term "bipolar" indicates that only electrons are used as current carriers.

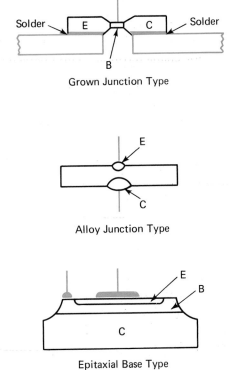

Grown Junction Type

Alloy Junction Type

Epitaxial Base Type

Fig. 5-9   Three transistor types.

59

8. The collector-base junction must be reverse-biased for proper transistor action. ⊤

9. An NPN transistor and a PNP transistor can be interchanged. ⊨

10. Even though the collector-base junction is reverse-biased, considerable current can flow in this part of the circuit. ⊤

11. The base of junction transistors is heavily doped with impurities. ⊬

12. When I_B is equal to zero, I_C will also be equal to zero. ⊤

13. Base current controls collector current. ⊤

14. Base current is much less than emitter current. ⊤

15. Transistor beta is measured in milliamperes. ⊨

16. All 2N2712 transistors have a current gain of 80 from the base to the collector. ⊨

17. In a PNP transistor, the emitter emits holes and the collector collects them. ⊤

18. Most circuits use PNP transistors. ⊨

5-3 CHARACTERISTIC CURVES

As with diodes, transistor characteristic curves can provide much information. There are many types of transistor characteristic curves. One of the more popular types is the *collector family* of curves. An example of this type is shown in Fig. 5-10. The vertical axis shows collector current (I_C) and is calibrated in milliamperes. The horizontal axis shows collector-emitter bias (V_{CE}) and is calibrated in volts. Figure 5-10 is called a *collector family* since several volt-ampere characteristic curves for the collector are presented.

Figure 5-11 shows a circuit that can be used to find the data points for a collector family of curves. Three meters are used to monitor base current I_B, collector current I_C, and collector-emitter voltage (V_{CE}). To develop a graph of three values, one value can be held constant as the other two vary. This produces one curve. Then, the constant value is set to a new level. Again, the other two values are changed and recorded. This produces the second curve. The process can be repeated as many times as required. For a collector family of curves, the constant value is the base current. The variable resistor in Fig. 5-11 is adjusted to produce the desired level of base current. Then, the adjustable source will be set to some value of V_{CE}. The collector current is recorded. Next, V_{CE} is changed to a new value. Again, I_C is recorded. These data points are plotted on a graph to produce a volt-ampere characteristic curve of I_C versus

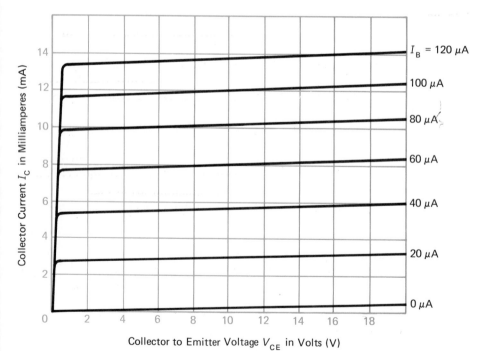

Fig. 5-10   A collector family of curves for an **NPN** transistor.

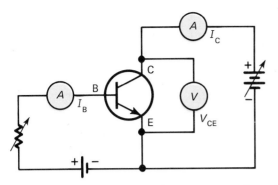

Fig. 5-11 Circuit for collecting transistor data.

values intersect at a base current of 60 μA. Now, β can be calculated:

$$\beta = \frac{I_C}{I_B} = \frac{8 \text{ mA}}{60 \text{ } \mu\text{A}}$$

$$= \frac{8 \times 10^{-3}}{60 \times 10^{-6}} = 133$$

Calculate β for the conditions of $V_{CE} = 16$ V and $I_C = 14$ mA. These values intersect at $I_B = 120$ μA. To find β,

$$\beta = \frac{I_C}{I_B} = \frac{14 \text{ mA}}{120 \text{ } \mu\text{A}}$$

$$= \frac{14 \times 10^{-3}}{120 \times 10^{-6}} = 117$$

These calculations reveal another fact about transistors. Not only does β vary from transistor to transistor; but it also varies with I_C and V_{CE}. Later, it will be shown that temperature also affects β.

It is standard practice to plot positive values to the right on the horizontal axis and up on the vertical axis. Negative values go to the left and down. A family of curves for a PNP transistor may be plotted on a graph as shown in Fig. 5-12. The collector voltage must be negative in a PNP transistor. Thus the curves go to the left. The collector current is in the opposite direction, compared to an NPN transistor. Thus the curves go down. Curves for a PNP transistor are often drawn up and to the right. Either method is equally useful for presenting the collector characteristics.

Some shops and laboratories are equipped with a device called a *curve tracer*. This device draws the characteristic curves on a cathode-ray tube or picture tube. This is far more convenient than collecting many data points and plotting the curves by hand. Curve tracers often show NPN curves in the first quadrant (as in Fig. 5-10) and PNP curves in the third quadrant (as in Fig. 5-12).

The collector family of curves can be used to show the safe operating area for a transistor. Figure 5-13 shows an example of this. A constant power curve has been added to the graph that clearly divides the curves into those operating points below 7.5 W and those above 7.5 W. This makes it very easy to find safe areas of operation for the transistor. For example, if the maximum safe transistor dissipation is 7.5 W, then no operation to the right of the power curve would be safe.

Base current

Collector current

Curve tracer

Constant power curve

V_{CE}. A very accurate curve can be produced by recording many data points. The next curve in the family is produced in exactly the same way but at a new level of base current.

The curves of Fig. 5-10 show some of the important characteristics of junction transistors. Notice that over most of the graph the collector-emitter voltage has very little effect on the collector current. Examine the curve for $I_B = 20$ μA. How much change in collector current can you see over the range from 2 to 18 V? This is a ninefold increase in voltage. Ohm's law tells us to expect the current to increase 9 times. It would increase if the transistor were a simple resistor. In a transistor, the base current has the major effect on collector current. If the collector voltage is very low, it can affect the collector current.

It is important to be able to convert the curves back into data points. For example, can you read the value of I_C when $V_{CE} = 10$ V and $I_B = 20$ μA? Refer to Fig. 5-10. First, locate 10 V on the horizontal axis. Project up from this point until you reach the 20-μA curve. Now, project from this point to the left and read the value of I_C on the vertical axis. You should obtain a value of 3 mA. Try another: find the value of I_B when $I_C = 10$ mA and $V_{CE} = 4$ V. These two data points cross on the 80-μA curve. The answer is 80 μA. It may be necessary to estimate a value. For example, what is the value of base current when $V_{CE} = 2$ V and $I_C = 4$ mA? The crossing of these two values occurs well away from any of the curves in the family. It is about halfway between the 20-μA curve and the 40-μA curve, so 30 μA is a good estimate.

The curves of Fig. 5-10 give enough information to calculate β. What is the value of β at $V_{CE} = 8$ V and $I_C = 8$ mA? The first step is to find the value of the base current. The two

Transistor
dissipation

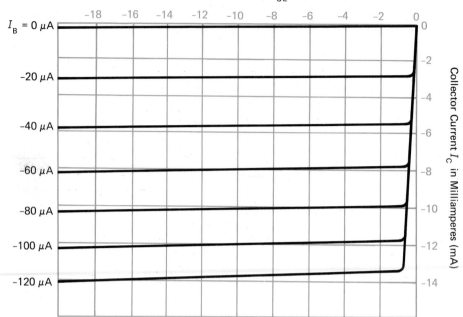

Collector to Emitter Voltage V_{CE} in Volts (V)

Fig. 5-12   A collector family of curves for a PNP transistor.

Transistor dissipation is usually calculated for the collector circuit. It is based on this power formula:

$$P = V \times I$$

Thus, collector dissipation is calculated by

$$P_C = V_{CE} \times I_C$$

Now, the power curve on Fig. 5-13 can be

verified. At $V_{CE} = 4$ V, the power curve crosses at a little less than 1.9 A on the I_C axis:

$$P_C = 4 \times 1.9 = 7.6 \text{ W}$$

At $V_{CE} = 8$ V, the power curve crosses a bit above 0.9 A on the I_C axis:

$$P_C = 8 \times 0.9 = 7.2 \text{ W}$$

All points along the power curve represent a

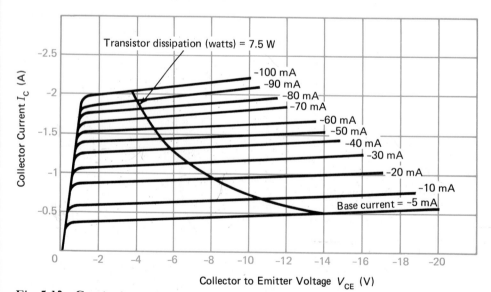

Fig. 5-13   Constant-power curve.

$\uparrow E = I \uparrow \bar{R}$

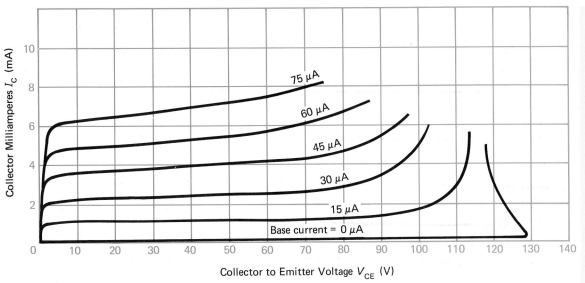

Fig. 5-14 Collector breakdown.

product of 7.5 W. The negative values need not be taken into account. They indicate the transistor is a PNP type. If negative values are used, the answers remain the same since multiplying a negative voltage by a negative current produces a positive power value.

If the collector characteristic curves are extended to include higher voltages, collector breakdown can be shown. Like diodes, transistors have limits as to the amount of reverse bias that can be applied. Transistors have two junctions, and their breakdown ratings are complicated. Figure 5-14 shows a collector family of curves where the horizontal axis is extended to 140 V. If collector voltage becomes very high, it begins to control collector current. This is generally undesirable. The base current is supposed to control the collector current. This is why transistors usually are not operated near their maximum voltage ratings. As can be seen from Fig. 5-14, collector breakdown is not a fixed point as it was in a diode circuit. It varies with the amount of base current. At 15 μA, the collector breakdown point is around 110 V. At 0 μA, it occurs near 130 V.

The *transfer characteristic curve* shown in Fig. 5-15 is another example of how curves can be used to show the electrical characteristics of a transistor. Curves of this type show how one transistor terminal (the base) affects another (the collector). This is why they are called transfer curves. It has been stated that base current controls collector current. It can be seen also from Fig. 5-15 that base-

emitter voltage controls collector current. This is because the base-emitter bias sets the level of base current.

Figure 5-15 also shows one of the important differences between silicon transistors and germanium transistors. Like diodes, germanium transistors turn on at a much lower voltage (approximately 0.2 V). The silicon device turns on near 0.6 V. These voltages are important to remember. They are reasonably constant and can be of great help in troubleshooting transistor circuits. They can also help determine if a transistor is made of silicon or germanium.

Germanium transistors are not nearly as popular as they once were. Most modern equipment utilizes silicon devices. Ger-

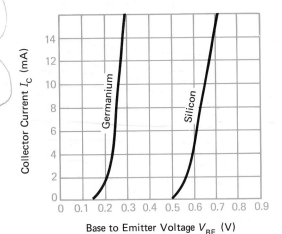

Fig. 5-15 Comparing silicon and germanium transistors.

Electronics:
Principles and
Applications
CHAPTER 5

Data sheet

Case style

manium does offer a few advantages for some applications, however. The carrier mobility is better in germanium. Some high-frequency transistors make the most of this advantage. Certain high-power transistors use germanium since it is a better conductor than silicon. The low turn-on voltage of germanium is also an advantage in some circuits. Silicon transistors are inexpensive and show much better high-temperature performance. These two reasons make them the logical choice for most applications.

Review Questions

Supply the missing word in each statement.

19. Refer to Fig. 5-10. Voltage $V_{CE} = 2$ V and current $I_C = 10$ mA. Therefore $I_B =$ _____?_____

20. Refer to Fig. 5-10. Current $I_B = 50$ μA and voltage $V_{CE} = 16$ V. Therefore, $I_C =$ _____?_____.

21. Refer to Fig. 5-10. Voltage $V_{CE} = 4$ V and current $I_C = 10$ mA. Therefore, beta = _____?_____

22. Refer to Fig. 5-10. Current $I_B = 100$ μA and voltage $V_{CE} = 10$ V. Therefore, $P_C =$ _____?_____

23. Germanium transistors turn on when V_{BE} reaches _____?_____ V.

24. Silicon transistors turn on when V_{BE} reaches _____?_____ V.

25. Of the two popular semiconductor materials, _____?_____ is the better conductor.

5-4 TRANSISTOR DATA

Solid-state device manufacturers provide data sheets covering the mechanical, thermal, and electrical characteristics of the parts they make. These data sheets are also bound into volumes. Data sheets and volumes are very useful to the electronics technician, but they are not always available. Often, a less detailed information source can be used.

One way for a technician to learn something about a particular transistor is to use substitution guides. These guides are not totally accurate, but they do provide a good, general idea about the device of interest. Another good source of information is a parts catalog. Figure 5-16 is a sample of transistor listings

Type	Case	Material Function	Maximum Ratings			Beta H_{FE}@Ic		f_T MHz
			Dissipation Watts	Col'r To Base Volts	Col'r Curr. mA	Min. Max.	mA	
2N2870/ 2N301	TO-3	GP AP	30C	80	3A	50-165	1A	.200
2N2876	TO-60	SN AV	17.5C	80	2.5A			.200
2N2894	TO-18	SP SH	1.2C	12	200	40-150	30	400
2N2895	TO-18	SN GP	.500	120	1A	60-150	1	120
2N5070	TO-60	SN AP	70C	65	3.3A	10-100	3A	100
2N5071	TO-60	SN AP	70C	65	3.3A	10-100	3A	100
2N5086	TO-92	SP GP	.310	50	50	150-	1	40
2N5087	TO-92	SP GP	.310	50	50	250-	1	40
2N5088	TO-92	SN GP	.310	35	50	350-	1	
2N5172	TO-98	SN GP	.200	25	100	100-	10	
2N5179	TO-72	SN AU	.200	20	50	25-	20	900
2N5180	TO-104	SN AU	.180	30		20-	2	650
2N5183	TO-104	SN GP	.500	18	1A	70-	10	62
2N5184	TO-104	SN GP	.500	120	50	10-	50	50

Material code: GP . . . . Germanium, PNP
SN . . . . Silicon, NPN
SP . . . . Silicon, PNP

Function code: AP . . . . Amplifier, power
AV . . . . Amplifier, VHF
SH . . . . Switch, high speed
GP . . . . General purpose
AU . . . . Amplifier, UHF

Fig. 5-16   Transistor catalog listings.

from a catalog. The prices have been deleted. These listings may even include some of the nonregistered device numbers. Notice that quite a bit of information is listed in the catalog for each transistor number. For example, a 2N5179 transistor is seen to use a TO-72 case style, it is a silicon NPN type, it is used as an ultra-high-frequency (UHF) amplifier, it can dissipate 0.2 W, and so on. Such parts catalogs are available for a small cost or often are free. It is a good idea to gather a collection of these catalogs and obtain new ones as they become available.

Figure 5-17 shows another example of the information that can be found in substitution guides and parts catalogs. Transistors are made in many case styles. The physical characteristics can be just as important as the electrical details. Figure 5-17 is only a sample of the many case styles used today. Notice that this material is also valuable because it can help you to identify the emitter, base, and collector leads. This information is not usually on the transistor case.

In replacing transistors, the best choice is an exact replacement. This would be one with the same number as the original. When this is not possible, a substitution guide can be very helpful. In some critical circuits, a substitute unit will not work well. This may happen in fewer than 10 percent of the cases. The most difficult situation for many technicians is when the part number cannot be

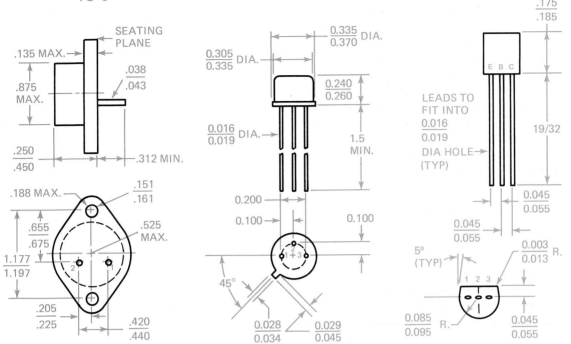

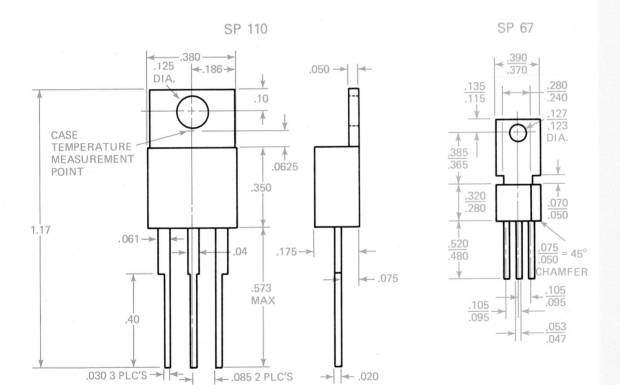

Fig. 5-17   Transistor case styles.

65

Dynamic test

Noise figure

Overlay-type
transistor

found in any of the available guides or on the original transistor. Often, a general type of unit can be used in these cases. It is very important to find out if the original is germanium or silicon and if it is NPN or PNP. The substitute should be of the same material and the same polarity. The circuit voltages can be inspected to give some idea of the voltage ratings that the new transistor should have. The power ratings can be established by inspecting the circuit current levels and voltages. Of course, the physical characteristics should be similar. Finally, knowing the function of the original unit is helpful in picking a substitute. Substitution guides and catalogs often list transistors as audio types, very-high-frequency (VHF) types, switching types, and in other descriptive ways.

Review Questions

Determine whether each statement is true or false.

26. Device manufacturers publish data sheets and data volumes for solid-state devices.

27. Almost all solid-state devices have the leads marked on the case.

28. Substitute transistors will not work in all cases.

29. It is possible to choose a replacement transistor by considering polarity, semiconductor material, voltage and current levels, and circuit function.

5-5 TRANSISTOR TESTING

One way to test transistors is to use a curve tracer. This technique is used often in the electronics industry. If a transistor shows normal characteristic curves, it is good. Slight problems can be detected on a curve tracer. These problems may not show up on

other testers. Many technicians do not have a curve tracer. Thus, other techniques are used more often.

A technique sometimes used in industry is to place the transistor in a special fixture or test circuit. This is a dynamic test because it makes the device operate with real voltages and signals. This method of testing is very valuable for very high and ultra-high-frequency amplifier transistors. Dynamic testing reveals power gain and noise figure under real conditions. *Noise figure* is a measure of a transistor's ability to amplify weak signals. Some transistors may make enough electrical noise to overpower a weak signal. These transistors are said to have a poor noise figure.

Since curve tracing and dynamic testing are rather limited techniques, most technicians need another way to check transistors.

A few transistor types may show a gradual loss of power gain. Radio-frequency (RF) power amplifiers, for example, may use overlay-type transistors. These transistors can have over 100 separate emitters. Such transistors can suffer base-emitter changes which can decrease power gain. Another problem is humid air. It can enter the transistor package and degrade the transistor.

For the most part, transistors fail suddenly and completely. One or both junctions may short-circuit. An internal connection can break loose or burn out from an overload. This type of failure is easy to check. Most bad transistors can be identified with some simple ohmmeter tests.

A good transistor has two PN junctions. Both can be checked with an ohmmeter. As shown in Fig. 5-18, a PNP transistor can be compared to two diodes with a common cathode connection. The base lead acts as the common cathode. Figure 5-19 shows an NPN transistor as two diodes with a common anode connection. If two good diodes can be

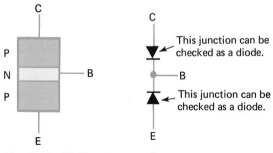

Fig. 5-18   PNP junction polarity.

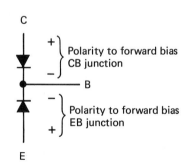

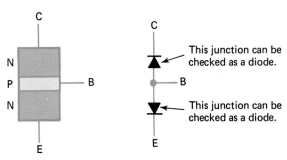

Fig. 5-19 NPN junction polarity.

verified by ohmmeter tests, the transistor is probably good.

The ohmmeter can also be used to identify the polarity (NPN or PNP) of a transistor and the three leads. This can be helpful when data are not available. The ohmmeter should be set to the $R \times 100$ range for testing most transistors. Germanium power transistors may be easiest to check by using the $R \times 1$ range. Power transistors are easy to recognize because they are large compared to small-signal transistors.

The first step in testing transistors is to connect the ohmmeter leads across two of the transistor leads, as shown in Fig. 5-20. If a low resistance is indicated, the leads are across one of the diodes or else the transistor is shorted. To decide which is the case, reverse the ohmmeter leads. If the diode is good, the ohmmeter will show a high resistance, as seen in Fig. 5-21. If you happen to connect across the emitter and collector leads of a good transistor, the ohmmeter will show high resistance in both directions. The reason is that two junctions are in the ohmmeter circuit. Study Figs. 5-18 and 5-19 and verify that with either polarity applied from emitter to collector, one of the diodes will be reverse-biased.

Once the emitter-collector connection is found, the base has been identified by the process of elimination. Now, connect the negative lead of the ohmmeter to the base lead. Touch the positive lead to one and then the other of the two remaining leads. If a low resistance is shown, the transistor is a PNP type. Connect the positive lead to the base

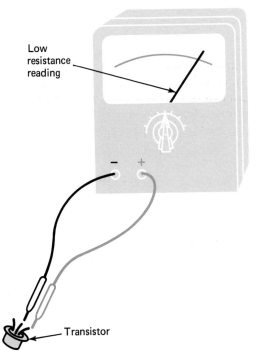

Fig. 5-20 Establishing a forward-biased junction.

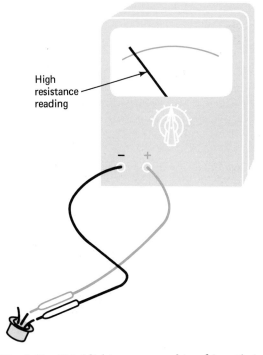

Fig. 5-21 Establishing a reverse-biased junction.

Checking
gain

lead. Touch the negative lead to one and then the other of the two remaining leads. If a low resistance is shown, then the transistor is an NPN type.

Thus far, you have identified the base lead and the polarity of the transistor. Now it is possible to check the transistor for gain and to identify the collector and emitter leads. All that is needed is a 100,000-Ω resistor and the ohmmeter. If you are checking a germanium power transistor on the $R \times 1$ range, use a 1000-Ω resistor.

The resistor will be used to provide the transistor with a small amount of base current. If the transistor has good current gain, the collector current will be much greater. The ohmmeter will indicate a resistance much lower than 100,000 Ω, and this proves that the transistor is capable of current gain. This check is made by connecting the ohmmeter across the emitter and collector leads at the same time that the resistor is connected across the collector and base leads. For an NPN transistor, the technique is shown in Fig. 5-22. If you guess wrong and have the positive lead to the emitter and the negative lead to the collector, a low resistance reading will not be seen. Just remember that the resistor must be connected from the positive lead to the base when testing for gain in an NPN transistor.

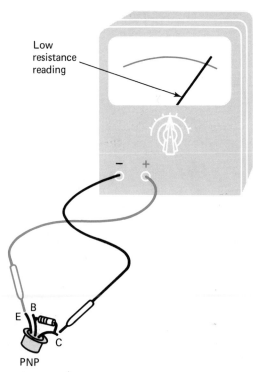

Fig. 5-23   Checking PNP gain.

The emitter-collector combination showing the most gain (lowest resistance) is the correct connection. When this is obtained, the ohmmeter leads will identify the collector and the emitter, just as in Fig. 5-22 for NPN transistors. You will also be sure that the transistor has gain because of the low resistance reading. Usually it will be much less than 100,000 Ω.

In checking a PNP transistor for gain, the connection for the lowest resistance reading is shown in Fig. 5-23. Remember that the resistor must be connected from the negative lead to the base when a PNP transistor is tested for gain. The combination that shows the best gain (lowest resistance) is just as in Fig. 5-23.

The entire process is more difficult to describe than to do. With some practice, it becomes very quick and easy. The only drawback to this technique is that it cannot be used on transistors in a circuit. A summary of the steps follows:

1. Use the $R \times 100$ range of your ohmmeter (or $R \times 1$ for germanium power transistors).
2. Find the two leads showing high resistance with both polarities applied. The remaining lead is the base lead.

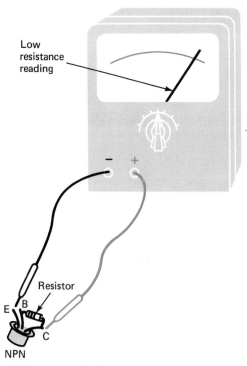

Fig. 5-22   Checking NPN gain.

3. With the positive lead on the base, a low resistance should be found to either of the two remaining leads if the transistor is NPN. For a PNP transistor, the negative lead will have to be on the base to obtain a low resistance.

4. With the ohmmeter across the emitter-collector combination, connect the resistor (100 kΩ or 1 kΩ) from the positive lead to the base terminal for an NPN unit. Reverse the emitter-collector combination. The lowest resistance is obtained when the positive lead is on the collector.

5. In checking a PNP transistor, the resistor goes from the negative lead to the base. The correct combination (lowest resistance) is when the negative lead is on the collector.

The process is easier to remember and less confusing if you know why it works. Figure 5-24 shows what is happening when an NPN transistor is being checked for gain. The positive lead of the ohmmeter is applied directly to the collector. This reverse-biases the collector as it should be. The positive lead of the ohmmeter is also connected to the base but through a high resistance. This forward-biases the base as it should be. However, the high value of the resistor keeps the base current very low. If the transistor has gain, the emitter-collector current will be greater. The current is supplied by the ohmmeter, and the ohmmeter shows a low resistance because of the increased current from emitter to collector.

Some ohmmeters may have reversed polarity. Some ohmmeters use a very low supply voltage to avoid turning on PN junctions. These characteristics of the ohmmeter must be known before being used to test a transistor.

Transistors have some leakage current. This is due to minority carrier action. One leakage current in a transistor is called I_{CBO}. (The symbol I stands for current, CB stands for the collector-base junction, and O tells us the emitter is open.) This is the current that flows across the collector-base junction under conditions of reverse bias and with the emitter lead open. Another transistor leakage current is I_{CEO}. (The symbol I stands for current, CE stands for the collector-emitter terminals, and O tells us the base terminal is open.) This is the largest leakage current. It is an

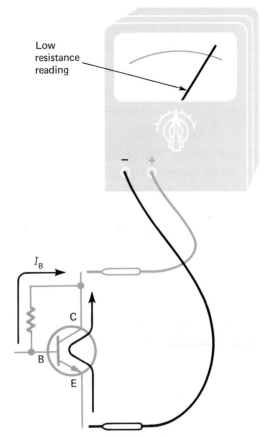

Low resistance reading

Leakage current

I_B

Collector current is β times larger and the ohmmeter shows low resistance

Fig. 5-24   How the ohmmeter test works.

amplified form of I_{CBO}:

$$I_{CEO} = \beta \times I_{CBO}$$

With the base terminal open, any current leaking across the reverse-biased collector-base junction will have the same effect on the base-emitter junction as an externally applied base current. With the base terminal open, there is no other place for the leakage current to go. The transistor amplifies this leakage just as it would any base current:

$$I_C = \beta \times I_B$$

You should recognize the above formula as a form of the β formula discussed earlier:

$$\beta = \frac{I_C}{I_B}$$

Silicon transistors have very low leakage currents. When ohmmeter tests are made,

69

Electronics:
Principles and
Applications
CHAPTER 5

In-circuit testing

Signal injection

Unipolar
transistor

Junction
field-effect
transistor (JFET)

N-channel JFET

the ohmmeter should show an infinite resistance when the junctions are reverse-biased. Anything less may mean the transistor is defective. Germanium transistors have much greater leakage currents. This will probably show up as a high, but not infinite, reverse resistance. It will be most noticeable when checking from the emitter to the collector terminal. This is because I_{CEO} is an amplified version of I_{CBO}. Some technicians use this test to tell the difference between a silicon transistor and a germanium transistor. It works, but remember that you could be confused by a leaky silicon transistor.

Quite a bit of information can be learned from ohmmeter tests. Unfortunately, the transistor must be removed from the circuit. Transistor testers exist that will check transistors in the circuit. These are very helpful. Most transistors are soldered to the circuit board. Removing them takes time and can cause damage. A good in-circuit tester is helpful for a technician whose job is to troubleshoot solid-state equipment.

In-circuit testing can be accomplished in other ways. When a transistor fails in a circuit, there may be voltage changes at the transistor terminals. These can be found with a voltmeter. Another in-circuit test uses an oscilloscope to check transistor input and output signals (*signal tracing*). A bad transistor may have an input signal but no output signal. Finally, in-circuit checking can be accomplished by *signal injection*. The technician will apply a test signal from a generator. If the signal goes through the rest of the circuit when applied at the output but not when applied at the input of an amplifier, it is fairly certain that something is wrong with that amplifier.

In summary, in-circuit transistor testing can be done in four ways (there are others too):

1. Using an in-circuit transistor tester
2. Using a voltmeter (voltage analysis)
3. Using an oscilloscope (signal tracing)
4. Using a signal generator (signal injection)

A good technician will use any of or all these techniques. Depending on certain conditions, one technique may prove to be quicker and easier in a given situation. Voltage analysis, signal tracing, and signal injection are covered in more detail in later chapters.

Review Questions

Determine whether each statement is true or false.

30. Transistor junctions can be checked with an ohmmeter. T

31. Junction failures account for most bad transistors. T

32. A good transistor should show a high resistance from emitter to collector, regardless of the ohmmeter polarity. T

33. It is not possible to locate the base lead of a transistor with an ohmmeter. F

34. Suppose that the positive lead of an ohmmeter is connected to the base of a good transistor. Also assume that touching either of the remaining transistor leads with the negative lead shows a low resistance. The transistor must be a PNP type. F

35. It is possible to verify transistor gain by using an ohmmeter and a resistor. T

36. Transistor testing with an ohmmeter is usually limited to out-of-circuit conditions. T

37. It is not possible to check transistors that are soldered into a circuit. F

5-6 OTHER TRANSISTOR TYPES

Bipolar junction transistors are used in most circuits. However, another transistor type is becoming more popular. This type is classified as *unipolar* devices. A unipolar (one-polarity) transistor uses only one type of current carrier. The junction field-effect transistor (JFET) is an example of a unipolar transistor. Figure 5-25 shows the structure and schematic symbol for an N-channel JFET.

The JFET can be made in two basic ways.

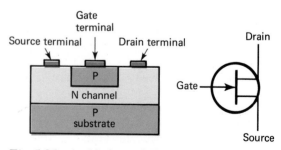

Fig. 5-25   An N-channel JFET.

The channel can be N-type material or P-type material. The schematic symbol of Fig. 5-25 is for an N-channel device. The symbol for a P-channel device will show the arrow on the gate lead pointing out. Remember, pointing iN means an N-channel device.

In a bipolar junction transistor, both holes and electrons are used to support conduction. In an N-channel JFET, only electrons are used. In a P-channel JFET, only holes are used.

The JFET operates in the *depletion mode*. A control voltage at the gate terminal can deplete (remove) the carriers in the channel. For example, the transistor of Fig. 5-25 will normally conduct from the source terminal to the drain terminal. The N channel contains enough free electrons to support the flow of current. If the gate is made negative, the free electrons can be pushed out of the channel. Like charges repel. This leaves the channel with fewer free carriers. The resistance of the channel is now much higher, and this tends to decrease the source and drain currents. In fact, if the gate is made negative enough, the device can be turned off—no current will flow.

Examine the curves of Fig. 5-26. Notice that as the negative voltage from gate to source (V_{GS}) increases, the drain current (I_D) decreases. Compare this operation with a bipolar junction transistor:

1. A bipolar junction transistor is off (there is no collector current) until base current is provided.
2. A JFET is on (drain current is flowing) until the gate voltage becomes high enough to remove the carriers from the channel.

These are important differences. The bipolar device is current-controlled. The unipolar device is voltage-controlled. The bipolar transistor is normally off. The JFET is normally on.

Will there be any gate current in the JFET? Check Fig. 5-25. The gate is made of P-type material. To control channel conduction, the gate is made negative. This reverse-biases the gate-channel diode. The gate current should be zero (there may be a very small leakage current).

There are also P-channel JFETs. They use P-type material for the channel and N-type

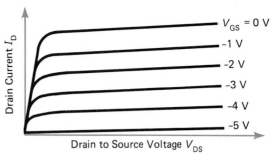

Fig. 5-26   Characteristic curves of a JFET.

material for the gate. The gate will be made positive to repel the holes in the channel. Again, this reverse-biases the gate-channel diode, and the gate current will be zero. Since the polarities are opposite, N-channel JFETs and P-channel JFETs are not interchangeable.

Field-effect transistors (FETs) do not require any gate current for operation. This means the gate structure can be completely insulated from the channel. Thus any slight leakage current resulting from minority carrier action is blocked. The gate can be made of metal. The insulation used is an oxide of silicon. This structure is shown in Fig. 5-27. It is called a metal oxide semiconductor field-effect transistor (MOSFET). The MOSFET can be made with P channels or N channels. Again, the arrow pointing iN tells us the channel is N-type material.

Early MOSFETs were very delicate devices. The thin oxide insulator was easily damaged by excess voltage. The static charge

P-channel JFET

Depletion mode

Metal oxide semiconductor field-effect transistor (MOSFET)

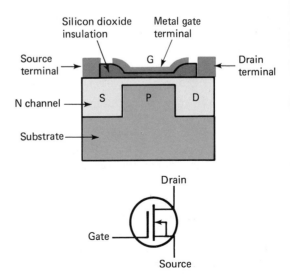

Fig. 5-27   An N-channel MOSFET.

71

Enhancement
mode

Unijunction
transistor (UJT)

Negative-
resistance
region

on a technician's body could easily break down the gate insulator. These devices had to be handled very carefully. Their leads were shorted together until the device was soldered into the circuit. Special precautions were needed to safely make measurements in a MOSFET circuit. Today, most MOSFET devices have built-in diodes to protect the gate insulator. If the gate voltage goes too high, the diodes turn on and safely discharge the potential. However, many manufacturers still advise careful handling of MOSFET devices.

The gate voltage in a MOSFET can be of either polarity since a diode junction is not used. This makes another mode of operation possible—the *enhancement* mode. An enhancement-mode device normally has no conductive channel from the source to the drain. It is a normally-off device. The proper gate voltage will attract carriers to the gate region and form a conductive channel. The channel is enhanced (aided by gate voltage). Figure 5-28 shows the schematic symbols used for enhancement-mode MOSFETs. Note that the line from source to drain is broken. This implies that the channel is not always present. Figure 5-29 shows a family of curves for an N-channel enhancement-mode device. As the gate is made more positive, more electrons are attracted into the channel area. This enhancement improves channel conduction, and the drain current increases. A JFET is not operated in the enhancement mode because the gate diode would be forward-biased and excess gate current would flow.

All field-effect transistors have some advantages over bipolar transistors that make them desirable for certain applications. Their gate terminal does not require any current. This is a good feature when an amplifier with high input resistance is needed. This is easy to understand by inspecting Ohm's law:

$$R = \frac{V}{I}$$

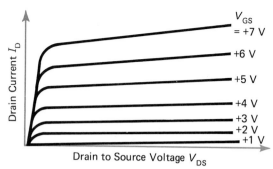

Fig. 5-29 Enhancement-mode characteristic curves.

Consider V to be a signal voltage supplied to an amplifier and I the current taken by the amplifier. In the equation, as I decreases, R increases. This means that an amplifier which draws very little current from a signal source has a high input resistance. Bipolar transistors are current-controlled. A bipolar amplifier must take a great deal more current from the signal source. As I increases, R decreases. Bipolar amplifiers have a low input resistance.

The final transistor type that will be covered is the unijunction transistor (UJT). This transistor is not used as an amplifier. It is used in timing and control applications. The structure can be seen in Fig. 5-30. The device has an N-type silicon structure with a tiny P-type zone near the center. This produces only one PN junction in the device ("uni" means one). The characteristic curve for the UJT is shown in Fig. 5-31. This curve has a unique region called the *negative-resistance* region. When a device shows decreasing voltage drop with increasing current, it can be said to have negative resistance. According to Ohm's law, V = IR. Thus, as current increases, the voltage drop is expected to increase also. If the reverse occurs, the resistance must be changing. It is easy to explain by using some

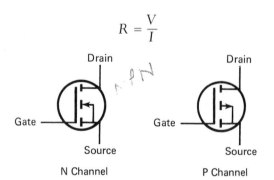

Fig. 5-28 Enhancement-mode MOSFETs.

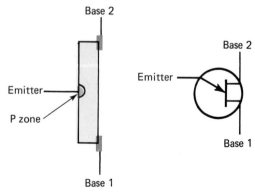

Fig. 5-30 The unijunction transistor.

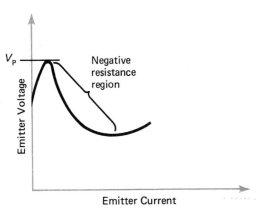

Fig. 5-31 Characteristic curve of a UJT.

tion of the transistor. The injected holes greatly improve the conductivity of this part of the N-type material. Greater conductivity means lower resistance. This sudden drop in resistance occurs with an increase of current. It can be used to trigger or turn on other devices. The trigger point (V_P) is predictable. This makes the UJT very useful in timing and control circuits. Remember, it is not useful as an amplifier.

numbers. Suppose the current through a device is 1 A and the resistance is 10 Ω. The voltage drop would be

$$V = I \times R = 1 \times 10 = 10 \text{ V}$$

Now, if the current increases to 2 A, we can expect the voltage drop to increase, too:

$$V = I \times R = 2 \times 10 = 20 \text{ V}$$

But, if the resistance drops to 2 Ω, then

$$V = I \times R = 2 \times 2 = 4 \text{ V}$$

Strictly speaking, negative resistance does not exist. It is simply a property that has been applied to a family of devices which show a sudden drop in resistance at some point on their characteristic curve.

The UJT is a member of the negative-resistance family. When the emitter voltage reaches a certain point (V_P on the curve of Fig. 5-31), the emitter diode becomes forward-biased. This causes holes to cross over from the P-type zone into the N-type silicon. These holes are injected into the region between the emitter and the Base 1 connec-

Review Questions

Determine whether each statement is true or false.

38. The JFET is a bipolar device. F

39. A depletion-mode transistor uses gate voltage to increase the number of carriers in the channel. voltage der.

40. Bipolar transistors are current amplifiers while unipolar transistors are voltage amplifiers. T

41. It is possible to turn off an N-channel JFET with negative gate voltage. T

42. In P-channel JFET circuits, the gate diode is normally forward-biased. F

43. It is a good idea to handle MOSFETs carefully to prevent breakdown of the gate insulator. T

44. It is possible to operate a MOSFET in the enhancement mode. T

45. The enhancement mode means that carriers are being pushed out of the channel by gate voltage. F

46. The FET makes a better high-input-resistance amplifier than the bipolar type. T

47. The UJT is a negative-resistance device. T

48. The UJT makes a good amplifier. F

Summary

1. Gain is the basic function of an amplifier.
2. Gain can be measured in units of voltage, current, or power.
3. Power gain is the product of voltage gain and current gain.
4. The term "voltage amplifier" is often used to describe a small-signal amplifier.
5. The term "power amplifier" is often used to describe a large-signal amplifier.

6. Bipolar junction transistors are manufactured in two polarities: NPN and PNP. The NPN types are more popular.
7. In a junction transistor, the emitter emits the carriers, the base is the control region, and the collector collects the carriers.
8. The schematic symbol of an NPN transistor shows the emitter lead arrow Not Pointing iN.

9. Normal operation of a junction transistor requires that the collector-base junction be reverse-biased and the base-emitter junction be forward-biased.

10. Most of the current carriers coming from the emitter cannot find carriers in the base region to combine with. This tends to make the base current much lower.

11. The base is very narrow, and the collector bias attracts the carriers coming from the emitter. This tends to make the collector current almost as high as the emitter current.

12. β, or h_{fe}, is the current gain from the base terminal to the collector terminal. A typical value is 100, but it varies considerably.

13. Base current controls collector current and emitter current.

14. Emitters of PNP transistors emit holes. Emitters of NPN transistors emit electrons.

15. A collector characteristic curve is produced by plotting a graph of I_C versus V_{CE} with I_B at some fixed value.

16. Collector voltage has only a small effect on collector current over most of the operating range.

17. A power curve can be plotted on the graph of the collector family to show the safe area of operation.

18. Collector dissipation is the product of collector-emitter voltage and collector current.

19. Germanium transistors require base-emitter bias of about 0.2 V for turn-on. Silicon units need about 0.6 V.

20. Silicon transistors are more widely used than germanium transistors.

21. Substitution guides provide the technician with needed information about solid-state devices.

22. The physical characteristics of a part can be just as important as the electrical characteristics.

23. Transistors can be tested with curve tracers, testers, and ohmmeters or by various in-circuit checks.

24. Most transistors fail suddenly and completely. One or both PN junctions may short or open.

25. An ohmmeter can check both junctions, identify polarity, identify leads, check gain, indicate leakage, and may even identify the transistor material.

26. Leakage current I_{CEO} is β times larger than I_{CBO}.

27. Bipolar transistors (NPN and PNP) use both holes and electrons for conduction.

28. Unipolar transistors (N-channel and P-channel types) use either electrons or holes for conduction.

29. A bipolar transistor is a normally-off device. It is turned on with base current.

30. A JFET is a normally-on device. It is turned off with gate voltage. This is called the depletion mode.

31. A MOSFET uses an insulated gate structure. It is available in both the depletion mode and the enhancement mode.

32. An enhancement-mode MOSFET is a normally-off device. It is turned on by gate voltage.

33. Field-effect transistors have a very high input resistance.

34. Unijunction transistors have one junction. They are not used as amplifiers.

35. The UJT belongs to the category of negative-resistance devices.

36. The UJT is useful in timing and control applications.

Chapter Review Questions

Supply the missing word in each statement.

5-1. An amplifier provides a voltage gain of 5 and a current gain of 11. Its power gain is _____?_____. 55

5-2. An amplifier must give an output signal of 25 V peak-to-peak. If its voltage gain is 10, the input signal to the amplifier must be _____?_____. 2.5 V

5-3. Small-signal amplifiers are often called ___*voltage*___ amplifiers.

5-4. Large-signal amplifiers are often called ___*power*___ amplifiers.

5-5. Bipolar junction transistors are made in two basic polarities: NPN and ___?___. *PNP*

5-6. Current flow in bipolar transistors involves two types of carriers: electrons and ___?___ *holes*

5-7. The base-emitter junction of a bipolar transistor shows a low resistance because it is forward-___?___ *Bias*.

5-8. The collector-base junction of a bipolar transistor shows a high resistance because it is reverse-___?___ *Bias*.

5-9. The smallest current in a bipolar transistor is the ___*Base*___ current.

5-10. A bipolar transistor has a base current of 100 μA and an emitter current of 10 mA. The collector current is ___?___ *9.9 mA*

5-11. The beta of the transistor described in question 5-10 is ___?___. *99*

5-12. For proper operation, the base terminal of a bipolar NPN transistor should be ___?___ with respect to the emitter terminal. *pos.*

5-13. For proper operation, the base terminal of a bipolar PNP transistor should be ___?___ with respect to the emitter terminal. *neg.*

5-14. The symbol h_{fe} represents a transistor's ___?___. *Beta*

5-15. The emitter of a PNP transistor produces ___?___ current. *hole*

5-16. The emitter of an NPN transistor produces ___?___ current. *electron*

5-17. Refer to Fig. 5-10 where V_{CE} = 10 V and I_B = 20 μA. Therefore, beta = ___?___. *150*

5-18. Refer to Fig. 5-12 where V_{CE} = − 16 V and I_c = −7 mA. Thus, I_B = ___?___. *−50 μA*

5-19. Refer to Fig. 5-12 where I_B = − 100 μA and V_{CE} = − 10 V. Thus P_c = ___?___. *120 mW*

5-20. Refer to Fig. 5-15. The transistor is silicon and V_{BE} = 0.4 V. Then I_c = ___?___ *zero*

5-21. In testing bipolar transistors with an ohmmeter, a good diode indication should be noted at the collector-base and ___?___ junctions. *Emitter Base*

5-22. Refer to Fig. 5-26. As V_{GS} becomes more negative, drain current ___?___. *decreases*

5-23. An N-channel JFET uses ___?___ to support the flow of current. *electrons*

5-24. A P-channel JFET uses ___?___ to support the flow of current. *holes*

5-25. Gate voltage in a JFET can remove carriers from the channel. This is known as the ___?___ mode. *depletion*

5-26. Gate voltage in a MOSFET can produce carriers in the channel. This is known as the ___?___ mode. *enhancement*

5-27. A JFET is not operated in the enhancement mode because the gate diode may become ___?___ biased. ~~reverse~~ *forward*

5-28. Once the firing voltage V_P is reached in a UJT, the resistance is expected to ___?___. *decrease*

.0100
.0001
.0099

$BETA = \dfrac{I_c}{I_B}$

$\beta = \dfrac{I_c}{I_B}$

$\dfrac{3 \text{ mA}}{20 \mu A} =$

$P_c = V_{CE} \times I_c$

-10×-12

Answers to Review Questions

1. T	17. T	33. F
2. T	18. F	34. F
3. F	19. 80 μA	35. T
4. F	20. 7 mA	36. T
5. F	21. 125	37. F
6. T	22. 120 mW	38. F
7. F	23. 0.2	39. F
8. T	24. 0.6	40. T
9. F	25. Germanium	41. T
10. T	26. T	42. F
11. F	27. F	43. T
12. T	28. T	44. T
13. T	29. T	45. F
14. T	30. T	46. T
15. F	31. T	47. T
16. F	32. T	48. F

Introduction to Small-Signal Amplifiers

This chapter details the concept of gain. *Gain* is the ability of an electronic circuit to increase the level of a signal. As you will see, gain can be measured as a ratio or as a logarithm of a ratio.

Transistors provide gain. This chapter will show you how they can be used with other components to make amplifier circuits. You will learn how to evaluate some of the popular amplifier circuits with some simple calculations.

This chapter is limited to small-signal amplifiers. As mentioned before, these are often called voltage amplifiers.

6-1 MEASURING GAIN

Gain is the basic function of all amplifiers. It is a comparison of the signal fed into the amplifier with the signal coming out of the amplifier. Because of the gain, we can expect the output signal to be greater than the input signal. Figure 6-1 shows how the measurements are used to calculate the voltage gain of an amplifier. For example, if the signal in is 1 V and the signal out is 10 V, the gain is

$$\text{Gain} = \frac{\text{signal out}}{\text{signal in}} = \frac{10 \text{ V}}{1 \text{ V}} = 10$$

Note that the units of voltage cancel and gain is a pure number. It is not correct to say the gain of the amplifier is 10 V.

Voltage gain is used to describe the operation of *small-signal amplifiers*. Power gain is used to describe the operation of *large-signal amplifiers*. If the amplifier of Fig. 6-1 were a power or large-signal amplifier, the gain would be based on watts rather than on volts. For example, if the signal in were 0.5 W and the signal out were 8 W, the gain would be

$$\text{Gain} = \frac{\text{signal out}}{\text{signal in}} = \frac{8 \text{ W}}{0.5 \text{ W}} = 16$$

Early work in electronics was in the communications area. The useful output of most circuits was audio for headphones or speakers. Thus, engineers and technicians needed a way to relate circuit performance to human hearing. The human ear is not *linear*. It does not recognize intensity or loudness in the way a linear device does. For example, if you are listening to a speaker with 0.1-W input and the power suddenly increases to 1 W, you will judge that the sound has become louder. Then, assume the power suddenly increases again to 10 W. Again, you will judge the sound to be louder. The interesting thing is

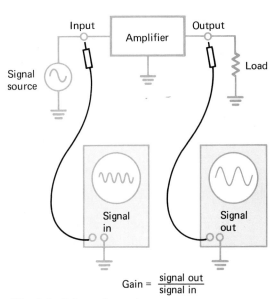

$$\text{Gain} = \frac{\text{signal out}}{\text{signal in}}$$

Fig. 6-1   Measuring gain.

From page 77:
Voltage gain

On this page:
Linear change

Logarithmic gain

Decibel

that you will probably rate the second increase in loudness as about equal to the first increase in loudness.

A linear device would indicate the second increase to be 10 times greater than the first. Let us see why:

- First, increase from 0.1 to 1 W which is a 0.9-W *linear change*.
- Second, increase from 1 to 10 W, which is a 9-W *linear change*.

Even though the second loudness change involved 10 times the linear change, your hearing judges the changes to be about the same. This shows that human hearing is nonlinear for loudness. The loudness response of human hearing is *logarithmic*. Logarithms are very useful to describe the performance of audio systems. We are more interested in the logarithmic gain of an amplifier than in its linear gain. Gain expressed in this way is very convenient, and practically all electronic amplifiers are described by their logarithmic performance. What started out as a convenience in audio work has now become the universal standard for amplifier performance.

Common logarithms are powers of 10. For example,

$$10^1 = 10$$
$$10^2 = 100$$
$$10^3 = 1000$$

The logarithm of 10 is 1. The logarithm of 100 is 2. The logarithm of 1000 is 3. All positive numbers can be described as a common logarithm. These logarithms can be found on slide rules, on scientific calculators, and in log tables. They are very useful in electronics.

Power gain is very often measured in *decibels*. A decibel is a logarithmic unit. It can be found by this formula:

$$\text{Decibels of power gain} = 10 \times \log_{10}\left(\frac{\text{power out}}{\text{power in}}\right)$$

Decibels are usually abbreviated dB. Gain in decibels is based on common logarithms. Common logarithms are based on 10. This is shown in the above equation as $\log_{10}$ (the base is 10). Hereafter the base 10 will be dropped, and "log" will be understood to mean $\log_{10}$.

Now, we can apply the equation to the example given previously:

First loudness increase

$$\text{dB power gain} = 10 \times \log\left(\frac{1\ \text{W}}{0.1\ \text{W}}\right)$$
$$= 10 \times \log 10$$

Remember, the logarithm of 10 (log 10) is 1.

$$\text{dB power gain} = 10 \times 1 = 10$$

Thus, the first increase in level or loudness was equal to 10 dB.

Second loudness increase

$$\text{dB power gain} = 10 \times \log\left(\frac{10\ \text{W}}{1\ \text{W}}\right)$$
$$= 10 \times \log 10$$
$$= 10 \times 1 = 10$$

Thus, the second increase was also equal to 10 dB. Since the decibel is a logarithmic unit and because your hearing is logarithmic, the two 10-dB increases sound about the same. The average human ear can detect a change as small as 1 dB. Any change smaller than 1 dB would be very difficult for most people to hear.

Why has the decibel, which is really an audio unit, come to be used in all areas of electronics where gain is important? The answer is that it is so convenient to work with. Figure 6-2 shows why. Five stages, or parts, of an electronic system are shown. Three of the stages show a gain (+ dB), and two show a loss (− dB). Loss means that more signal goes in than comes out. The ratio of signal out to

Fig. 6-2 Gain and loss in decibels.

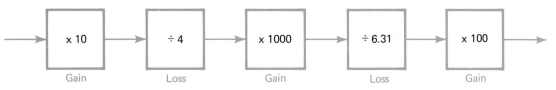

Fig. 6-3 Gain and loss in ratios.

signal in will be less than 1. Logarithms for numbers less than 1 are negative. To evaluate the overall performance of the system shown in Fig. 6-2, it is necessary only to add the numbers:

Overall gain = + 10 − 6 + 30 − 8 + 20
            = + 46 dB

When the gain or loss of individual parts of a system is given in decibels, it is very easy to evaluate the overall performance. This is why the decibel has come to be so widely used in electronics.

Figure 6-3 shows the same system without the decibel gain or loss given. Now, it is not so easy to evaluate the overall performance. The overall performance will be given by

Overall gain = (10 ÷ 4) × (1000 ÷ 6.31) × 100
            = 39,619.65

Notice that it is necessary to multiply for the gain stages and divide for the loss stages. When stage performance is given in decibels, gains are added and losses subtracted. Which do you think is easier?

Figures 6-2 and 6-3 are the same system. One has gain and loss specified in decibels, and the other in ratios. One has an overall gain of + 46 dB, and the other has an overall gain of 39,619.65. To check this:

$$dB = 10 \times \log 39{,}619.65$$
$$dB = 10 \times 4.60$$
$$dB = 46$$

The decibel is based on the ratio of the power output to the power input. It can also be used to describe the ratio of two voltages. The equation for finding voltage gain is slightly different from the one used for finding power gain:

$$dB \text{ voltage gain} = 20 \times \log \left(\frac{\text{voltage out}}{\text{voltage in}}\right)$$

Notice that the logarithm is multiplied by 20 in the above equation. This is because power varies as the square of the voltage:

$$\text{Power} = \frac{V^2}{R}$$

Power gain can be specified as follows:

$$\text{Power gain} = \frac{(V_{out})^2/R_{out}}{(V_{in})^2/R_{in}}$$

If R_{out} and R_{in} happen to be equal, they will cancel. Now power gain reduces to

$$\text{Power gain} = \frac{(V_{out})^2}{(V_{in})^2}$$

Since $\log V^2 = 2 \times \log V$, the logarithm can be multiplied by 2 to remove the need for squaring the two voltages:

$$dB \text{ voltage gain} = 10 \times 2 \times \log \left(\frac{V_{out}}{V_{in}}\right)$$
$$= 20 \times \log \left(\frac{V_{out}}{V_{out}}\right)$$

It is important to remember that when decibels are used to express gain in a voltage amplifier, the output resistance R_{out} is assumed equal to the input resistance R_{in}. This is because the decibel is really a unit of power gain and can be adapted to voltage gain only by using this assumption.

The technician needs to have a *feeling* for gain and loss expressed in decibels. Generally, a quick estimate is all that is required. Table 6-1 contains all the information needed for making good estimates. The following examples show the use of the table:

Example 6-1

A power amplifier has a gain of 15 dB. What is the ratio of power out to power in?

Voltage gain

79

Table 6-1 Ratios for finding gain and loss in decibels

POWER	
GAIN	LOSS
1 dB = 1.26	–1 dB = 0.79
3 dB = 2.00	–3 dB = 0.50
10 dB = 10.00	–10 dB = 0.10
VOLTAGE	
GAIN	LOSS
1 dB = 1.12	–1 dB = 0.89
3 dB = 1.41	–3 dB = 0.71
10 dB = 3.16	–10 dB = 0.32
20 dB = 10.00	–20 dB = 0.10

From Table 6-1,

$$10 \text{ dB} = 10$$
$$3 \text{ dB} = 2$$
$$1 \text{ dB} = 1.26$$

The value 15 dB can be split into parts:

$$10 \text{ dB} + 3 \text{ dB} + 1 \text{ dB} + 1 \text{ dB}$$

Now, the power ratios can be multiplied:

$$10 \times 2 \times 1.26 \times 1.26 = 31.75$$

The power output of the amplifier is 31.75 times greater than the power input to the amplifier.

Example 6-2

A power meter is used to measure the output of a transmitter. It is 1000 W. The power meter is then moved to the antenna, and the measurement there is 400 W. What is the loss in decibels in the line between the transmitter and the antenna?

From the table,

$$-3 \text{ dB} = 0.5$$

This would cause a loss of half the power. The loss in this case must be greater. Since -1 dB $= 0.79$, -4 dB would represent a total loss of

$$0.5 \times 0.79 = 0.4$$

Thus the antenna power would be

$$0.4 \times 1000 \text{ W} = 400 \text{ W}$$

The loss in the line is -4 dB.

Example 6-3

An audio amplifier is specified as capable of producing an output of 50 W, ± 1 dB. What is the actual range of power output from this amplifier?

$$1 \text{ dB} = 1.26$$

and

$$-1 \text{ dB} = 0.79$$
$$50 \times 1.26 = 63 \text{ W}$$

and

$$50 \times 0.79 = 39.5 \text{ W}$$

The actual power output can range between 39.5 and 63 W.

Example 6-4

A technician is using an oscilloscope to measure a high-frequency waveform. The oscilloscope specifications show a -3-dB loss at this frequency. If the true voltage is 10 V, what will be seen on the screen?

From the table,

$$-3 \text{ dB} = 0.71$$
$$10 \times 0.71 = 7.1 \text{ V}$$

The voltage on the screen will be 7.1 V.

Example 6-5

The response of a low-pass filter is supposed to be -6 dB at 5 kHz. A technician measures the output of the filter as 1 V at 1 kHz and finds that it drops to 0.5 V at 5 kHz. Is the filter working properly?

At -3 dB, the voltage loss is 0.71. Therefore, -6 dB would be

$$0.71 \times 0.71 = 0.50$$

The output voltage is, therefore,

$$1 \text{ V} \times 0.50 = 0.5 \text{ V}$$

The filter is working properly.

Example 6-6

It takes a 50-V video signal to produce a high-contrast picture in a television receiver. The signal at the tuner is only 100 μV. What must the overall gain ot the receiver be in decibels?

First, we must find the voltage ratio:

$$\text{Voltage ratio} = \frac{50}{100 \times 10^{-6}} = 500,000$$

A 20-dB voltage gain is represented by a ratio of 10. How many times can a ratio of 10 go into 500,000?

$500,000 \div 10 = 50,000$	(20 dB)
$50,000 \div 10 = 5000$	(20 dB)
$5000 \div 10 = 500$	(20 dB)
$500 \div 10 = 50$	(20 dB)
$50 \div 10 = 5$	(20 dB)
	100 dB

Thus, the gain must be in excess of 100 dB. To get an exact answer, the remaining 5 must be evaluated. The 5 can be split into these parts:

$5 \div 3.16 = 1.58$	(10 dB)
$1.58 \div 1.41 = 1.12$	(3 dB)
$1.12 \div 1.12 = 1$	(1 dB)
	14 dB

The gain of the receiver must be

$$100 + 14 = 114 \text{ dB}$$

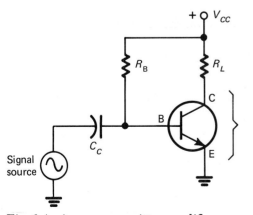

Fig. 6-4   A common-emitter amplifier.

Answer the following.

1. A two-stage amplifier has a voltage ratio of 200 in the first stage and a voltage ratio of 80 in the second stage. What is the overall voltage ratio of the amplifier? *16080*

2. A two-stage amplifier has a voltage gain of 46 dB in the first stage and 38 dB in the second stage. What is the overall gain of the amplifier in decibels? *84 Db*

3. A police receiver needs about 2 V of audio input to the speaker for good volume. If the receiver sensitivity is specified at 1 μV, what will the overall gain of the receiver have to be in decibels? *126 Db*

4. A 100-W audio amplifier is specified at −3 dB at 20 Hz. What power output can be expected at 20 Hz? *50w*

5. A transmitter produces 10 W of output power. An amplifier is added with a specified gain of 6 dB. What is the output power from the amplifier? *40w*

6. A transmitting station runs 100 W of output power into an antenna with a 9-dB gain. What is the effective radiated power? *800w*

6-2 COMMON-EMITTER AMPLIFIER

Figure 6-4 shows a common-emitter amplifier. It is so named because the emitter of the transistor is common to both the input circuit and the output circuit. The input signal is applied across ground and the base circuit of the transistor. The output signal appears across ground and the collector of the transistor. Since the emitter is connected to ground, it is common to both signals, input and output.

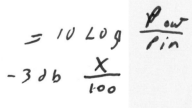

= 10 Log $\frac{P_{out}}{P_{in}}$

−3db $\frac{X}{100}$

81

Static state

There are two resistors in the circuit of Fig. 6-4. One is a base bias resistor R_B, and the other is a collector load resistor R_L. The base bias resistor is selected to limit the base current to some low value. The collector load resistor makes it possible to develop a voltage swing across the transistor (from collector to emitter). This voltage swing becomes the output signal.

Capacitor C_c in Fig. 6-4 is the coupling capacitor. It is required if the signal source has a dc component or if the signal source conducts. Without the capacitor, the transistor may be short-circuited by the signal source. Figure 6-5 shows why. The direct connection provides a dc path through the signal source, through R_B, and on to the power supply V_{cc}. This condition may turn off the transistor. The current through R_B is supposed to come from the base of the transistor, as shown in Fig. 6-6. There must be some base current for the transistor to be on.

Figure 6-6 has enough information to show how the amplifier operates. First the base current will be calculated. Two parts can limit current flow: R_B and the base-emitter junction. Resistor R_B has a very high resistance so it will limit the flow quite a bit. The base-emitter junction is forward-biased so its resistance is low. Thus, R_B is the only significant component from which the base current will be determined:

$$I_B = \frac{V_{cc}}{R_B}$$

$$= \frac{12 \text{ V}}{100 \times 10^3 \ \Omega}$$

$$= 120 \times 10^{-6} \text{ A}$$

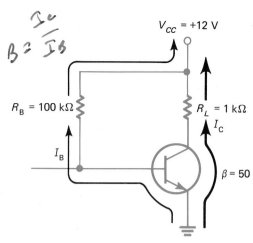

Fig. 6-6 Transistor circuit currents.

It can be seen that the base current is very small. It is only 120 μA. Since β is given, the collector current can now be found:

$$I_c = \beta \times I_B = 50 \times 120 \times 10^{-6} = 6 \times 10^{-3}$$

The collector current will be 6 mA. This current flows through the load resistor R_L. The voltage drop across R_L will be

$$V_L = I_c \times R_L = 6 \times 10^{-3} \times 1 \times 10^3 = 6 \text{ V}$$

With a 6-V drop across R_L, the drop across the transistor will be

$$V_{CE} = V_{cc} - V_{R_L} = 12 - 6 = 6 \text{ V}$$

The calculations show the condition of the amplifier at its *static*, or resting, state. The input signal can cause the static conditions to change. Figure 6-7 shows why. As the signal source goes positive with respect to ground, the base current will increase. The positive-going signal causes additional base current to flow onto the plate of the coupling capacitor. This is shown in Fig. 6-7(*a*). Figure 6-7(*b*) shows the input signal going negative. Current flows off the capacitor plate and up through R_B. This decreases the base current.

As the base current increases and decreases, so does the collector current. This is because base current controls collector current. As the collector current increases and decreases, the voltage drop across the load resistor also increases and decreases. This means that the voltage drop across the transistor must also be changing. It does not remain constant at 6 V.

Figure 6-8 shows how the output signal is produced. A transistor can be thought of as a

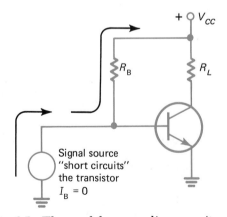

Fig. 6-5 The need for a coupling capacitor.

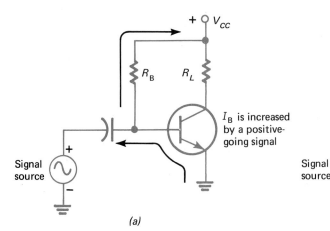

Fig. 6-7 The effect of the input signal on I_B.

resistor from its collector terminal to its emitter terminal. The better the transistor conducts, the lower this resistor is in value. The poorer it conducts, the higher this resistor is in value. Transistor conduction does change as base current changes. So, we can assume that an input signal will change the collector-emitter resistance of the transistor.

In Fig. 6-8(a) the amplifier is at its static state. The supply voltage is divided equally between R_L and R_{CE}. Resistor R_L is the load, and R_{CE} represents the resistance of the transistor. The time graph shows that the output V_{CE} is a steady 6 V.

Now, the input signal goes negative. This *decreases* the base current and, in turn, decreases the collector current. The transistor is now offering more resistance to current flow. Figure 6-8(b) shows what happens. Resistance R_{CE} has increased to 2 kΩ. The voltages do not divide equally:

$$V_{CE} = \frac{R_{CE}}{R_{CE} + R_L} \times V_{cc}$$

$$= \frac{2\ \text{k}\Omega}{2\ \text{k}\Omega + 1\ \text{k}\Omega} \times 12$$

$$= 8\ \text{V}$$

Thus the output signal has increased to 8 V. The time graph shows this change. Another way to solve for the voltage drop across the transistor would be to solve for the current:

$$I = \frac{V_{cc}}{R_L + R_{CE}}$$

$$= \frac{12}{1\ \text{k}\Omega + 2\ \text{k}\Omega}$$

$$= 4 \times 10^{-3}\ \text{A}$$

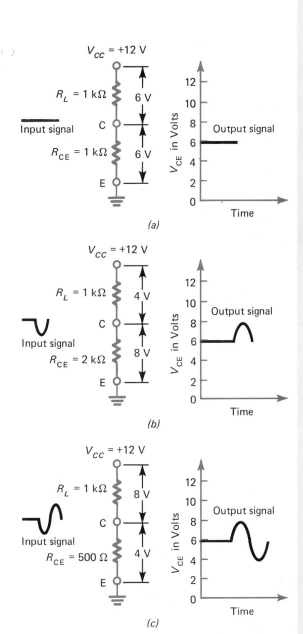

Fig. 6-8 How the output signal is created.

Phase inversion

Distortion

Clipping

Load line

Now, this current can be used to calculate the voltage across the transistor:

$$V_{CE} = I \times R_{CE}$$
$$= 4 \times 10^{-3} \times 2 \times 10^3$$
$$= 8 \text{ V}$$

This agrees with voltage found by using the ratio technique.

Figure 6-8(c) shows what happens in the amplifier circuit when the input signal goes positive. The base current increases. This makes the collector current increase. The transistor is conducting better, so its resistance has decreased. The output voltage V_{CE} is now

$$V_{CE} = \frac{0.5 \text{ k}\Omega}{0.5 \text{ k}\Omega + 1 \text{ k}\Omega} \times 12 = 4 \text{ V}$$

The time graph shows this change in output voltage.

Note that in Fig. 6-8 the input signal is 180° out of phase with the output signal. When the input goes negative [Fig. 6-8(b)], the output goes in a positive direction. When the input goes positive [Fig. 6-8(c)], the output goes in a negative direction (less positive). This is called *phase inversion*. It is an important characteristic of the common-emitter amplifier.

The output signal should be a good replica of the input signal. If the input is a sine wave,

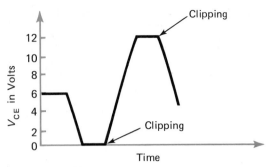

Fig. 6-9 A clipped sine wave.

the output should be a sine wave. When this is achieved, the amplifier is *linear*. One thing that can make an amplifier nonlinear is too much input signal. This will cause the output signal to show distortion, as shown in Fig. 6-9. The waveform is clipped. The output V_{CE} cannot exceed 12 V. This means that the output signal will approach this limit and then suddenly stop increasing. Note that the positive-going part of the sine wave has been clipped off. The negative-going part of the signal also has been clipped. The output signal is distorted. Such distortion would cause speech or music to sound bad in an audio amplifier. This is what happens when the volume control is turned up too far. The clipping makes the audio very distorted.

Clipping can be avoided by keeping the input signal small and by operating the amplifier at the proper static point. This is best shown by drawing a load line. Figure 6-10

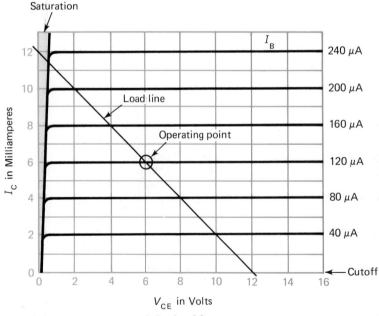

Fig. 6-10 Transistor amplifier load line.

shows a load line drawn on the collector family of characteristic curves. To draw a load line, it is necessary to know the supply voltage V_{CC} and the value of the load resistor R_L. Voltage V_{CC} sets the lower end of the load line. If V_{CC} is equal to 12 V, one end of the load line is found at 12 V on the horizontal axis. The other end of the load line is set by the *saturation current*. This is the current that will flow if the collector resistance drops to zero. Under this condition, only R_L will limit the flow. Ohm's law is used to find the saturation current:

$$I_{sat} = \frac{V_{CC}}{R_L}$$
$$= \frac{12\ V}{1\ k\Omega}$$
$$= 12 \times 10^{-3}\ A,\ or\ 12\ mA$$

This value of current is found on the vertical axis. It is the other end of the load line.

As shown in Fig. 6-10, the load line for the amplifier runs between 12 mA and 12 V. These two values are the circuit limits. No matter what the input signal is, the collector current cannot exceed 12 mA and the output voltage cannot exceed 12 V. If the input signal is too large, the output will clip at these points.

It is possible to operate the amplifier at any point along the load line. The best operating point is usually in the center of the load line. This point is circled in Fig. 6-10. Notice that the operating point is the intersection of the load line and the 120-μA curve. This is the same value of base current as was calculated before. Project straight down from the operating point. The voltage V_{CE}, is 6 V. This also agrees with the previous calculation. Project to the left from the operating point. The current I_C is 6 mA. Again there is agreement with the previous calculations.

The load line does not really provide new information. It is helpful because it provides a visual or graphical view of circuit operation. For example, examine Fig. 6-11(a). The operating point is in the center of the load line. Notice that as the input changes, the output signal can be projected with no clipping. Look at Fig. 6-11(b). The operating point is near the saturation end of the load line. The output signal is now clipped on the negative-going portion. Figure 6-11(c) shows the operating point near the cutoff end of the

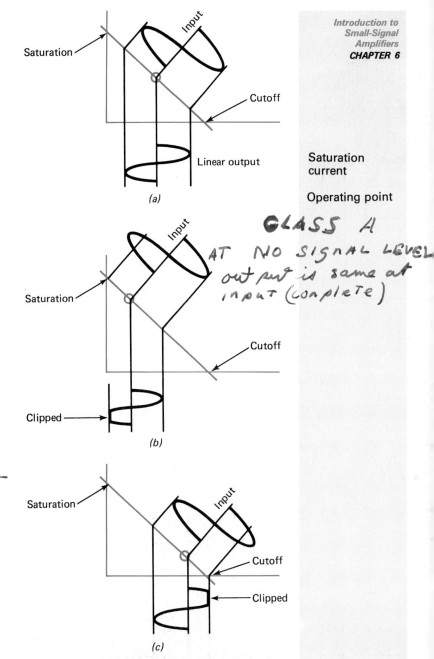

Fig. 6-11 Comparing amplifier operating points.

load line. The signal is clipped on the positive-going portion. It should be obvious now why the best operating point is near the center of the load line. It allows the most output before clipping occurs.

Usually saturation and cutoff are both avoided in linear amplifiers. They cause clipping and distortion of the signal. Three possible conditions for an amplifier are shown in Fig. 6-12. Figure 6-12(a) indicates that a saturated amplifier is similar to a closed switch. The current is maximum because the tran-

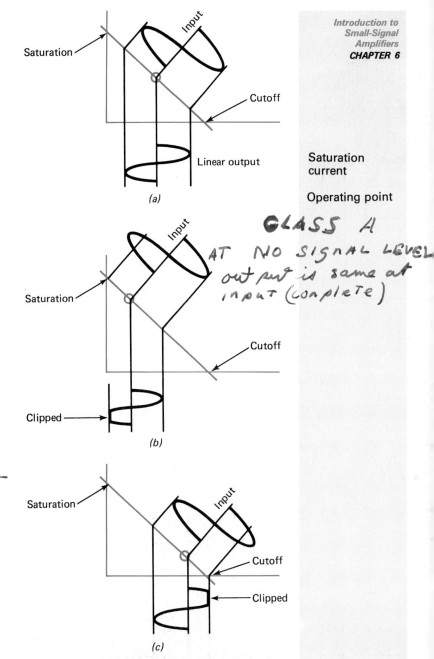

Saturation current

Operating point

CLASS A
AT NO SIGNAL LEVEL
output is same at
input (complete)

85

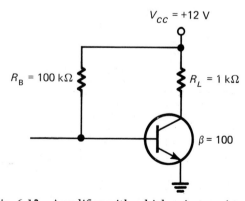

$+\circ\ V_{CC}$

R_L

Saturation

$V_{CE} = 0$ and I_C is maximum

C

E

(a)

$+\circ\ V_{CC}$

R_L

Cutoff

$V_{CE} = V_{CC}$ and $I_C = 0$

C

E

(b)

$+\circ\ V_{CC}$

R_L

Active

$V_{CE} \approx \frac{1}{2} V_{CC}$ and $I_C = \frac{1}{2}$ maximum

C

E

(c)

Fig. 6-12   Three possible amplifier conditions.

[handwritten margin notes:]
$I_B = 240\,uA$
$I_C = 12\ mA$
$V_L = 12v$
$V_{CE} = 0$
SATURATION

sistor is turned on so hard. It is as if the transistor were a closed switch. No voltage drops across a closed switch, so $V_{CE} = 0$. Figure 6-12(b) shows the amplifier at cutoff. The transistor is turned off, and no current flows. All the voltage drops across the open switch, so V_{CE} is equal to the supply voltage. An active transistor amplifier lies in between the two extremes. The transistor is partly on. It can be represented as a resistor. The current is about half of the maximum, or saturation, value; and V_{CE} is about half of the supply voltage. The conditions of Fig. 6-12 should be memorized. They are very useful when troubleshooting.

Review Questions

Answer the following.

7. Refer to Fig. 6-6. Change the value of R_B to 300 kΩ. What is the base current now?

8. With R_B changed to 300 kΩ in Fig. 6-6, what is the new value of collector current? *2 mA*

9. What is the new value of the voltage drop across the load resistor for Fig. 6-6? *2 ✓*

10. What is the new value of the voltage drop across the transistor for Fig. 6-6? *10 ✓*

11. Refer to Fig. 6-10. Find the new operating point on the load line. Project down to the voltage axis. How does this answer agree with your answer to question 10? *It agrees*

12. Refer to Fig. 6-10. Project to the left from the new operating point. How does this answer agree with your answer to question 8? *yes*

13. Refer to Fig. 6-6. Change the value of R_B to 50 kΩ. Calculate the new base current, the new collector current, the new voltage drop across the load resistor, and the voltage drop across the transistor. Is the transistor in saturation, in cutoff, or in the linear range?

14. Refer to Figs. 6-6 and 6-10. If the supply voltage V_{CC} is changed to 10 V, where will the new load line run? *run on V_{CE} LINE*

6-3 STABILIZING THE AMPLIFIER

Figure 6-13 shows a common-emitter amplifier that is identical to the one shown in Fig. 6-6 with one important exception. The transistor has a β of 100. If we analyze the circuit for base current, we get

$$I_B = \frac{V_{CC}}{R_B}$$

$$= \frac{12}{100 \times 10^3}$$

$$= 120\ \mu A$$

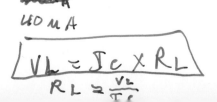

Fig. 6-13   Amplifier with a high-gain transistor.

[handwritten at bottom:]

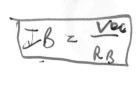

$I_C = \beta \times I_B$

$\frac{\square}{\beta} = \frac{I_C}{\beta}$

$I_B = \frac{V_{CC}}{R_B}$

40 uA

$V_L = I_C \times R_L$

$R_L = \frac{V_L}{I_C}$

This is the same value of base current that was calculated before. The collector current is

$$I_C = \beta \times I_C$$
$$= 100 \times 120 \times 10^{-6}$$
$$= 12 \text{ mA}$$

This is twice the collector current of Fig. 6-6. Now, we can solve for the voltage drop across R_L:

$$V_{R_L} = I_C \times R_L$$
$$= 12 \times 10^{-3} \times 1 \times 10^3$$
$$= 12 \text{ V}$$

Finally, the voltage drop across the transistor is

$$V_{CE} = V_{CC} - V_{R_L} = 12 - 12 = 0 \text{ V}$$

There is no voltage across the transistor. The transistor amplifier is in saturation. A saturated transistor is not capable of linear amplification. The circuit of Fig. 6-13 would produce severe clipping and distortion.

The only change from Fig. 6-6 to Fig. 6-13 is in the β of the transistor. The value of β does vary widely among transistors with the same part number. This means that the amplifier circuit is not practical. It is too sensitive to β, which varies from transistor to transistor and with temperature. We need a circuit that is not as sensitive to β.

Figure 6-14 shows a practical common-emitter amplifier. This circuit is not nearly so sensitive to β. It is also much more stable over a wide temperature range. It is a very popular circuit and is used quite often in elec-

tronic equipment. The operating point for this circuit is found by the following process:

1. Calculate the voltage drop across R_{B_2}. This is called the base voltage, or V_B. Resistors R_{B_1} and R_{B_2} form a voltage divider across V_{CC}.
2. Assume a 0.7-V drop from base to emitter for silicon transistors and a 0.2-V drop for germanium transistors. Thus $V_E = V_B - 0.7$ (for silicon transistors) and $V_E = V_B - 0.2$ (for germanium transistors).
3. Calculate the emitter current I_E, using Ohm's law.
4. Assume a collector current equal to the emitter current: $I_C = I_E$.
5. Calculate the voltage drop across the load resistor using Ohm's law.
6. Calculate the voltage drop across the transistor using Kirchhoff's voltage law:

$$V_{CE} = V_{CC} - V_{R_L} - V_E$$

The six-step process is not exact. It is accurate enough for most practical purposes. The first step ignores the base current that flows through R_{B_1}. This current is usually only about one-tenth the divider current. Thus, a small error is made by using only the resistor values to compute V_B. The second step is based on what we know about silicon and germanium junctions. The junctions have a very predictable voltage drop when forward-biased. The third step is Ohm's law, and no error is introduced here. The fourth step introduces a small error. We expect the emitter current to be a few percent greater than the collector current. This few percent difference will not cause a large error. The last two steps are circuit laws, and no error is introduced. All in all, the procedure can provide very useful answers. Notice that β is not used anywhere in the six steps.

Using this procedure to solve the circuit in Fig. 6-14, we get

1. $$V_B = \frac{R_{B_2}}{R_{B_1} + R_{B_2}} \times V_{CC}$$
    $$= \frac{2.2 \text{ k}\Omega}{18 \text{ k}\Omega + 2.2 \text{ k}\Omega} \times 12$$
    $$= 1.307 \text{ V}$$

2. $$V_E = V_B - 0.7 = 1.307 - 0.7 = 0.607 \text{ V}$$

3. $$I_E = \frac{V_E}{R_E} = \frac{0.607}{100} = 6.07 \text{ mA}$$

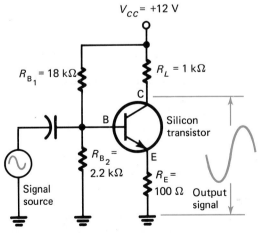

$V_{CC} = +12 \text{ V}$

$R_{B_1} = 18 \text{ k}\Omega$

$R_L = 1 \text{ k}\Omega$

C

B

Silicon transistor

$R_{B_2} = 2.2 \text{ k}\Omega$

E

$R_E = 100 \Omega$

Signal source

Output signal

Fig. 6-14   A practical common-emitter circuit.

87

4. $I_C = I_E = 6.07$ mA

5. $V_{R_L} = I_C \times R_L = 6.07 \times 10^{-3} \times 1 \times 10^3$
    $= 6.07$ V

6. $V_{CE} = V_{CC} - V_{R_L} - V_E$
    $= 12 - 6.07 - 0.607$
    $= 5.32$ V

Since the collector-emitter voltage is near half the supply voltage, we can assume the circuit will make a good linear amplifier. The circuit will work well with any reasonable value of β and will be stable over a wide temperature range.

The next step is to evaluate the amplifier for voltage gain. Junction transistors of the bipolar type are actually current amplifiers. It is the changing base current that causes the output signal. However, a changing voltage at the input to such an amplifier will cause a changing current. This makes it possible to discuss and calculate for voltage gain even though the transistor is current-controlled.

The first step in calculating voltage gain is to estimate the ac resistance of the emitter of the transistor. This resistance is determined by the emitter current. Its symbol is r_E. The ac resistance r_E is found by

$$r_E = \frac{25}{I_E} \qquad I_E \text{ is in milliamperes}$$

We have already solved the circuit of Fig. 6-14 for the emitter current, so we can now estimate r_E:

$$r_E = \frac{25}{6.07} = 4.12 \ \Omega$$

In actual circuits, r_E can be much higher. It can be as high as

$$r_E = \frac{50}{I_E}$$

So, for the circuit of Fig. 6-14, r_E could be as high as

$$r_E = \frac{50}{6.07} = 8.24 \ \Omega$$

Now, it is possible to find the voltage gain of the circuit by using the equation

$$A_V = \frac{R_L}{R_E + r_E}$$

Notice that A_V is the symbol for voltage gain. The voltage gain of the amplifier of Fig. 6-14 will be

$$A_V = \frac{1000}{100 + 4.12} = 9.6$$

The amplifier will have a voltage gain of 9.6. If the input signal is 1 V peak-to-peak, then the output signal should be 9.6 V peak-to-peak. If the input signal is 2 V peak-to-peak, then the output will be clipped. It is not possible to exceed the supply voltage in peak-to-peak output in this type of amplifier. The calculated gain will hold true only if the amplifier is operating in a linear fashion.

Sometimes, a much higher gain is needed. It is possible to improve the gain quite a bit by adding an emitter bypass capacitor. Figure 6-15 shows this capacitor added to the amplifier. The capacitor is chosen to have a low reactance at the frequency of operation. It acts as a short circuit for the ac signal. This means that the ac signal is bypassed around R_E. Current will take the path of least opposition. Since R_E has been bypassed, the voltage gain equation no longer shows it:

$$A_V = \frac{R_L}{r_E}$$

The voltage gain for the circuit of Fig. 6-15 is

$$A_V = \frac{1000}{4.12} = 243$$

But, remember, r_E might be as high as 8.24 Ω. The gain might be somewhat less:

$$A_V = \frac{1000}{8.24} = 121$$

Thus, the voltage gain of the amplifier of Fig. 6-15 will range between 121 and 243. It is possible to make accurate predictions for voltage gain. The procedure is more involved, and it is necessary to know more about the transistor. For practical work, the procedure used here is good.

Without the bypass capacitor, the voltage gain was calculated at 9.6. With the bypass

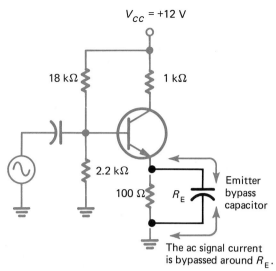

V_{CC} = +12 V

18 kΩ 1 kΩ

2.2 kΩ

100 Ω R_E

Emitter
bypass
capacitor

The ac signal current
is bypassed around R_E.

Fig. 6-15 Adding an emitter bypass capacitor.

capacitor, the voltage gain will be at least 121. This is almost an amazing increase in gain for simply adding a capacitor. The gain comes at a price. The capacitor may be a costly item. It must have a low reactance at its frequency of operation. If the amplifier is to operate at very low frequencies, the capacitor may have to be quite large. It is usually chosen to have one-tenth the reactance in ohms as compared to the emitter resistor in ohms. In Fig. 6-15, the emitter resistor is 100 Ω. This means that the capacitor must not have a reactance greater than 10 Ω at the lowest operating frequency. An audio amplifier may have to operate down to 15 Hz. The capacitor reactance equation can now be used to select the bypass capacitor:

$$X_C = \frac{1}{6.28fC} \quad \text{or} \quad C = \frac{1}{6.28fX_C}$$

Solving for C, we get

$$C = \frac{1}{6.28 \times 15 \times 10} = 1061 \ \mu F$$

This is, indeed, a large capacitor. The voltage rating can be very low since the emitter voltage is only 0.61 V in this circuit. However, the capacitor will still be expensive and physically large.

Emitter bypassing is more profitable in high-frequency circuits. The higher frequencies require a much lower value of capacitance for effective bypassing.

Review Questions

*Introduction to
Small-Signal
Amplifiers*
CHAPTER 6

Input impedance

Answer the following.

Questions 15 through 21 refer to Fig. 6-14 with these changes:
R_{B_2} = 3.3 kΩ and R_L = 470 Ω.

15. Calculate V_B.
16. Calculate I_E.
17. Calculate V_{R_L}.
18. Calculate V_{CE}.
19. Is the amplifier operating in the center of the load line?
20. Calculate A_V.
21. Calculate the range for A_V if an emitter bypass capacitor is added to the circuit.

6-4 OTHER CONFIGURATIONS

The common-emitter amplifier is a very popular circuit. It serves as the basis for most linear amplifiers. However, under some circuit conditions, other types of amplifiers may be a better choice.

Amplifiers have many characteristics. Among these is input impedance. The input impedance of an amplifier is the loading effect it will present to a signal source. Figure 6-16 shows that when a signal source is connected to an amplifier, that source sees a load, not an amplifier. The load seen by the source is the input impedance of the amplifier.

Signal sources vary widely. An antenna is the signal source for a radio receiver. An antenna might have an impedance of 50 Ω. A microphone is the signal source for a public address system. A microphone might have an impedance of 100,000 Ω. Every signal source has a characteristic impedance.

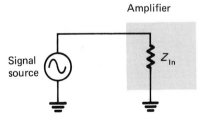

Amplifier

Signal
source

Z_{In}

The signal source is loaded by Z_{In},
the input impedance of the amplifier

Fig. 6-16 Amplifier loading effect.

89

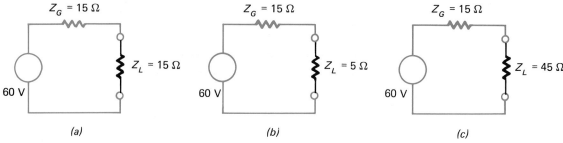

(a)   *(b)*   *(c)*

Fig. 6-17 The need for impedance matching.

Impedance
matching

Common-collector
amplifier

The situation can be stated simply. For best efficiency, the source impedance should equal the amplifier input impedance. This is called *impedance matching*. Figure 6-17 shows why impedance matching gives the best efficiency. In Fig. 6-17(a) a 60-V signal source has an impedance of 15 Ω. This impedance Z_G is drawn as an external resistor in series with the generator (signal source). Since the generator impedance does act in series, the circuit is a good model, as shown. The load impedance in Fig. 6-17(a) is also 15 Ω. Thus, we have an impedance match. Let us see how efficient the circuit is:

$$I = \frac{V}{Z} = \frac{60}{15 + 15} = 2 \text{ A}$$

Now, we can solve for the load dissipation since all loads are resistive:

$$P = I^2 \times Z = 2^2 \times 15 = 60 \text{ W}$$

This proves that a 15-Ω load will dissipate 60 W. Figure 6-17(b) shows the same circuit, but the load is now 5 Ω. Solving this circuit gives

$$I = \frac{60}{15 + 5} = 3 \text{ A}$$
$$P = 3^2 \times 5 = 45 \text{ W}$$

The dissipation is *less* when the load impedance is less than the source impedance. Figure 6-17(c) shows the load impedance at 45 Ω. Solving this circuit gives

$$I = \frac{60}{15 + 45} = 1 \text{ A}$$
$$P = 1^2 \times 45 = 45 \text{ W}$$

The dissipation is less when the load impedance is greater than the source impedance.

Maximum load power will occur only when the impedances are matched.

The common-emitter amplifier has an input impedance of around 1000 Ω. The actual value depends on both the transistor used and the other parts in the amplifier. This may or may not be a good value. It depends on the signal source.

Figure 6-18 shows a common-collector amplifier. It is so named because the collector terminal is common to the input and the output circuits. This may be a bit difficult to recognize at first. The circuit of Fig. 6-18 may not appear very different from the common-emitter configuration. There are two very important differences:

1. The collector is bypassed to ground with a capacitor. This capacitor has very low reactance at the signal frequency. As far as signals are concerned, the collector is grounded.
2. The load resistor is in the emitter circuit. The output signal is across this load. Thus, the emitter is now the output terminal. The collector was the output terminal in the common-emitter configuration.

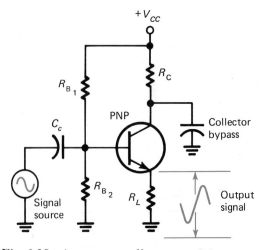

Fig. 6-18 A common-collector amplifier.

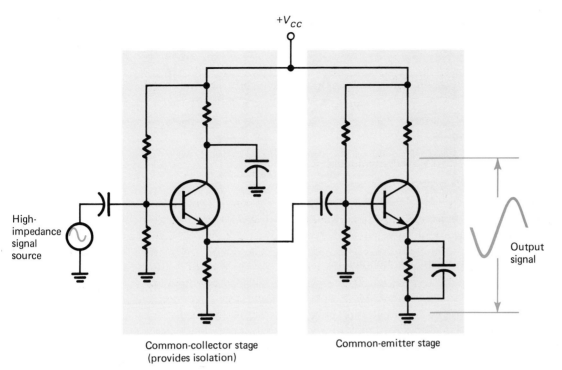

Isolation
amplifier

Emitter follower

Common-base
amplifier

Fig. 6-19   A two-stage amplifier.

The common-collector amplifier has a very high input impedance.   It is several hundred thousand ohms.   If a signal source has a very high characteristic impedance, the common-collector amplifier may prove to be the best choice.   The stage following the common collector could be a common-emitter configuration.   Figure 6-19 shows this arrangement. The common-collector stage is sometimes referred to as an *isolation amplifier.*   Its high input impedance loads the signal source very lightly.   Only a very small signal current will flow.   Thus, the signal source has been isolated from the loading effects of the rest of the circuit.

In addition to a high input impedance, the common-collector amplifier has some other important characteristics.   It is not capable of giving any voltage gain.   The output signal will always be less than the input signal as far as voltage is concerned.   The current gain is very high.   There is also a moderate power gain.   There is no phase inversion in the common-collector amplifier.   As the signal source drives the base terminal in a positive direction, the output (emitter terminal) also goes in a positive direction.   The fact that the emitter follows the base has led to a second name for this amplifier.   It is often called an *emitter follower.*

The last configuration to be discussed is the *common-base amplifier.*   This amplifier has its base terminal common to both the input and output signals.   It has a very low input impedance, on the order of 50 Ω.   It is useful with low-impedance signal sources.   It is also a good performer at very high frequencies. Very high frequencies are often called radio frequencies.   This means that the common-base amplifier makes a good radio frequency (RF) amplifier when the source impedance is low.

Figure 6-20 is a schematic diagram of a common-base RF amplifier.   It is designed to amplify weak radio signals from the antenna circuit.   The antenna impedance is low, on the order of 50 Ω.   This makes a good impedance match from the antenna to the amplifier.   The base is grounded at the signal frequency by C_4.   The signal is fed into the emitter terminal of the transistor, and the amplified output signal is taken from the collector.   Circuits L_1C_2 and L_2C_5 are tuning circuits.   They are resonant at the desired frequency of operation.   This helps the receiver reject other frequencies that could cause interference.   Coil L_2 and capacitor C_5 make the collector load for the amplifier. This load will be a high impedance at the resonant frequency.   This makes the voltage

91

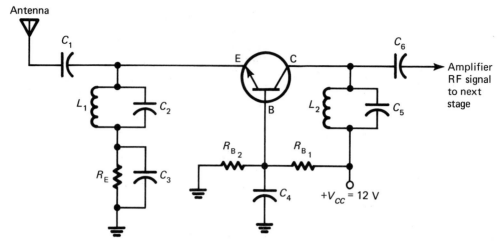

Fig. 6-20   A common-base RF amplifier.

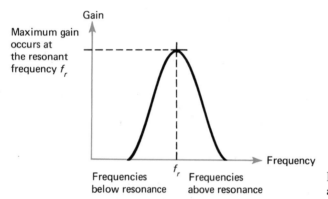

Maximum gain
occurs at
the resonant
frequency f_r

Gain

Frequency

Frequencies
below resonance

f_r Frequencies
above resonance

Fig. 6-21   Frequency response of the tuned **RF**
amplifier.

Table 6-2   Summary of amplifier configurations

	Common Base	Common Collector	Common Emitter
Basic circuit (Showing signal source and load R_L)			
Power gain	Yes	Yes	Yes (highest)
Voltage gain	Yes	No (less than 1)	Yes
Current gain	No (less than 1)	Yes	Yes
Input impedance	Lowest ($\approx 50\ \Omega$)	Highest ($\approx 300\ k\Omega$)	Medium ($\approx 1\ k\Omega$)
Output impedance	Highest ($\approx 1\ M\Omega$)	Lowest ($\approx 300\ \Omega$)	Medium ($\approx 50\ k\Omega$)
Phase inversion	No	No	Yes
Application	Used mainly as an RF amplifier	Used mainly as an isolation amplifier	Universal—works best in most applications

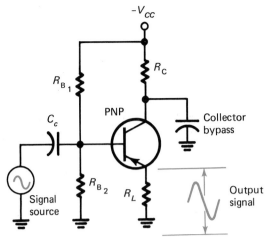

Fig. 6-22 A PNP transistor amplifier.

gain very high for this frequency. Other frequencies will have less gain through the amplifier. Figure 6-21 shows the gain performance for a tuned RF amplifier of this type.

The common-base amplifier is not capable of providing any current gain. The input current will always be more than the output current. This is because the emitter current is always highest in bipolar junction transistors. The amplifier is capable of a very large voltage gain. It is also capable of power gain. As with the emitter follower, it does not invert the signal (no phase inversion).

Only the common-emitter amplifier provides all three forms of gain: voltage, current, and power. It provides the best power gain of any of the configurations. It is easily the most useful of the three. Table 6-2 summarizes the important details for the three amplifier con-figurations. Study this information. It is very important for understanding and trouble-shooting linear electronic devices.

So far, only NPN transistor amplifiers have been shown. Everything that applies to these NPN circuits will also apply to PNP circuits. The only major difference will be the *polarity* of the supply voltage. Figure 6-22 shows a PNP amplifier. Note that the supply V_{CC} is negative.

Review Questions

Answer the following.

22. A signal source has an impedance of 300 Ω. It develops an output of 1 V. Calculate the power transfer from this source into each of the following amplifiers:

    A. Amplifier A has an input impedance of 100 Ω.

    B. Amplifier B has an input impedance of 300 Ω.

    C. Amplifier C has an input impedance of 900 Ω.

23. Refer to Fig. 6-18. The transistor is silicon; $R_{B_1} = 47$ kΩ, and $R_{B_2} = 68$ kΩ. If R_L is a 470-Ω resistor, what is the dc emitter current I_E?

24. Refer to Fig. 6-20. The transistor is silicon; $R_{B_1} = 10$ kΩ, $R_{B_2} = 2.2$ kΩ, and $R_E = 270$ Ω. Assume both coils have zero resistance. What is the collector current I_C?

25. Refer to Fig. 6-22. What is the configuration of this amplifier?

Summary

1. Amplifier gain is determined by dividing the output by the input.

2. Gain is specified as a voltage ratio, a power ratio, or a logarithm of a ratio.

3. When each part of a system is specified in decibel gain or loss, the overall performance can be obtained by simply adding all gains and subtracting all losses.

4. When each part of a system is specified in ratios, the overall performance is obtained by multiplying all gains and dividing by all losses.

5. The decibel is based on power gain or loss. It can be adapted to voltage gain or loss by assuming the resistance to be constant.

6. In a common-emitter amplifier, the emitter of the transistor is common to both the input signal and the output signal.

7. The collector load resistor allows the output voltage swing to be developed.

8. The base bias resistor limits base current to the desired steady or static level.

9. The input signal causes the base current,

the collector current, and the output voltage to change.

10. The common-emitter amplifier produces a 180° phase inversion.

11. One way to show amplifier limits is to draw a load line. Linear amplifiers usually are operated in the center of the load line.

12. A saturated transistor can be compared with a closed switch. The voltage drop across it will be zero.

13. A cutoff transistor can be compared with an open switch. The voltage drop across it will be equal to the supply voltage.

14. A transistor set up for linear operation should be between saturation and cutoff. The voltage drop across it should be about half the supply voltage.

15. For a transistor amplifier to be practical, it cannot be too sensitive to β.

16. A practical and stable circuit uses a voltage divider to set the base voltage and a resistor in the emitter lead.

17. The voltage gain of a common-emitter amplifier is set by the load resistance and the emitter resistance.

18. The common-emitter amplifier is the most popular of the three possible configurations.

19. The most efficient transfer of signal into an amplifier occurs when the source impedance matches the amplifier input impedance.

20. The common-collector, or emitter-follower, amplifier has a very high input impedance.

21. The common-collector amplifier has a voltage gain of less than 1.

22. Because of its high input impedance, the common-collector amplifier makes a good isolation amplifier.

23. The common-base amplifier has a very low input impedance.

24. The common-base amplifier works well at very high frequencies. It makes a good RF amplifier.

25. The common-emitter amplifier is the only one that gives both voltage and current gain. It has the best power gain.

26. Any of the three amplifier configurations can use either NPN or PNP transistors. The major difference is in polarity.

Chapter Review Questions

Supply the missing word in each statement.

6-1. The signal fed into an amplifier is 100 mV, and the output signal is 5 V. The voltage gain of the amplifier is ____?____.

6-2. The sensitivity of human hearing to loudness is not linear but ____?____.

6-3. An amplifier with a power gain of 10 dB develops an output signal of 20 W. The signal input power is ____?____.

6-4. A 100-W transmitter is fed into a coaxial cable with a 3-dB loss. The power reaching the antenna is ____?____.

6-5. A two-stage amplifier has a gain of 10 in the first stage and a gain of 18 in the second stage. The overall gain is ____?____.

6-6. A two-stage amplifier has a gain of 18 dB in the first stage and 10 dB in the second stage. The overall gain is ____?____.

6-7. An oscilloscope has a frequency response that is − 3 dB at 25 MHz. A 10-V peak-to-peak, 25-MHz signal is fed into the oscilloscope. The oscilloscope will show a voltage of ____?____.

6-8. In a common-emitter amplifier, the signal is fed to the base circuit of the transistor, and the output is taken from the ____?____.

6-9. Refer to Fig. 6-4. The component that prevents the signal source from shorting the base-emitter junction is ____?____.

6-10. Refer to Fig. 6-4. The component that allows the amplifier to develop an output voltage signal is ___?___.

6-11. Refer to Fig. 6-4. If R_B opens (infinite resistance), then the transistor will operate in ___?___.

6-12. Refer to Fig. 6-4. Assume $R_B = 100$ kΩ and $V_{CC} = 20$ V. The base current will be ___?___.

6-13. Refer to Fig. 6-6. Assume that $\beta = 80$. The voltage from collector to emitter will be ___?___.

6-14. Refer to Fig. 6-6. As an input signal drives the base in a positive direction, the collector will change in a ___?___ direction.

6-15. Refer to Fig. 6-6. As an input signal drives the base in a positive direction, the collector current should ___?___.

6-16. Clipping can be avoided by controlling the input signal and by operating the amplifier at the ___?___ of the load line.

6-17. Refer to Fig. 6-10. Assume the base current is 200 μA. The collector current will be ___?___.

6-18. Refer to Fig. 6-10. The base current is 40 μA. The voltage from collector to emitter will be ___?___.

6-19. Refer to Fig. 6-10. The base current is zero. The amplifier will be in ___?___.

6-20. A technician is troubleshooting an amplifier and measures V_{CE} to be near 0 V. Voltage V_{CC} is normal. The transistor is operating in ___?___.

6-21. Refer to Fig. 6-13. This amplifier circuit is not practical because it is too sensitive to temperature and ___?___.

6-22. Refer to Fig. 6-14. Assume the transistor to be germanium. The emitter current will be ___?___.

6-23. Refer to Fig. 6-14. Assume the load resistor is 860 Ω. Voltage V_{CE} will be ___?___.

6-24. Refer to Fig. 6-15. Assume the emitter current to be 1 mA. The voltage gain could be as high as ___?___.

6-25. A signal source has an impedance of 50 Ω. For best efficiency, an amplifier designed for this signal source should have an input impedance of ___?___.

6-26. Refer to Fig. 6-18. As the base is driven in a positive direction, the emitter will go in a ___?___ direction.

6-27. Refer to Fig. 6-18. This configuration is noted for a high input impedance and a ___?___ output impedance.

6-28. An amplifier is needed with a low input impedance for a radio frequency application. The best choice is probably the common ___?___ configuration.

6-29. An amplifier is needed with a moderate input impedance and the best possible power gain. The best choice is probably the common ___?___ configuration.

6-30. An amplifier is needed to isolate a signal source from any loading effects. The best choice is probably the common ___?___ configuration.

Answers to Review Questions

1. 16,000
2. 84 dB
3. 126 dB
4. 50 W
5. 40 W
6. 800 W
7. 40 μA
8. 2 mA
9. 2 V
10. 10 V
11. It agrees

12. It agrees
13. I_B = 240 μA
    I_C = 12 mA
    V_R = 12 V
    V_{CE} = 0 V
    Saturation
14. 10 V on V_{CE} axis
    10 mA on I_C axis
15. 1.86 V
16. 11.6 mA
17. 5.45 V

18. 5.39 V
19. Very near the center
20. 4.6
21. 109 to 218
22. 0.625 mW to A
    0.833 mW to B
    0 625 mW to C
23. 13.6 mA
24. 5.4 mA
25. Common collector

Advanced Small-Signal Amplifiers

- This chapter treats some advanced concepts concerning small-signal amplifiers. These amplifiers are often multistage circuits. You will learn several methods for getting the signal from one stage to another. You will also learn the characteristics for each method.

  This chapter also covers field-effect transistor (FET) amplifiers. The FET has certain advantages that make it attractive for some amplifier applications.

  The final section of this chapter introduces some of the characteristics of negative feedback. You will learn how feedback can be used to improve certain amplifier specifications.

7-1 AMPLIFIER COUPLING

A single stage of gain may not be enough. Two, three, or more stages may be needed. *Coupling* refers to the method used to transfer the signal from one stage to the next. There are three basic types of amplifier coupling.

Capacitive coupling is useful when the signals are ac. The coupling capacitors are selected to have a low reactance at the lowest signal frequency. This ensures good performance over the frequency range of the amplifier. Any dc component in the signal will be blocked by the capacitor. This is because good capacitors have a nearly infinite reactance for 0 Hz. The frequency of dc is 0 Hz.

Figure 7-1 shows why it is important to block the dc component. Transistor Q_1 serves as the first stage of gain. Its static collector voltage is 7 V. This is measured from ground to the collector terminal. It is very important to remember that terminal voltages in electronic circuits are measured with respect to ground. The only exception to this rule is when two terminals are specified, such as V_{CE}. The CE means the voltage is to be measured from the collector to the emitter. Transistor Q_2 in Fig. 7-1 shows a static base voltage of 3 V. This is measured from ground to the base terminal. Because the grounds are common, it is easy to calculate the voltage drop across the coupling capacitor from the collector of transistor Q_1 to the base of Q_2:

$$V = 7 - 3 = 4 \text{ V}$$

There is a 4-V drop across the capacitor.

What would happen in Fig. 7-1 if the coupling capacitor were replaced by a wire? The wire has very low resistance. We could expect almost no voltage drop across the wire. This means that the two voltages would have to be the same. The collector of Q_1 and the base of Q_2 would show the same voltage with respect to ground. This would greatly change the operating point of Q_2. The base voltage of Q_2 would be much higher than 3 V. This extra base voltage would drive Q_2 into saturation. It would no longer be capable of linear operation.

Coupling capacitors used in transistor circuits are often of the electrolytic type. This is especially true in low-frequency amplifiers. High values of capacitance are needed to pass the signals with little loss. Polarity is an important factor when working with electrolytic capacitors. Again, refer to Fig. 7-1. The collector of Q_1 is 4 V more positive than the base of Q_2. The capacitor must be installed with the polarity shown.

Capacitive coupling is very good for many uses. Some applications, however, require operation down to dc. Electronic instruments, such as oscilloscopes and meters, may

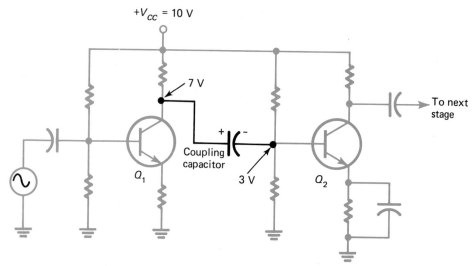

Fig. 7-1 A capacitively coupled amplifier.

From page 97:
Coupling

Capacitive coupling

Saturation

Electrolytic capacitor

On this page:
Direct coupling

Temperature sensitivity

Darlington connection

Darlington transistor

need to perform at dc. The amplifiers in such instruments cannot use capacitive coupling.

Direct coupling does perform down to 0 Hz (dc). A direct-coupled amplifier uses wire or some other direct connection between stages. Figure 7-2 shows a direct-coupled amplifier. Notice that the signal is transferred by direct connections. An amplifier of this type will have to be designed so that the terminal voltages are not upset by the direct connections. In Fig. 7-2, the emitter voltage of Q_1 will have to be the same as the base voltage of Q_2. Every stage will require operating points based on common voltages. When many stages are direct-coupled, this can be difficult to achieve.

Another problem in direct-coupled amplifiers is temperature sensitivity. Temperature does affect transistors. As temperature goes up, β increases and leakage current increases.

Fig. 7-2 A direct-coupled amplifier.

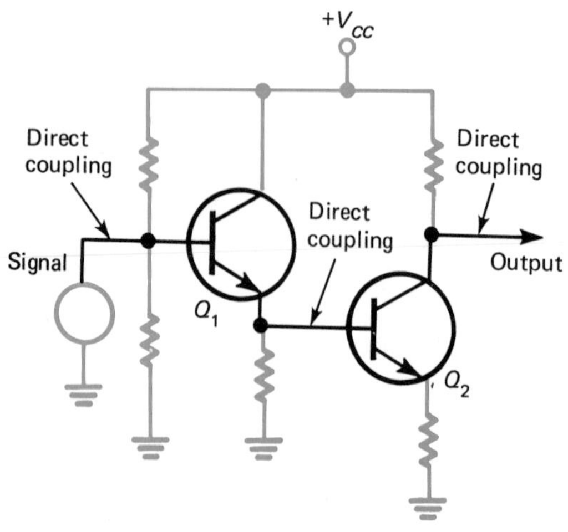

This will tend to shift the operating point. All the stages following will amplify the shift of the operating point. In Fig. 7-2, assume the temperature has gone up. This tends to make Q_1 conduct more. More current will flow through the emitter resistor of Q_1, increasing the voltage drop across the resistor. The base of Q_2 now sees more voltage, and it is turned on harder. If a third and fourth stage follow, the slight shift in the operating point of Q_1 may cause the fourth stage to be driven out of the linear range of operation.

Direct-coupling a few stages is not difficult. It may be the least expensive way to get the gain needed. Direct coupling may be used in sections of an audio amplifier where the lowest frequency is around 20 Hz. Response to dc is not needed. Direct coupling is used in audio work when it is less expensive than another coupling method.

Figure 7-3 shows a popular way to direct-couple two transistors. This is called the *Darlington connection*. Two transistors connected in this way might also be called a Darlington pair. The pair is also manufactured in a single case with only three leads. This is called a Darlington transistor. The Darlington circuit provides an extremely high current gain. The gain is approximately equal to the *product* of the individual transistor gains:

$$\text{Current gain} = A_I \simeq \beta_1 \times \beta_2$$

If each transistor has a gain of 100, then

$$A_I \simeq 100 \times 100 = 10{,}000$$

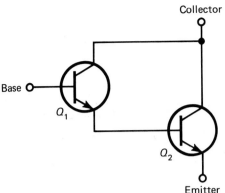

Fig. 7-3 A Darlington amplifier.

The Darlington circuit is a good choice when a high current gain or a high input impedance is required. Since the circuit has such high current gain, it requires a small-signal current. A small-signal current means that the source is lightly loaded by the Darlington amplifier. This is especially true in a Darlington emitter follower such as the one shown in Fig. 7-4.

Figure 7-5 shows a transformer-coupled amplifier. The transformer serves as the collector load for the transistor and as the coupling device to the amplifier load. The advantage of transformer coupling is easy to understand if we examine its impedance matching property. The turns ratio of a transformer is given by

$$\text{Turns ratio} = \frac{N_P}{N_S}$$

where N_P = number of primary turns
N_S = number of secondary turns

The impedance ratio is given by

$$\text{Impedance ratio} = (\text{turns ratio})^2$$

If the transformer in Fig. 7-5 has 100 primary turns and 10 secondary turns, its turns ratio will be

$$\text{Turns ratio} = \frac{100}{10} = 10$$

Its impedance ratio will be

$$\text{Impedance ratio} = 10^2 = 100$$

This means that the load seen by the collector of the transistor will be 100 times the impedance of the actual load. If the load is 10 Ω, the collector will see a load that appears to be $10 \times 100 = 1000$ Ω.

The output impedance of most common-emitter amplifiers is much higher than 10 Ω. When the amplifier must deliver signal energy to such a low impedance, a matching transformer greatly improves the circuit efficiency. The collector load of 1000 Ω in Fig. 7-5 is high enough to provide good voltage gain. If we assume the emitter current is 5 mA, we have enough information to calculate the gain. First, we must estimate the ac resistance of the emitter:

$$r_E = \frac{25}{I_E} = \frac{25}{5} = 5 \ \Omega$$

The emitter resistor is bypassed, so the voltage gain will be given by

$$A_V = \frac{R_L}{r_E}$$

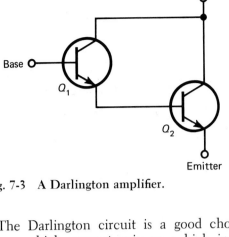

Fig. 7-4 A Darlington emitter follower.

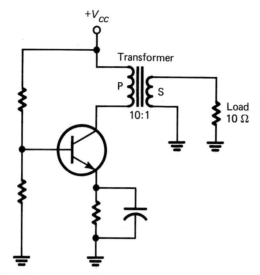

Fig. 7-5   A transformer-coupled amplifier.

There is no load resistor in Fig. 7-5, but there is an external load of 10 Ω. This impedance is transformed to 1000 Ω by the transformer and will set the voltage gain:

$$A_V = \frac{1000}{5} = 200$$

Does the external 10-Ω load see 200 times the signal voltage sent to the base of the transistor? No, because the transformer is a 10:1 step-down device. The load will see only

$$\frac{200}{10} = 20 \text{ times the base signal voltage}$$

This is still better than we can do without the coupling transformer. If the 10-Ω load were

connected directly in the collector circuit as a load resistor R_L, the gain would be

$$A_V = \frac{10}{5} = 2$$

Obviously, the transformer does quite a bit to improve the voltage gain of the circuit.

Audio coupling transformers present problems. If high quality is needed, the transformers will be expensive. They will need good cores, and this means the size and weight of the transformers will be another problem. Transformer coupling is not used too often in high-power audio work. It is more popular in low- and medium-quality audio devices where some shortcuts are acceptable.

At radio frequencies the transformers become less expensive. The cores can be smaller and lighter. Transformer coupling is more popular in RF amplifiers.

Tuned transformers are used often in RF amplifiers to provide selective gain. *Selective gain* means that only those frequencies near the tuned frequency will receive gain. Figure 7-6 shows a tuned, transformer-coupled, RF amplifier. The resonant frequency of T_1 will find a high impedance in the collector circuit of Q_1. This frequency will receive the greatest voltage gain. Transistor Q_2 is also tuned by T_2. If both transformers are tuned to the same frequency, the amplifier will be quite selective. The gain will drop sharply for those frequencies above and below the resonant frequency.

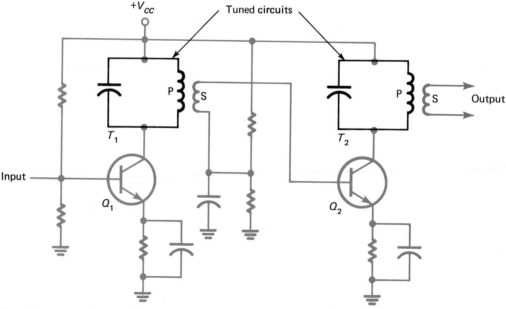

Fig. 7-6   A tuned RF amplifier.

Table 7-1 Summary of coupling methods

	Capacitor Coupling	Direct Coupling	Transformer Coupling
Response to dc	No	Yes	No
Provides impedance match	No	No	Yes
Advantages	Easy to use. Terminals at different dc levels can be coupled.	Simplicity when a few stages are used.	High efficiency. Can be tuned to make a selective amplifier.
Disadvantages	May require high values of capacity for low-frequency work.	Difficult to design for many stages. Temperature sensitivity.	Cost, size, and weight can be a problem.

The transformers in Fig. 7-6 also provide an impedance match. Transformer T_1 normally will have more primary turns than secondary turns. This matches the high collector impedance of Q_1 to the lower input impedance of Q_2.

The secondary of T_1 delivers the signal to the base of Q_2. It also provides the base bias to put Q_2 in the linear operating range. Note in Fig. 7-6 that the base divider network feeds into the bottom of the secondary. The secondary will have a low resistance, so that the voltage at the junction of the two resistors will also appear on the base of Q_2. The capacitor grounds the bottom of the secondary for the ac signal. Without this capacitor, the signal would flow in the divider network and much of it would be lost.

Three methods of coupling have been presented. Table 7-1 summarizes some of the important points for each method discussed.

Review Questions

Answer the following.

1. Refer to Fig. 7-1. A coupling capacitor should present no more than *one-tenth* the impedance of the load it is working into. If amplifier Q_2 has an input impedance of 1000 Ω and the circuit must amplify frequencies as low as 20 Hz, what is the minimum value for the coupling capacitor?

2. Refer to Fig. 7-1. Assume the coupling capacitor develops a short and the base voltage of Q_2 increases to 6 V. Also assume that Q_2 has a load resistor of 1200 Ω and an emitter resistor of 1000 Ω. Can you prove that Q_2 goes into saturation?

3. Refer to Fig. 7-2. Will the output signal from Q_2 be in phase or 180° out of phase with the signal input to Q_1?

4. What is the configuration of Q_1 in Fig. 7-2?

5. Refer to Fig. 7-3. If Q_1 has a β of 90 and Q_2 has a β of 75, what is the current gain from the base terminal to the emitter terminal?

6. Refer to Fig. 7-4. The transistors are silicon. What will the emitter current of Q_2 be? Do not forget to take into account *both* base-emitter drops.

7. Refer to Fig. 7-5. Assume the turns ratio is 5:1 (primary to secondary). What load does the collector of the transistor see?

7-2 FIELD-EFFECT TRANSISTOR AMPLIFIERS

The bipolar transistor is the workhorse of modern electronic circuitry. Its cost and performance make it the best choice for most applications. Field-effect transistors do, however, offer certain advantages that make them the best choice in some circuits. Some of these advantages are as follows:

1. They are voltage-controlled amplifiers. This makes their input impedance very high.
2. They have a low noise output. This makes them useful as preamplifiers where the noise must be very low because of high gain in following stages.
3. They have better linearity. This makes them attractive where distortion must be minimized.
4. They have low interelectrode capacity. At very high frequencies, interelectrode capacitance can make an amplifier work poorly. This makes the FET desirable in RF amplifier work.
5. They can be manufactured with two gates. The second gate is useful for gain control or the application of a second signal.

Common-source
amplifier

Operating point

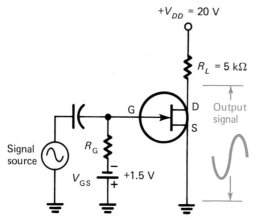

Fig. 7-7   An FET amplifier using fixed bias.

Figure 7-7 shows an FET common-source amplifier. The source terminal is common to both the input and the output signals. This circuit is similar to the bipolar common-emitter configuration. The supply voltage V_{DD} is positive with respect to ground. The current will flow from ground, through the N channel, through the load resistor, and into the positive end of the power supply. Note that a bias supply V_{GS} is applied across the gate-source junction. The polarity of this bias supply is arranged to *reverse-bias* the junction. Therefore, we may expect the gate current to be zero.

The gate resistor R_G in Fig. 7-7 will normally be a very high value (around 1 MΩ). It will not drop any dc voltage because there is no gate current. Using a large value of R_G keeps the input impedance high. If $R_G = 1$ MΩ, the signal source sees an impedance of 1 MΩ. At very high frequencies, other effects can lower this impedance. At low frequencies, the amplifier input impedance is equal to the value of R_G.

The load resistor in Fig. 7-7 allows the circuit to produce a voltage gain. Figure 7-8 shows the characteristic curves for the transistor and the load line. As before, one end of the load line is equal to the supply voltage, and the other end of the load line is set by Ohm's law:

$$I_{sat} = \frac{V_{DD}}{R_L} = \frac{20 \text{ V}}{5 \text{ k}\Omega} = 4 \text{ mA}$$

Thus, the load line for Fig. 7-7 runs from 20 V on the voltage axis to 4 mA on the current axis. This is shown in Fig. 7-8.

In linear work, the operating point should be near the center of the load line. The operating point is set by the gate-source bias voltage V_{GS} in a FET amplifier. The bias voltage is equal to -1.5 V in Fig. 7-7. In Fig. 7-8

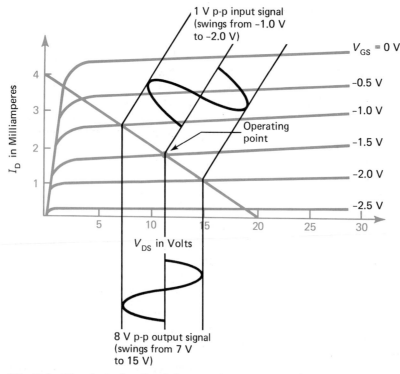

Fig. 7-8   The drain family of characteristic curves.

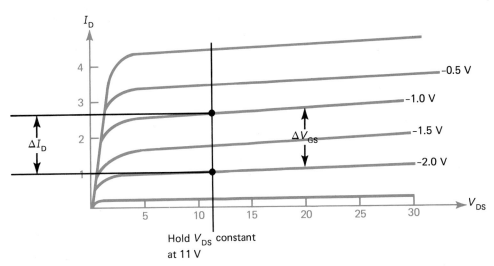

Fig. 7-9 Calculating forward transfer admittance.

the operating point is shown on the load line. By projecting down from the operating point, it is seen that the resting voltage across the transistor will be about 11 V. This is roughly half the supply voltage, so the transistor is operating in its linear range.

The signal will drive the amplifier above and below its operating point. As shown in Fig. 7-8, a 1-V peak-to-peak input signal will swing the output about 8 V peak to peak. The voltage gain is

$$A_V = \frac{\text{signal out}}{\text{signal in}} = \frac{8}{1} = 8$$

As with the common-emitter circuit, the common-source configuration produces a 180° phase shift. Inspect Fig. 7-8. As the input signal shifts the operating point from -1.5 to -1.0 V (positive direction), the output signal swings from about 11 to 7 V (negative direction). Common-drain and common-gate FET amplifiers do not produce this phase inversion.

There is a second way to calculate voltage gain for the common-source amplifier. It is based on a characteristic of the transistor called the *forward transfer admittance* Y_{fs}:

$$Y_{fs} = \frac{\Delta I_D}{\Delta V_{GS}} \bigg| V_{DS}$$

where ΔV_{GS} = change in gate-source voltage
ΔI_D = change in drain current with V_{DS} held constant
V_{DS} = drain-source voltage

(The vertical bar to the left of V_{DS} means that it is to be held constant.)

Figure 7-8 has enough information to calculate Y_{fs} for the transistor. In Fig. 7-9 the drain-source voltage V_{DS} is held constant at 11 V. The change in gate-source voltage ΔV_{GS} is from -1.0 to -2.0 V. The change is $(-1.0) - (-2.0) = 1$ V. The change in drain current ΔI_D is from 2.6 to 1 mA, or a change of $2.6 - 1 = 1.6$ mA. We can now calculate the forward transfer admittance of the transistor:

$$Y_{fs} = \frac{1.6 \times 10^{-3} \text{ A}}{1 \text{ V}}$$

$$= 1.6 \times 10^{-3} \text{ siemens (S)}$$

The siemen is the unit for conductance (although some older textbooks may still use the former unit, the mho). It is abbreviated by the letter S. Conductance is the reciprocal of resistance:

$$G = \frac{1}{R}$$

Conductance is a dc characteristic. Admittance is an ac characteristic equal to the reciprocal of impedance. They share the unit siemen.

The voltage gain of a common-source FET amplifier is given by

$$A_V \simeq Y_{fs} \times R_L$$

For the circuit of Fig. 7-7, the voltage gain will be

103

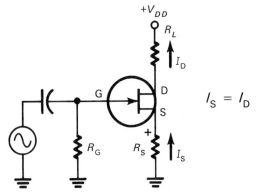

$+V_{DD}$

$I_S = I_D$

Fig. 7-10   An FET amplifier using source bias.

$$A_V = 1.6 \times 10^{-3} \times 5 \times 10^3 = 8$$

This agrees with the graphical solution of Fig. 7-8. The nice thing about the voltage-gain equation is that it makes it easy to calculate the voltage gain for different values of load resistance. It will not be necessary to draw another load line. If the load resistance is changed to 8.2 kΩ, the voltage gain will be

$$A_V = 1.6 \times 10^{-3} \times 8.2 \times 10^3 = 13.12$$

With a load resistance of 5000 Ω, the circuit gives a voltage gain of 8. With a load resistance of 8200 Ω, the circuit gives a voltage gain slightly over 13. This shows that gain is directly related to load resistance. This was also the case in the common-emitter bipolar amplifier circuits. Remember this concept because it is valuable when troubleshooting amplifiers.

Figure 7-10 shows an improvement in the common-source amplifier. The bias supply V_{GS} has been eliminated. Instead, we find resistor R_S in the source circuit. As the source current flows through this resistor, voltage will drop across it. This voltage drop will serve to bias the gate-source junction of the transistor. If the desired bias voltage and current are

known, it is an easy matter to calculate the value of the source resistor. Since the drain current and the source current are equal,

$$R_S = \frac{V_{GS}}{I_D}$$

If we assume the same operating conditions as in the circuit of Fig. 7-7, the gate bias voltage should be −1.5 V. The source resistor should be

$$R_S = \frac{1.5}{1.9 \times 10^{-3}} = 790 \ \Omega$$

Note that the value of current used above is about half the saturation current. Check Fig. 7-8 and verify that this value of drain current is near the center of the load line.

The circuit of Fig. 7-10 is called a *source bias circuit*. The bias voltage is produced by source current flowing through a resistor from ground to source. The drop across the resistor makes the source terminal positive with respect to ground. The gate terminal is at ground potential. There is no gate current and therefore no drop across the gate resistor R_G. Thus, the source is positive with respect to the gate. To say it another way, the gate is negative with respect to the source. This accomplishes the same purpose as the separate supply V_{GS} in Fig. 7-7.

Source bias is much simpler than using a separate bias supply. The source resistor is a very inexpensive part. The voltage gain does suffer, however. To see why, examine Fig. 7-11. As the input signal drives the gate in a positive direction, the source current increases. This makes the voltage drop across the source resistor increase. The source terminal is now more positive with respect to ground. This makes the gate more negative with respect to the source. The overall effect is that much of the input signal is canceled.

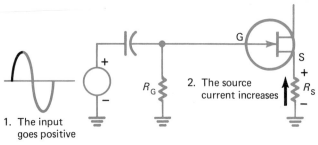

3. The voltage drop across R_S increases

2. The source current increases

4. Increase voltage drop across R_S makes the gate more negative with respect to the source

1. The input goes positive

5. Some of the input signal is cancelled

Fig. 7-11   Source feedback.

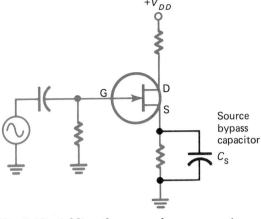

Fig. 7-12 Adding the source bypass capacitor.

When an amplifier develops a signal that interacts with the input signal, the amplifier is said to have *feedback*. Figure 7-11 shows one way feedback can affect an amplifier. In this example, the feedback is acting to cancel part of the effect of the input signal. When this happens, the feedback is said to be *negative*.

Negative feedback *decreases* the amplifier's gain. It is also capable of increasing the frequency range of an amplifier. Negative feedback may be used to decrease an amplifier's distortion. So, negative feedback is not good or bad—it is a mixture. If maximum voltage gain is required, the negative feedback will have to be eliminated. In Fig. 7-12 the source bias circuit has a source bypass capacitor. This capacitor will cancel the negative feedback and restore the gain. It is selected to have low reactance at the signal frequency. It will prevent the source terminal voltage from swinging with the increases and decreases in source current. It has pretty much the same effect as the emitter bypass capacitor in the common-emitter amplifier studied before.

When we studied bipolar junction transistors, we found them to have an unpredictable β. This made it necessary to design a circuit that was not sensitive to β. Field-effect transistors are also somewhat unpredictable. Again, we find it necessary to design circuits that will work within a reasonable range of transistor characteristics.

The circuit of Fig. 7-7 is called *fixed bias*. This circuit will work only if the transistor is close to the expected characteristics. The fixed-bias circuit usually is not a good choice. The circuit of Fig. 7-10 is called *source bias*. It is much better and allows the transistor characteristics to vary. If, for example, the

transistor tended toward more current, the source bias would automatically increase. More bias would reduce the current. Thus, it can be seen that the source resistor stabilizes the circuit.

The greater the source resistance, the more stability we can expect in the operating point. But, too much source resistance could create too much bias, and the circuit will operate too close to cutoff. If there were some way to offset this effect, a better circuit would result. Figure 7-13 shows a way. This circuit uses *combination bias*. The bias is a combination of a fixed positive voltage applied to the gate terminal and source bias. The positive voltage is a divided portion of V_{DD}. The divider network is made up of R_{G_1} and R_{G_2}. These resistors will usually be very high in value to keep the input impedance high.

The combination-bias circuit can use a high value for R_S, the source resistor. The bias voltage V_{GS} will not be excessive because a positive, fixed voltage is applied to the gate. This fixed, positive voltage will reduce the effect of the voltage drop across the source resistor.

A few calculations will show how the combination-bias circuit works. Assume in Fig. 7-13 that the desired source current is to be 1.9 mA. The voltage drop across R_S is

$$V_{R_S} = 1.9 \times 10^{-3} \text{ A} \times 2.2 \times 10^3 \ \Omega = 4.18 \text{ V}$$

Next, the voltage divider drop across R_{G_2} is

$$V_G = \frac{2.2 \text{ M}\Omega}{2.2 \text{ M}\Omega + 15 \text{ M}\Omega} \times 20 \text{ V} = 2.56 \text{ V}$$

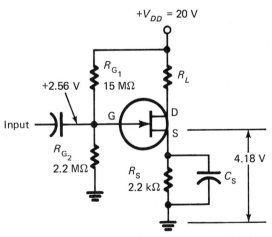

Fig. 7-13 An FET amplifier using combination bias.

105

Electronics:
Principles and
Applications
CHAPTER 7

Dual-gate
MOSFET
amplifier

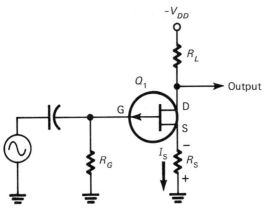

Fig. 7-14 A P-channel JFET amplifier.

Both of the above voltages are positive with respect to ground. The source voltage is more positive. The gate must be negative with respect to the source because

$$V_{GS} = 4.18 \text{ V} - 2.56 \text{ V} = 1.62 \text{ V}$$

In measuring bias voltage in circuits such as Fig. 7-13, remember that V_G and V_{GS} are different. The gate voltage V_G is measured with respect to ground. The gate-source voltage V_{GS} is measured with respect to the source.

Figure 7-14 shows a P-channel junction field-effect transistor amplifier. The supply voltage is negative with respect to ground. Note the direction of the source current I_S. The voltage drop produced by this current will reverse-bias the gate-source diode. This is proper, and no gate current will flow through R_G.

It is possible to have linear operation with zero bias, as shown in Fig. 7-15. You should recognize the transistor as a MOSFET. The gate is insulated from the source in this type of transistor. This prevents gate current

regardless of the gate-source polarity. As the signal goes positive, the drain current will increase. As the signal goes negative, the drain current will decrease. This type of transistor can operate in both the enhancement and depletion modes. This is not true of other FET types. Thus, the circuit of Fig. 7-15 is restricted to depletion-mode MOSFETs for linear work.

Figure 7-16 shows the schematic for a *dual-gate* MOSFET amplifier. The circuit uses tuned transformer coupling. Good gain is possible at frequencies near the resonant point of the transformers. The signal is fed to gate 1 (G_1) of the FET. The output signal appears in the drain circuit. The gain of this amplifier can be controlled over a large range by gate 2 (G_2). The graph of Fig. 7-17 shows a typical gain range for this type of amplifier. Note that maximum power gain, 20 dB, occurs when the gate 2 is positive with respect to the source by about 3 V. At zero bias, the gain is only about 5 dB. As gate 2 goes negative with respect to the source, the gain continues to drop. The minimum gain is about -28 dB. Of course, -28 dB represents a large loss.

The total range of gain for the amplifier is from $+20$ to -28 dB, or 48 dB. This means a power ratio of nearly 100,000:1. Thus, with the proper control voltage applied to G_2, the circuit of Fig. 7-16 can keep a constant output over a tremendous range of input levels.

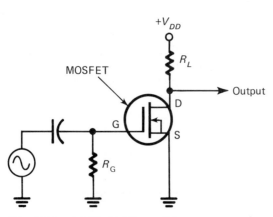

Fig. 7-15 A MOSFET amplifier.

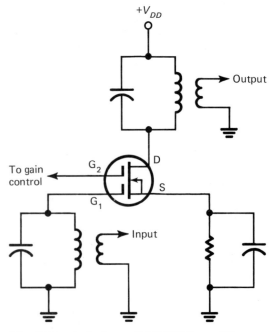

Fig. 7-16 A dual-gate MOSFET amplifier.

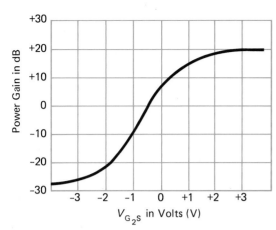

Fig. 7-17 The effect of V_{G_2} on gain.

Amplifiers of this type are useful in fields such as communications where a wide range of signal levels is expected.

Field-effect transistors have some very good characteristics. However, bipolar transistors usually cost less and give much better voltage gains. This makes the bipolar transistor the best choice for most applications.

Field-effect transistors are used when some special feature is needed. For example, they are a good choice if a very high input impedance is required. When used, FETs are generally found in the first stage or two of a linear system.

Examine Fig. 7-18. Transistor Q_1 is an FET, and Q_2 is a bipolar transistor. Thanks to the FET, the input impedance is very high. Thanks to the bipolar device, the power gain is good and the cost is reasonable. This is typical of the way circuits are designed to have the best performance for the least cost.

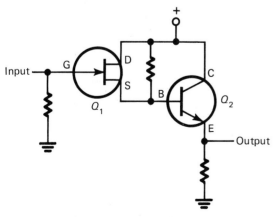

Fig. 7-18 Combining FET and bipolar devices.

Answer the following.

8. Refer to Fig. 7-7. Assume $I_D = 2.5$ mA. Find V_{DS}.

9. Refer to Fig. 7-8. Assume $V_{GS} = 0$ V. Where would the transistor be operating?

10. Refer to Fig. 7-8. Assume $V_{GS} = -3.0$ V. Where would the transistor be operating?

11. Refer to Fig. 7-10. If $I_D = 2$ mA and $R_S = 1000$ Ω, what is the value of V_{GS}?

12. Refer to Fig. 7-13. Assume $V_{DD} = 15$ V and $I_D = 2$ mA. What is the value of V_{GS}? What is the gate polarity with respect to the source?

13. Refer to Fig. 7-18. In what circuit configuration is Q_1 connected? Q_2?

14. Refer to Fig. 7-18. If the input is going in a positive direction, in what direction will the output go?

7-3 NEGATIVE FEEDBACK

Figure 7-19 shows the basic idea of feedback. Some of the output signal is returned to the input circuit. When this feedback signal is returned with a phase or polarity that cancels some of the input signal, the feedback is negative. Negative feedback lowers the voltage gain of an amplifier. When the emitter resistor is unbypassed in a common-emitter amplifier, the voltage gain is much lower. An ac signal is developed across this resistor that acts to cancel much of the effect of the input signal. An unbypassed source resistor in a common-source amplifier also produces negative feedback. Again, the voltage gain is reduced.

Negative feedback produces other effects. It increases the bandwidth of an amplifier. The amplifier will be able to give gain over a wider frequency range. Figure 7-20 shows why. The amplifier gain is best without negative feedback. At higher frequencies, the

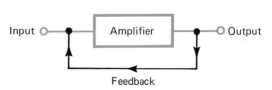

Fig. 7-19 The basic idea of feedback.

107

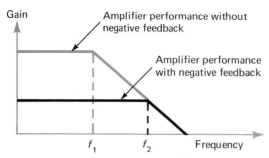

Gain

Amplifier performance without
negative feedback

Amplifier performance
with negative feedback

f_1　　f_2　Frequency

Fig. 7-20　The effect of feedback on gain and
bandwidth.

gain begins to drop off. This begins to occur
at f_1 in Fig. 7-20. The decrease in gain is due
to the performance of the transistor and circuit capacitance. All transistors show less
gain as the frequency increases. The reactance of a capacitor decreases as the frequency increases. This lower reactance loads
the amplifier, and the gain drops.

Now, look at the performance of the amplifier in Fig. 7-20 with negative feedback. The
gain is much less at the lower frequencies.
The gain does not begin dropping off until f_2 is
reached. Frequency f_2 is much higher than
f_1. With negative feedback, the amplifier provides less, but more uniform, gain over a wider
range of frequencies. The loss in gain can be
overcome with another stage of amplification.

Negative feedback also reduces noise and
distortion. Suppose the signal picks up some
noise or distortion in the amplifier. This appears on the output signal. Some of the output signal is fed back to the input. The noise
or distortion will be placed on the input signal
in an opposite way. Remember, the feedback
is out of phase with, or opposite to, the input.
Much of the noise and distortion is canceled
in this way. By intentionally distorting the
input signal, opposite to the way the amplifier
distorts it, the circuit becomes more linear.

Figure 7-21 shows the schematic diagram
for a stereo tape player. Only one channel is
shown. The other channel is the same. The
input signal comes from the tape head at the
left. It is amplified by five stages, and the output drives a loudspeaker. The other channel
uses a separate tape head and loudspeaker.

Figure 7-21 uses several types of negative
feedback. Transistors Q_1 and Q_2 are direct-
coupled. The emitter resistor of Q_2 produces
a voltage drop. This drop across R_3 is divided
by R_1 and R_2 to bias the base of Q_1. This is
negative feedback. It is dc feedback, how-

ever. It is not used to decrease gain, increase
bandwidth, or improve linearity. It stabilizes
the operating point for Q_1 and Q_2.

Suppose in Fig. 7-21 that Q_1 begins conducting too much current. This will cause an
increase in the voltage drop across R_4, its load
resistor. This will mean less positive voltage
at the base of Q_2, since they are direct-
coupled. Transistor Q_2 will now conduct less.
Less current will flow through R_3. This will
decrease the drop across R_3. Less forward
bias will be available for the base of Q_1. This
tends to reduce the conduction in Q_1. The dc
feedback has stabilized the operating point.

There is also some ac feedback from Q_2 to
Q_1. Note in Fig. 7-21 that the collector of Q_2
feeds back to the emitter of Q_1 through a resistor and a capacitor. This acts to cancel
part of the input signal coming from the tape
head. If, for example, the tape head is going
in a positive direction, then the collector of Q_1
will be driven in a negative direction. Do not
forget the phase inversion in a common-
emitter amplifier. The collector of Q_2 will go
in a positive direction for the same reason.
This positive-going signal is sent back to the
emitter of Q_1. The original input signal drives
the base in a positive direction, and the feedback signal drives the emitter in a positive
direction. This decreases or cancels part of
the base-emitter signal in Q_1.

The ac feedback from Q_2 to Q_1 in Fig. 7-21
will decrease the voltage gain. It will not improve the bandwidth of the amplifier, however. The feedback capacitor is chosen to
only couple the higher audio frequencies.
The capacitor has a high reactance at the
lower frequencies. This action reduces the
gain for the higher frequencies only. In tape
recording, the high frequencies are boosted.
This boost helps reduce tape hiss and other
high-frequency noises. During playback, the
high-frequency gain must be cut down to
offset the way the tape was recorded. The
high-frequency feedback from Q_2 to Q_1 does
this.

Refer again to Fig. 7-21. Notice that Q_3,
Q_4, and Q_5 are all direct-coupled. Again, dc
feedback is needed to stabilize the operating
point. The collector of Q_5 supplies the bias
through R_7. Resistors R_5 and R_6 form a voltage divider for the base of Q_3. Suppose that
Q_3 conducts too much current. This will
cause its collector to go less positive. The collector of Q_4 will then go more positive. Since

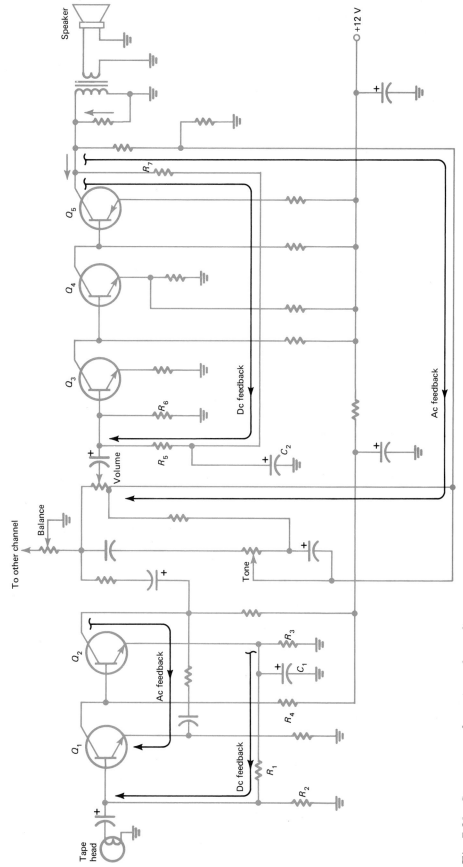

Fig. 7-21 Stereo tape player (one channel).

Q_5 is a PNP transistor, a positive-going change at its base will decrease its conduction. The decrease in conduction will lower the voltage drop across the primary circuit of the output transformer. Now, less positive voltage is fed back to Q_3, and this reduces the conduction of Q_3. The dc feedback stabilizes the circuit.

Capacitors C_1 and C_2 in Fig. 7-21 prevent signal feedback in the dc feedback circuits. They are large electrolytic capacitors chosen for low reactance at the signal frequencies. This low reactance effectively shorts the ac component to ground.

Figure 7-21 shows an ac feedback from the collector circuit of Q_5 to the volume control circuit. This negative signal feedback increases the linearity of the amplifier and improves its frequency response. It also tends to cancel any noise or disturbance on the +12-V supply line. A much better sound results from this negative feedback.

You might be wondering about the PNP transistor Q_5. Note that its collector does not go to the +12-V supply line. Rather, its

emitter does. This properly biases the transistor. Compare this with the way the other transistors are connected. They are NPN devices. Their emitters go to ground, and their collectors go to the +12-V line. As seen in this circuit, NPN and PNP devices can be used in the same circuit.

Review Questions

Answer the following.

15. Refer to Fig. 7-21. Assume the signal at the base of Q_1 is negative-going. What will be the signal at the collector of Q_2?

16. What component in Fig. 7-21 prevents any ac signal from being fed back to the base of Q_1?

17. All the transistors in Fig. 7-21 are operating in the same configuration. What is that configuration?

18. Is ac or dc feedback used to stabilize the operating point of an amplifier? Is it negative or positive feedback?

Summary

1. There are three basic types of amplifier coupling: capacitive, direct, and transformer coupling.
2. Capacitor coupling is useful in ac amplifiers because a capacitor will block dc and allow the ac signal to be coupled.
3. Electrolytic coupling capacitors must be installed with the correct polarity.
4. Direct coupling provides dc gain. It can be used only when the dc terminal voltages are compatible.
5. A Darlington amplifier is an example of a direct-coupled transistor. It has a very high current gain.
6. Transformer coupling gives the advantage of impedance matching.
7. The impedance ratio of a transformer is the square of the turns ratio.
8. Radio-frequency transformers can be tuned to give selectivity. Only those frequencies near the resonant frequency will receive gain.
9. Most electronic amplifiers use bipolar transistors.
10. It is necessary to reverse-bias the gate-source junction for linear operation with junction field-effect transistors.

11. Because there is no gate current, the input impedance of FET amplifiers is very high.
12. The voltage gain in an FET amplifier is approximately equal to the product of the load resistance and the forward transfer admittance of the transistor.
13. More load resistance means more voltage gain.
14. Fixed bias, in FET amplifiers, is not desirable because the characteristics of the transistor vary quite a bit from unit to unit.
15. Source bias tends to stabilize an FET amplifier and make it more immune to the characteristics of the transistor.
16. Combination bias uses fixed and source bias to make the circuit even more stable.
17. A common-source FET amplifier using source bias must use a source bypass capacitor to realize maximum voltage gain.
18. The dual-gate MOSFET amplifier is capable of a tremendous range of gain by applying a control voltage to the second gate.
19. When the feedback tends to cancel the effect of the input to an amplifier, that feedback is negative.
20. Dc negative feedback can be used to stabilize the operating point of an amplifier.

21. Dc negative feedback is especially valuable in direct-coupled amplifiers.
22. Ac negative feedback decreases the voltage gain of an amplifier.
23. Ac negative feedback increases the bandwidth and linearity of an amplifier. It can also reduce amplifier noise.
24. Ac negative feedback can also be used to reduce the high-frequency gain of an amplifier. This is useful in tape players.

Chapter Review Questions

For questions 7-1 through 7-18 determine whether each statement is true or false.

7-1. When electrolytic capacitors are used as coupling capacitors, they may be installed with any polarity.

7-2. Capacitive coupling provides response to 0 Hz.

7-3. Temperature sensitivity may be a problem in a direct-coupled amplifier.

7-4. A Darlington amplifier will not provide gain at 0 Hz.

7-5. A transformer can match a high-impedance collector circuit to a low-impedance load.

7-6. Refer to Fig. 7-6. This amplifier will provide gain at 0 Hz.

7-7. Refer to Fig. 7-8. If $V_{GS} = -2.5$ V, the negative-going portion of the signal will be severely clipped.

7-8. Refer to Fig. 7-10. Increasing the value of R_L should decrease the voltage gain of the amplifier.

7-9. Refer to Fig. 7-12. The effect of the capacitor C_S is to increase the voltage gain.

7-10. Refer to Fig. 7-14. Transistor Q_1 is a P-channel JFET.

7-11. Refer to Fig. 7-15. Voltage V_{GS} is 0 V.

7-12. Refer to Figs. 7-16 and 7-17. To decrease the gain of the amplifier, G_2 must be made more positive with respect to the source.

7-13. Refer to Fig. 7-18. The input terminal and the output terminal should be 180° out of phase.

7-14. Negative feedback tends to cancel the input signal.

7-15. Negative feedback increases the voltage gain of the amplifier but at the expense of reduced bandwidth.

7-16. Negative feedback predistorts the signal so as to reduce the overall distortion of the amplifier.

7-17. Negative dc feedback can be used to stabilize the amplifier operating point.

7-18. Refer to Fig. 7-21. If Q_1 tends to increase in conduction, R_1 will supply less forward bias and correct the drift.

For questions 7-19 through 7-30, answer the following.

7-19. Refer to Fig. 7-2. The base voltage at Q_1 measures 4.2 V. What should the emitter voltage measure at Q_2?

7-20. Refer to Fig. 7-3. Each transistor has a beta of 200. What is the overall current gain of the pair?

7-21. Refer to Fig. 7-5. The secondary has 10 turns, and the primary has 200 turns. What is the turns ratio of the transformer?

7-22. Refer to Fig. 7-5. The collector signal is 6 V peak-to-peak. What signal will appear across the 10-Ω load resistor?

7-23. Refer to Fig. 7-5. The transformer turns ratio is 7:1 from primary to secondary. What signal load does the collector see?

7-24. Refer to Fig. 7-6. The inductance of the transformer primaries is 100 μH. The capacitors across the primaries are both 470 pF. At what frequency will the gain of the amplifier be the greatest?

7-25. Refer to Fig. 7-7. If $R_L = 1000\ \Omega$, where will the load line terminate on the vertical axis?

7-26. Refer to Fig. 7-7. If $V_{DD} = 12$ V, where will the load line terminate on the horizontal axis?

7-27. A field-effect transistor drain swings 2 mA with a gate swing of 1 V. What is the forward transfer admittance for this FET?

7-28. An FET has a forward transfer admittance of 4×10^{-3} S. It is to be used in the common-source configuration with a load resistor of 4700 Ω. What voltage gain can be expected?

7-29. Refer to Fig. 7-10. Assume a source current of 10 mA and a source resistor of 100 Ω. What is the value of V_{GS}?

7-30. Refer to Fig. 7-12. It is desired that $V_{GS} = -2.0$ V at $I_D = 8$ mA. What should be the value of the source resistor?

Answers to Review Questions

1. 80 μF
2. Yes, $I_E = 5.4$ mA, $V_R = 6.48$ V, and the sum of the drops across the emitter resistor and the load resistor is greater than V_{CC}.
3. Out of phase
4. Common collector or emitter follower
5. 6750
6. 2.43 mA
7. 250 Ω
8. 7.5 V

9. Saturation
10. Cutoff
11. -2.0 V
12. 2.48 V, negative
13. Q_1 is common drain (source follower) and Q_2 is common collector (emitter follower).
14. Positive
15. Negative-going
16. C_1
17. Common emitter
18. Dc feedback, negative

Large-Signal Amplifiers

• This chapter introduces the idea of efficiency in an amplifier. An efficient amplifier delivers a large part of the power it receives from the supply as a useful output signal. Efficiency is most important when large amounts of signal power are required.

It will be shown that amplifier efficiency is related to how the amplifier is biased. It is possible to make large improvements in efficiency by moving the operating point away from the center of the load line.

8-1 AMPLIFIER CLASS

All amplifiers are power amplifiers. However, those operating in the early stages of the signal processing system deal with small signals. These early stages are designed to give good voltage gain. Since voltage gain is the most important function of these amplifiers, they are called voltage amplifiers. Figure 8-1 is a block diagram of a simple audio amplifier. The microphone produces a very small signal, in the millivolt range. The first two stages amplify this audio signal, and it becomes larger. The last stage produces a much larger signal. It is called a power amplifier.

A power amplifier is designed for good power gain. It must handle large voltage and current swings. These high voltages and currents make the power high. It is very important to have good efficiency in a power amplifier. An efficient power amplifier delivers the most signal power for the dc power it takes from the supply. Look at Fig. 8-2. Note that the job of the power amplifier is to change dc

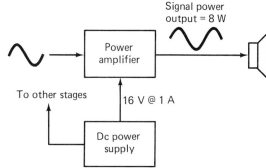

Fig. 8-2 Comparing the output signal to the dc power input.

power into signal power. Its efficiency is given by

$$\% \text{ Efficiency} = \frac{\text{signal power output}}{\text{dc power input}} \times 100$$

The power amplifier of Fig. 8-2 produces 8 W of signal power output. Its power supply develops 16 V and the amplifier draws 1 A. The

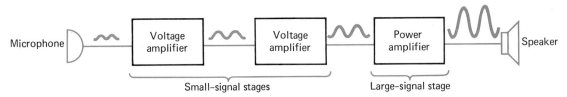

Fig. 8-1 Block diagram of an amplifier.

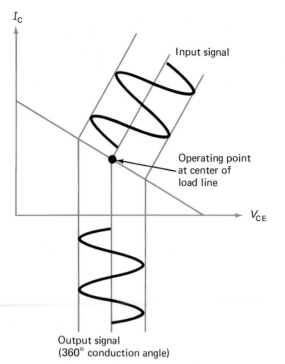

Fig. 8-3   Class A operating point.

dc power input to the amplifier is

$$P = V \times I = 16 \times 1 = 16 \text{ W}$$

The efficiency of the amplifier is

$$\% \text{ Efficiency} = \frac{8 \text{ W}}{16 \text{ W}} \times 100 = 50\%$$

Efficiency is very important in high-power systems.   For example, assume that 100 W of

audio power is required in a music amplifier. Also assume that the power amplifier is only 10 percent efficient.   What kind of a power supply would be required? The power supply would have to deliver 1000 W to the amplifier! A 1000-W power supply is a large, heavy, and expensive item.   Heat would be another problem in this music amplifier.   Of the 1000-W input, 900 W will become heat.   This system would probably need a cooling fan.

All the amplifier circuits discussed to this point have been *class* A.   Class A amplifiers operate at the center of the load line.   Refer to Fig. 8-3.   The operating point is class A. This gives the best possible output swing. The output signal is a good replica of the input signal.   This means that distortion is low. This is the greatest advantage of class A operation.

Figure 8-4 shows another class of operation. The operating point is at cutoff on the load line.   This is done by applying zero bias across the base-emitter junction of the transistor. Zero bias means that only half the input signal will be amplified.   Only that half of the signal which can turn on the base-emitter diode will produce any output signal.   The transistor conducts for half of the input cycle.   A *class B* amplifier is said to have a conduction angle of 180°.   Class A amplifiers conduct for the entire input cycle.   They have a conduction angle of 360°.

What is to be gained by operating in class B? Obviously, we have a distortion problem that was not present in class A.   In spite of the distortion, class B is useful because it gives better

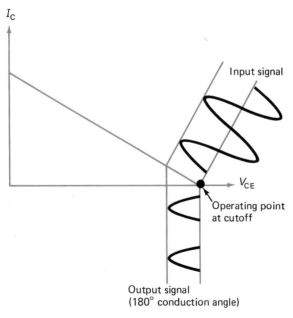

Fig. 8-4   Class B operating point.

efficiency. Biasing an amplifier at cutoff saves power.

Class A wastes power. This is especially true at very low signal levels. The class A operating point is in the center of the load line. This means that about half the supply voltage is dropped across the transistor. The transistor is conducting half the saturation current. This voltage drop and current produce a power loss in the transistor. This power loss is constant in class A. There is a drain on the power supply even if no signal is being amplified.

The class B amplifier operates at cutoff. The transistor current is zero. Zero current means 0 W. There is no drain on the supply until a signal is being amplified. The larger the amplitude of the signal, the larger the drain on the supply. The class B amplifier eliminates the fixed drain from the power supply and is therefore more efficient.

The better efficiency of class B is very important in high-power applications. Some of the distortion can be eliminated by using two transistors: each will amplify half the signal. Such circuits are a bit more complicated, but the improved efficiency is worth the effort.

There are also class AB and class C amplifiers. Again, it is a question of bias. Bias controls the operating point, the conduction angle, and the class of operation. Table 8-1 summarizes the important features of the various amplifier classes. Study this table now and refer to it after completing later sections in this chapter.

It is easy to become confused when studying amplifiers for the first time. There are so many categories and descriptive terms. Table 8-2 has been prepared to help you organize your thinking. You may wish to refer to this table from time to time.

Class AB amplifier

Class C amplifier

Review Questions

Determine whether each statement is true or false.

1. A voltage amplifier or small-signal amplifier gives no power gain.

2. The efficiency of a class A amplifier is greater than that of a class B amplifier.

3. The conduction angle for a class A power amplifier is 270°.

4. Refer to Fig. 8-3. With no input signal, the power taken from the supply will be 0 W.

5. Refer to Fig. 8-4. With no input signal, the power taken from the supply will be 0 W.

Table 8-1 Summary of amplifier classes

	Class A	Class AB	Class B	Class C
Efficiency	50%*	between classes A and B	78.5%*	100%*
Conduction angle	360°	between classes A and B	180°	small (approx. 90°)
Distortion	low	moderate	high	extreme
Bias (emitter-base)	forward (center of load line)	forward (near cutoff)	zero (at cutoff)	reverse (beyond cutoff)
Applications	Practically all small-signal amplifiers. A few moderate power amplifiers in audio applications.	High-power stages in both audio and radio-frequency applications	High-power stages — generally not used in audio applications due to distortion.	Generally limited to radio-frequency applications. Tuned circuits remove much of the extreme distortion.

*Theoretical maximums. Cannot be achieved in actual practice.

Table 8-2   Amplifier characteristics

	Explanations and Examples
Voltage amplifiers	Voltage amplifiers are small-signal amplifiers. They can be found in early stages in the signal system. They are often designed for good voltage gain. An audio preamplifier would be a good example of a voltage amplifier.
Power amplifiers	The power amplifiers are large-signal amplifiers. They can be found late in the signal system. They are designed to give power gain and reasonable efficiency. The output stage of an audio amplifier would be a good example of a power amplifier.
Configuration	The configuration of an amplifier tells how the signal is fed to and taken from the amplifying device. For bipolar transistors, the configurations are common-emitter, common-collector and common-base. For field-effect transistors, the configurations are common-source, common-drain and common-gate.
Coupling	How the signal is transferred from stage to stage. Coupling can be capacitive, direct, or transformer.
Applications	Amplifiers may be categorized according to their use. Examples are audio amplifiers, video amplifiers, RF amplifiers, dc amplifiers, band-pass amplifiers, and wideband amplifiers.
Classes	This category refers to how the amplifying device is biased. Amplifiers can be biased for class A, B, AB, or C operation. Voltage amplifiers are usually biased for class A operation. For improved efficiency, power amplifiers may use class B, AB, or C operation.

6. Bias controls the amplifier operating point, the conduction angle, the class of operation, and the efficiency.

Answer the following.

7. Refer to Fig. 8-2. Suppose the power supply is rated at 12 V. What is the efficiency of the power amplifier?

8. A certain amplifier is producing an output power of 150 W. Its efficiency is 60 percent. How much power will the amplifier take from the supply?

8-2 CLASS A POWER AMPLIFIERS

The class A power amplifier operates near the center of the load line. It is not highly efficient, but it does offer low distortion. It is also the most simple design.

Figure 8-5 shows a class A power amplifier. We will use a load line to see how much signal power can be produced. The load line will be set by the supply voltage V_{CC} and the saturation current:

$$I_{sat} = \frac{V_{CC}}{R_{load}} = \frac{16\ V}{80\ \Omega} = 0.2\ A,\ or\ 200\ mA$$

The load line will run from 16 V on the horizontal axis to 200 mA on the vertical axis.

Next, we must find the operating point for the amplifier. Solving for the base current, we get

$$I_B = \frac{V_{CC}}{R_B} = \frac{16\ V}{16 \times 10^3\ \Omega} = 1\ mA$$

The transistor has a β of 100. The collector current will be

$$I_C = \beta \times I_B = 100 \times 1 = 100\ mA$$

The load line can be seen in Fig. 8-6 with the 100-mA operating point.

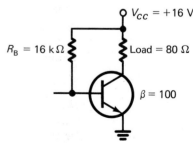

Fig. 8-5   Class A power amplifier.

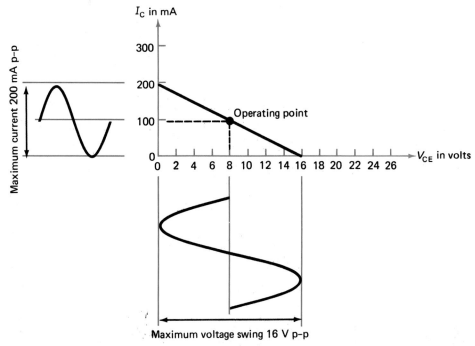

Fig. 8-6 Class A load line.

The amplifier can be driven to the load-line limits before clipping occurs. The maximum voltage swing will be 16 V peak-to-peak. The maximum current swing will be 200 mA peak-to-peak. Both of these maximums are shown in Fig. 8-6.

We now have enough information to calculate the signal power. The peak-to-peak values must be converted to rms values. This is done by

$$V_{rms} = \frac{V_{p-p}}{2} \times 0.707 = \frac{16}{2} \times 0.707 = 5.66 \text{ V}$$

Next, the rms current is

$$I_{rms} = \frac{I_{p-p}}{2} \times 0.707 = \frac{200}{2} \times 0.707$$
$$= 70.7 \text{ mA}$$

Finally, the signal power will be given by

$$P = V \times I = 5.66 \times 70.7 \times 10^{-3} = 0.4 \text{ W}$$

The maximum power (sine wave) is 0.4 W.

How much dc power is involved in producing this signal power? The answer is found by looking at the power supply. The supply voltage is 16 V. The current taken from the supply must also be known. The base current is small enough to ignore. The average collector current is 100 mA. Therefore, the average power is

$$P = V \times I = 16 \times 100 \times 10^{-3} = 1.6 \text{ W}$$

The amplifier needs 1.6 W from the power supply to produce a signal power of 0.4 W. The efficiency of the amplifier is

$$\% \text{ Efficiency} = \frac{P_{ac}}{P_{dc}} \times 100 = \frac{0.4 \text{ W}}{1.6 \text{ W}} \times 100$$
$$= 25\%$$

The Class A amplifier shows an efficiency of 25 percent. This occurs when the amplifier is driven to its maximum output. The efficiency is much less when the amplifier is not driven hard. With no drive, the efficiency drops to zero. An amplifier of this type would be a poor choice for high-power applications. The power supply would have to produce 4 times the required signal power. Three-fourths of this power would be wasted as heat in the load and the transistor. The transistor would probably require a large heat sink.

One reason that the class A amplifier is so wasteful is that dc power is wasted in the load. A big improvement is possible by removing the load from the dc circuit. Figure 8-7 shows how to do this. The transformer will couple

117

Ac load line

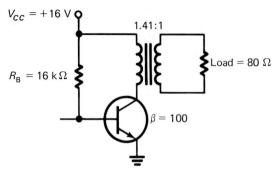

Fig. 8-7 Class A power amplifier with transformer coupling.

the signal power to the load. No dc current flow will appear in the load. Transformer coupling will allow the amplifier to produce twice the signal power.

Figure 8-7 shows the same supply voltage, the same bias resistor, the same transistor, and the same load as in Fig. 8-5. The only difference is the coupling transformer. The dc conditions are now quite different. The transformer primary will have very low resistance. This means that all the supply voltage will drop across the transistor at the operating point.

The dc load line for the transformer-coupled amplifier is shown in Fig. 8-8. It is vertical. The operating point is still at 100

mA. This is because the base current and β have not changed. The change is the absence of the 80-Ω dc resistance in series with the collector. All the supply voltage now drops across the transistor.

Actually, the load line will not be perfectly vertical. The transformer and even the power supply always have a little resistance. However, the dc load line is very steep. We cannot show any output swing from this load line.

There is a second load line in a transformer-coupled amplifier. It is the result of the ac load in the collector circuit. It is therefore called the *ac load line*. The ac load will not be 80 Ω for Fig. 8-7. The transformer is a step-down type. Remember, the impedance ratio is equal to the square of the turns ratio. Therefore, the ac load in the collector circuit will be

$$Ac\ load = (1.41)^2 \times 80 = 160\ \Omega$$

Notice that the ac load line in Fig. 8-8 runs from 32 V to 200 mA. This satisfies an impedance of

$$Z = \frac{V}{I} = \frac{32\ V}{200 \times 10^{-3}\ A} = 160\ \Omega$$

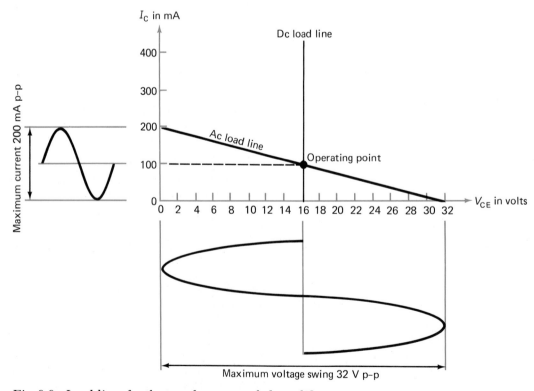

Fig. 8-8 Load lines for the transformer-coupled amplifier.

Also notice that the ac load line passes through the operating point. The dc load line and the ac load line must always pass through the same operating point.

How does the ac load line extend to 32 V? This is twice the supply voltage! There are two ways to explain this. First, it has to extend to 32 V if it is to pass through the operating point and satisfy a slope of 160 Ω. Second, a transformer is a type of inductor. When the field collapses, a voltage is generated. This voltage adds in series with the supply voltage. Thus, V_{CE} can swing to twice the supply voltage in a transformer-coupled amplifier.

Compare Fig. 8-8 with Fig. 8-6. The output swing doubles with transformer coupling. It is safe to assume the output power also doubles. The dc power has not changed. The supply voltage is still 16 V, and the average current is still 100 mA. Transformer-coupling the class A amplifier has given us twice as much signal power for the same dc power input. The efficiency of the transformer-coupled amplifier is

$$\% \text{ Efficiency} = \frac{P_{ac}}{P_{dc}} \times 100 = \frac{0.8}{1.6} = 50\%$$

Remember, however, that this efficiency is reached only at maximum signal level. The efficiency is much less for smaller signals.

Now, an efficiency of 50 percent is not too bad. This makes the class A power amplifier attractive for medium-power applications (up to 5 W or so). However, the transformer can be an expensive component. For example, in a high-quality audio amplifier, the output transformer may cost more than all the other amplifier parts combined! So for high-power and high-quality amplifiers, something other than class A may be a better choice.

The examples used have ignored some losses. First, we have ignored the saturation voltage across the transistor. In actual practice, the signal cannot swing down to 0 V. A power transistor might show a saturation of 0.7 V. This would have to be subtracted from the output swing. Second, we ignored transformer loss in the transformer-coupled amplifier. Transformers are not 100 percent efficient. Small, inexpensive transformers may be only 75 percent efficient at low audio frequencies. The calculated efficiencies of 25 and 50 percent are theoretical maximums. They are not realized in actual circuits.

Another problem with the class A circuit is the fixed drain on the power supply. Even when no signal is being amplified, the drain on the supply is fixed at 1.6 W. Most power amplifiers must handle signals that change in level. An audio amplifier, for example, will handle a broad range of volume levels. The class A amplifier will show a very poor efficiency when the signal is small.

Review Questions

Answer the following.

9. Refer to Fig. 8-5. The current gain of the transistor is 80. Calculate the power dissipated in the transistor with no input signal.

10. Refer to Fig. 8-6. The operating point is at $V_{CE} = 12$ V. Calculate the power dissipated in the transistor with no input signal.

11. Refer to Fig. 8-7. The transformer has a turns ratio from primary to secondary of 2:1. What load does the collector of the transistor see?

12. Refer to Fig. 8-7. The transformer has a turns ratio of 4:1. An oscilloscope shows a collector sinusoidal signal of 12 V peak-to-peak. What will the amplitude of the signal be across the 80-Ω load? What will be the rms signal power delivered to the load?

Determine whether each statement is true or false.

13. Transformer-coupling the output does not improve the efficiency of a class A amplifier.

14. Refer to Fig. 8-8. The dc load line is very steep because the dc resistance of the output transformer primary winding is so low.

15. In practice, it is possible to achieve 50 percent efficiency in class A by using transformer coupling.

8-3 CLASS B POWER AMPLIFIERS

The class B amplifier is biased at cutoff. No current will flow until an input signal provides the bias to turn on the transistor. This eliminates the large fixed drain on the power supply. The efficiency is much better. Only

119

Push-pull power amplifier

Driver transformer

Output transformer

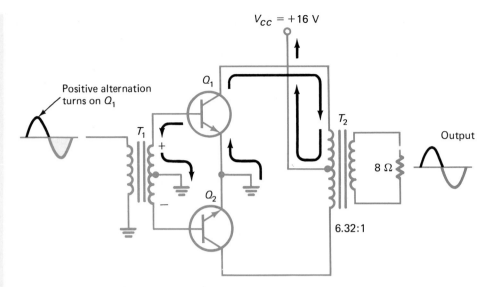

Fig. 8-9 Class B push-pull power amplifier with Q_1 turned on.

half the input signal is amplified, however. This produces extreme distortion. A single class B transistor would not be useful in audio work. The sound would be horrible.

Two transistors can be operated in class B. One can be arranged to amplify the positive-going portion of the input. The other will amplify the negative-going portion. Combining the two halves, or portions, will reduce much of the distortion. Two transistors operating in this way are said to be in *push-pull*.

Figure 8-9 shows a class B push-pull power amplifier. Two transformers are used. Transformer T_1 is called the *driver* transformer. It provides Q_1 and Q_2 with signal drive. Transformer T_2 is called the *output* transformer. It combines the two signals

and supplies the output to the load. Notice that both transformers are center-tapped.

With no signal input, there will not be any current flow in Fig. 8-9. Both Q_1 and Q_2 are cut off. There is no dc supply to turn on the base-emitter junctions. When the input signal drives the secondary of T_1 as shown, Q_1 is turned on. Current will flow through half of the primary of T_2. The positive signal alternation appears across the load.

When the signal reverses polarity, Q_1 is cut off and Q_2 turns on. This is shown in Fig. 8-10. Current will flow through half of the primary of T_2. This time the current is flowing up through the primary. When Q_1 was on, the current was flowing down the primary. This reversal produces the negative al-

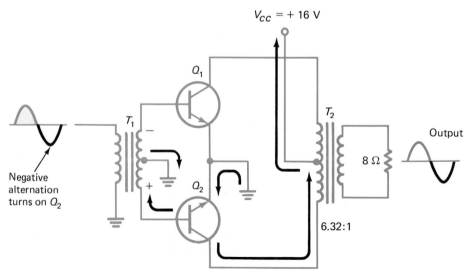

Fig. 8-10 Class B push-pull power amplifier with Q_2 turned on.

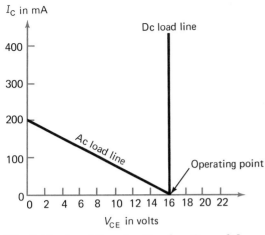

Fig. 8-11 Load lines for the class B amplifier.

ducting at any time. Therefore, only half the turns ratio will be used to calculate the impedance ratio:

$$\frac{6.32}{2} = 3.16$$

Now, the collector load will be equal to the square of half the turns ratio times the load resistance:

$$\text{Ac load} = (3.16)^2 \times 8 = 80\ \Omega$$

Each transistor sees an ac load of 80 Ω. The load line of Fig. 8-11 runs from 16 V to 200 mA. This satisfies a slope of 80 Ω:

$$R = \frac{V}{I} = \frac{16\ \text{V}}{0.2\ \text{A}} = 80\ \Omega$$

Figure 8-11 is correct but will show only one transistor. There is a better way to plot the graph of a push-pull circuit.

Figure 8-12 shows the graph for push-pull operation. This allows the entire output swing to be shown. The output voltage swings 32 V peak-to-peak. This must be converted to an rms value:

$$V_{\text{rms}} = \frac{V_{\text{p-p}}}{2} \times 0.707 = \frac{32}{2} \times 0.707$$

$$= 11.31\ \text{V}$$

ternation across the load. By operating two transistors in push-pull, much of the distortion has been eliminated. The circuit amplifies the entire input signal.

We can use graphs to show the output swing and efficiency for the class B push-pull amplifier. Figure 8-11 shows the dc and ac load lines for the push-pull circuit. The dc load line is vertical. There is very little resistance in the collector circuit. The ac load line is the transformed load in the collector circuit.

Transformer T_2 shows a turns ratio of 6.32:1. This turns ratio determines the ac load that will be seen in the collector circuit. Only half the transformer primary is con-

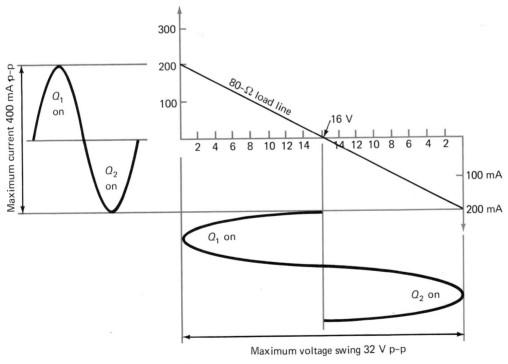

Fig. 8-12 Load lines for push-pull operation.

Electronics:
Principles and
Applications
CHAPTER 8

Crossover
distortion

Next, the rms current is

$$I_{rms} = \frac{I_{p-p}}{2} \times 0.707 = \frac{400}{2} \times 0.707$$

$$= 141.4 \text{ mA}$$

Finally, the signal power is

$$P = V \times I = 11.31 \times 141.4 \times 10^{-3} = 1.6 \text{ W}$$

To find the efficiency of the class B push-pull circuit, we will need the dc input power. The supply voltage is 16 V. The supply current varies from 0 to 200 mA. As in class A, the average collector current is what we need:

$$I_{av} = I_p \times 0.636 = 200 \times 0.636 = 127.2 \text{ mA}$$

The average input power is

$$P = V \times I = 16 \times 127.2 \times 10^{-3} = 2.04 \text{ W}$$

Our amplifier takes 2.04 W from the supply to give a signal output of 1.6 W. The efficiency is

$$\% \text{ Efficiency} = \frac{P_{ac}}{P_{dc}} \times 100 = \frac{1.6}{2.04} \times 100$$

$$= 78.5\%$$

The best efficiency for class A is 50 percent. The best efficiency for class B is 78.5 percent. This is a good improvement and makes the class B push-pull circuit attractive for high-power applications. For smaller signals, the class B amplifier takes less from the power supply. This means that its efficiency does not fall off as it did for class A. Thus, the improvement in class B is even more noticeable at smaller signal levels.

Class A power transistors need a high wattage rating. The reason is the poor efficiency of the transistor and the fact that it is always on. For example, to build a 100-W class A amplifier, the transistor will need at least a 200-W rating. This is based on

$$\% \text{ Efficiency} = \frac{P_{ac}}{P_{dc}} \times 100 = \frac{100}{200} \times 100$$

$$= 50\%$$

Look at the above equation: 200 W goes into

the transistor; 100 W comes out as signal power. The 100 W lost heats the transistor. What if the signal input is zero? The signal output is zero, yet 200 W still goes into the transistor. All 200 W is lost as heat in the transistor.

The wattage rating needed for class B is only one-fifth that needed for class A. To build a 100-W amplifier requires 200 W in class A. In class B

$$\frac{200}{5} = 40 \text{ W}$$

Two 20-W transistors operating in push-pull would handle it. Two 20-W transistors cost much less than one 200-W transistor. This is a great advantage of class B over class A.

The size of the heat sink is another factor. A transistor rating is based on some safe temperature. In high-power work, the transistor is mounted on a device which carries off the heat. A class B design will need only one-fifth the heat sink capacity for a given amount of power.

There is a very strong case for using class B in high-power work. However, there is still too much distortion for some applications. The push-pull circuit eliminates quite a bit of distortion, but some remains. The problem is called *crossover distortion*.

The base-emitter junction of a transistor behaves much as a diode. It is very nonlinear near the turn-on point. Figure 8-13 shows the characteristic curve for a typical silicon diode. Note the curvature near the 0.6-V forward-bias region. As one transistor is turning off and the other is coming on in a push-pull design, this curvature distorts the output signal.

The effect of crossover distortion on the output signal is seen in Fig. 8-14(*a*). It happens as the signal is *crossing over* from one transistor to the other. Crossover distortion

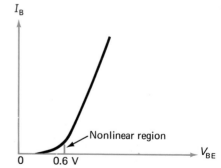

Fig. 8-13 Characteristic curve for a silicon PN junction.

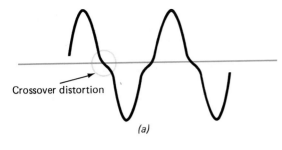

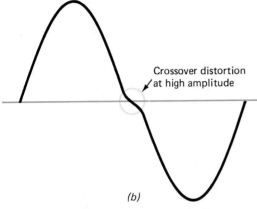

(a)

(b)

Fig. 8-14 Crossover distortion in the output signal.

is very noticeable when the signal is small. As shown in Fig. 8-14(*b*), the distortion is less noticeable for large signals. This can be a valuable clue when troubleshooting.

Review Questions

Determine whether each statement is true or false.

16. Refer to Fig. 8-9. Transistors Q_1 and Q_2 are a complementary pair.

17. Refer to Fig. 8-10. Transistors Q_1 and Q_2 will never be on at the same time.

18. Crossover distortion is due to the nonlinearity of the base-emitter junctions in the transistors.

Answer the following.

19. Refer to Fig. 8-10. Transformer T_2 has a turns ratio of 10:1. What is the load seen by the collector of Q_1? Q_2?

20. A class A power amplifier is designed to deliver 10 W of power. What is dissipated in the transistor at zero signal level?

21. A class B push-pull amplifier is designed to deliver 10 W of power. What is the most power that must be dissipated by each transistor?

22. Refer to Fig. 8-12. Assume the amplifier is being driven to only half its maximum swing. Calculate the rms power output.

23. Refer to Fig. 8-12. Assume the amplifier is driven to half its maximum swing. Calculate the average power input.

24. Refer to Fig. 8-12. Assume the amplifier is driven to half its maximum swing. Calculate the efficiency of the amplifier.

8-4 CLASS AB POWER AMPLIFIERS

The solution to the crossover distortion problem is to provide some forward bias for the base-emitter junctions. The forward bias will prevent the base-emitter voltage V_{BE} from ever reaching the nonlinear part of its curve. This is shown in Fig. 8-15. The forward bias is small and results in a class AB amplifier. It has characteristics between class A and class B. The operating point for class AB is shown in Fig. 8-16. Note that class AB operates near cutoff.

Figure 8-17 is a class AB push-pull power amplifier. Resistors R_1 and R_2 form a voltage divider to forward-bias the base-emitter junctions. The bias current flows through both halves of the secondary of T_1. Capacitor C_1 grounds the center tap for ac signals. It prevents R_2 from loading the drive signal.

A class AB amplifier will not have as high an efficiency as a class B amplifier. Its efficiency will be better than that of a class A design. It is a compromise design to achieve minimum distortion and reasonable efficiency. It is the most popular class for high-power audio work.

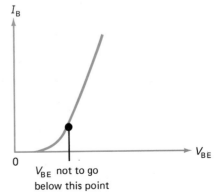

Fig. 8-15 Minimum value of V_{BE} to prevent crossover distortion.

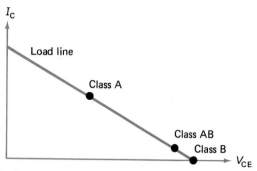

Fig. 8-16 Class AB operating point.

Complementary
symmetry
amplifier

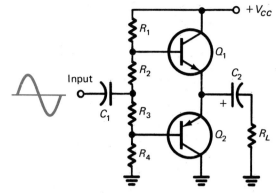

Fig. 8-18 A complementary symmetry amplifier.

Amplifiers such as the one shown in Fig. 8-17 are popular in portable radios and tape recorders.

The one big problem in this circuit is the transformers. For high-power and high-quality work, the transformers will be too expensive. The transformers, however, can be eliminated.

Driver transformers can be eliminated by using a combination of transistor polarities. A positive-going signal applied to the base of an NPN transistor tends to turn it on. A positive-going signal applied to the base of a PNP transistor tends to turn it off. This means that push-pull operation can be obtained without a center-tapped transformer.

Output transformers can be eliminated by using a different amplifier configuration. The emitter-follower (common-collector) amplifier is noted for its lower output impedance. This allows good matching to low-impedance loads such as loudspeakers.

Figure 8-18 shows an amplifier design that eliminates the transformers. Transistor Q_1 is an NPN transistor, and Q_2 is a PNP transistor. Push-pull operation will be obtained without a center-tapped driver transformer. Notice

that the load is capacitively coupled to the emitter leads of the transistors. The transistors are operating as emitter followers.

The circuit of Fig. 8-18 is known as a *complementary symmetry* amplifier. The transistors are complements. One is an NPN, and the other is a PNP. The curves in Fig. 8-19 show the symmetrical characteristics of NPN and PNP transistors. Good matching of characteristics is important in the complementary symmetry amplifier. For this reason, transistor manufacturers offer NPN-PNP pairs having good symmetry.

Figure 8-20 follows the output signal in a complementary symmetry amplifier when the input signal goes positive. Transistor Q_1, the NPN transistor, is turned on. Transistor Q_2, the PNP transistor, is turned off. Current flows through the load, through C_2, and through Q_1 into the power supply. This current charges C_2 as shown. Notice that there is no phase inversion in the amplifier. This is to be expected in an emitter follower.

When the input signal goes negative, the signal flow is as shown in Fig. 8-21. Now Q_1 is

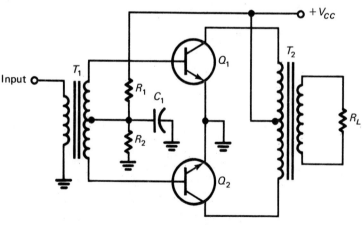

Fig. 8-17 Class AB push-pull power amplifier.

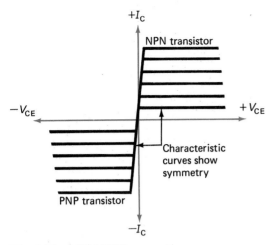

Fig. 8-19   NPN-PNP symmetry.

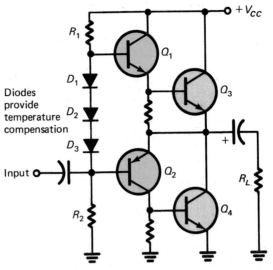

off and Q_2 is on.   This causes C_2 to discharge as shown.   Again, the output is in phase with the input.   Capacitor C_2 is usually a large capacitor (1000 μF or so).   This is necessary for good low-frequency response with low values of R_L.

Another possibility is shown in Fig. 8-22.

Fig. 8-22   A quasi-complementary symmetry amplifier.

This is known as a *quasi*-complementary symmetry amplifier.   The output transistors Q_3 and Q_4 are not complementary.   They are both NPN types.   The driver transistors Q_1

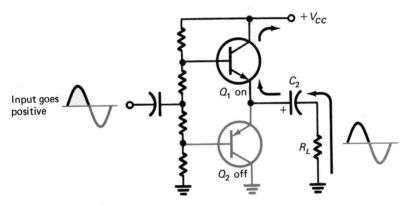

Fig. 8-20   A positive-going signal in a complementary symmetry amplifier.

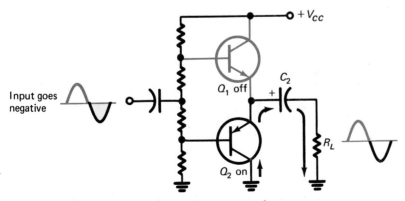

Fig. 8-21   A negative-going signal in a complementary symmetry amplifier.

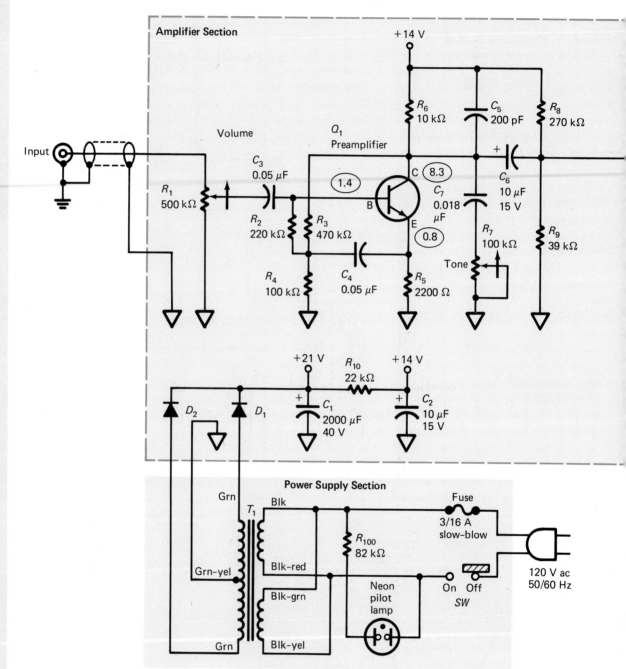

Fig. 8-23 An audio amplifier. (Courtesy of Heath Company.)

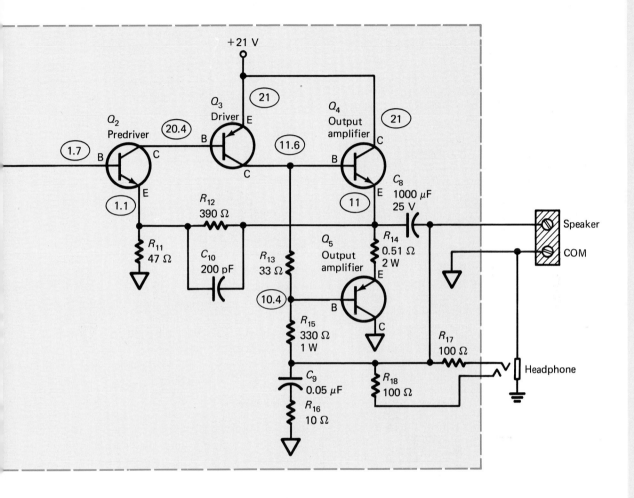

NOTES:

1. All resistors are 1/2–W unless marked otherwise.

2. This symbol indicates a positive dc voltage measurement, taken with an 11-MΩ input voltmeter, from the point indicated to chassis ground, with all controls in counterclockwise position, and speaker connected to the amplifier. Voltages may vary ±20%.

3. Indicates clockwise rotation of control.

4. Indicates common circuit board foil ground connections.

Temperature
compensation

Bootstrap circuit

Class C amplifier

Tank circuit

and Q_2 are complementary. A positive-going input signal will turn on Q_1, the NPN driver. It will turn off Q_2, the PNP driver. This results in a push-pull action since the drivers supply the base current for the output transistors. Again, no center-tapped transformer is necessary in the output.

Notice the diodes used in the bias network of Fig. 8-22. The diodes provide *temperature compensation*. Transistors tend to conduct more current as temperature goes up. This means that the operating point of an amplifier tends to change with temperature. This is undesirable. The drop across a conducting diode decreases with an increase in temperature. If the diode drop is part of the bias voltage for the amplifier, compensation results. The decreasing voltage across the diode will lower the amplifier current. Thus, the operating point is more stable with this arrangement.

Figure 8-23 is the schematic diagram for an audio amplifier. It is rated at 3.25-W output power. Transistors Q_4 and Q_5 are connected in a complementary symmetry, emitter-follower output circuit. The driver transistors Q_2 and Q_3 are direct-coupled to the output transistors. Resistor R_{12} and capacitor C_{10} apply both ac and dc negative feedback. This feedback reduces distortion and stabilizes the operating point. Note that the emitter of Q_4 is operating at 11 V. This is about half the supply voltage. The output point in an amplifier of this type should always operate at about half the supply voltage. This is valuable information for troubleshooting this amplifier.

Transistor Q_1 in Fig. 8-23 is the preamplifier. Capacitor C_4 feeds back signal from the emitter to R_2. The other end of R_2 is the input circuit. This is known as a *bootstrap* circuit. The feedback raises the input impedance of the amplifier. The input impedance of the amplifier is rated at 125,000 Ω. Without the bootstrap circuit, the input impedance would be about 46,000 Ω, too low for many uses.

Resistor R_7 and capacitor C_7 make up the tone control circuit in Fig. 8-23. When R_7 is set for minimum resistance, C_7 will bypass the higher audio frequencies to ground. This decreases the gain of the amplifier for the high frequencies. Setting R_7 to 0 Ω decreases the gain at 15 kHz by 22 dB.

Review Questions

Answer the following.

25. Is the efficiency of class AB better than that of class A but not as good as that of class B?

26. Refer to Fig. 8-17. Assume that R_1 opens (infinite resistance). In what class will the amplifier operate?

27. Refer to Fig. 8-17. Assume that Q_1 and Q_2 are running very, very hot. Could C_1 be shorted? Why or why not?

28. Refer to Fig. 8-17. Assume that Q_1 and Q_2 are running very, very hot. Could R_2 be open? Why or why not?

29. Refer to Fig. 8-18. An input signal drives C_1 in a positive direction. In what direction will the top of R_L be driven?

30. Refer to Fig. 8-18. An input signal drives C_1 in a positive direction. Which transistor is turning off?

31. Refer to Fig. 8-18. Voltage $V_{CC} = 14$ V. With no input signal, what should the voltage be at the emitter of Q_1? At the base of Q_1? At the base of Q_2? (*Hint:* The transistors are silicon.)

8-5 CLASS C POWER AMPLIFIERS

Class C amplifiers are biased beyond cutoff. Figure 8-24 is a class C amplifier with a negative voltage V_{BB} applied to the base circuit. This negative voltage reverse-biases the base-emitter junction of the transistor. The transistor will not conduct until the input signal overcomes the bias. This happens for only a small part of the input cycle. The transistor conducts for about 90° of the input waveform.

As shown in Fig. 8-24, the collector-current waveform is not a whole sine wave. It is not even half a sine wave. This extreme distortion means the class C amplifier *cannot* be used for audio work. Class C amplifiers are used at radio frequencies.

Figure 8-24 shows a *tank circuit* in the collector circuit of the class C amplifier. This tank circuit restores the sine wave input signal. Note that a sine wave is shown across R_L. Tank circuits can restore sine wave signals but not square waves or complex audio waves.

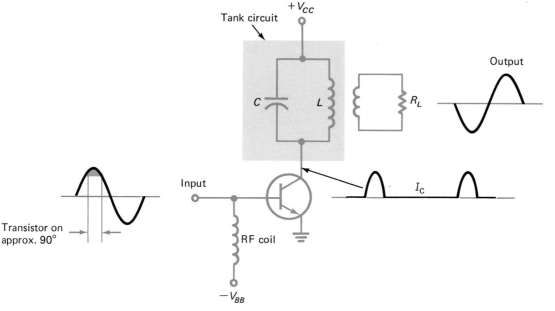

Damped sine wave

Resonance

Doubler

Tripler

Fig. 8-24   A class C amplifier.

Tank circuit action is explained in Fig. 8-25. The collector-current pulse charges the capacitor [Fig. 8-25(a)]. After the pulse, the capacitor discharges through the inductor [Fig. 8-25(b)]. Energy is stored in the field around the inductor. When the capacitor discharges to zero, the field collapses and keeps the current flowing [Fig. 8-25(c)]. This charges the capacitor again, but note that the polarity is opposite. After the field has collapsed, the capacitor again begins discharging through the inductor [Fig. 8-25(d)]. Note that current is now flowing in the opposite direction and the inductor field is expanding. Finally, the inductor field begins to collapse, and the capacitor is charged again to its original polarity [Fig. 8-25(e)].

Tank circuit action is due to a capacitor discharging into an inductor which later discharges into the capacitor and so on. Both the capacitor and the inductor are energy storage devices. As the energy goes from one to the other, a sine wave is produced. Circuit loss (resistance) will cause the sine wave to decrease gradually. This is shown in Fig. 8-26(a) and is called a *damped* sine wave. By pulsing the tank circuit every cycle, the sine wave can be constant. This is shown in Fig. 8-26(b). In a class C amplifier, the tank circuit is recharged by a collector-current pulse every cycle. This makes the sine wave constant.

The values of inductance and capacitance are very important in a class C amplifier tank circuit. They must *resonate* at the frequency of the input signal. The equation for resonance is

$$f_r = \frac{1}{6.28 \sqrt{LC}}$$

What is the resonant frequency of a tank circuit that has a 100-pF capacitance and 1-μH inductance? Substituting the values into the equation, we get

$$f_r = \frac{1}{6.28 \sqrt{1 \times 10^{-6} \times 100 \times 10^{-12}}}$$
$$= 15.9 \times 10^6 \text{ Hz}$$

The resonant frequency is 15.9 MHz.

In some cases, the tank circuit is tuned to resonate at 2 or 3 times the frequency of the input signal. This produces an output signal that will be 2 or 3 times the frequency of the input signal. Such circuits are called *doublers* or *triplers*. They are commonly used where high-frequency signals are needed. For example, suppose that a 150-MHz police transmitter is being designed. It is easier to start off at a much lower frequency for good stability. Figure 8-27 shows the block diagram for such a transmitter.

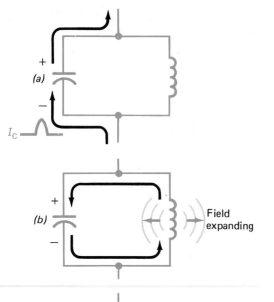

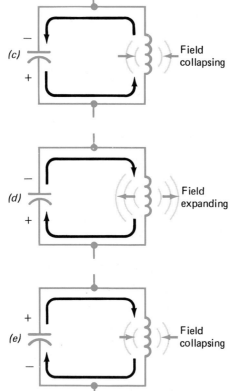

Fig. 8-25 Tank circuit action.

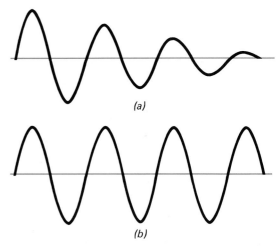

Fig. 8-26 Tank circuit waveforms.

The class C amplifier is the most efficient of all the amplifier classes. This high efficiency is shown by the waveforms of Fig. 8-28. The top waveform is the input signal V_{BE}. You can see that only the positive peak can forward-bias the base-emitter junction. This occurs at 0.6 V in a silicon transistor. Most of the input signal falls below this value because of the negative bias. The second waveform shown is the collector current I_C. It shows that collector current is in the form of narrow pulses. The last waveform shown is the collector-emitter voltage V_{CE}. It is sinusoidal because of tank circuit action. Note that the collector-current pulses occur when V_{CE} is near zero. This means that little power will be dissipated in the transistor:

$$P_C = V_{CE} \times I_C = 0 \times I_C = 0\ \text{W}$$

If no power is dissipated in the transistor, it must all become signal power. This leads to the conclusion that the class C amplifier is 100 percent efficient. Actually, there is power dissipated in the transistor. Voltage V_{CE} is

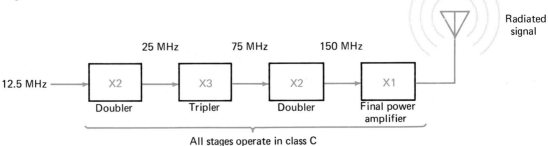

Fig. 8-27 Block diagram for a high-frequency transmitter.

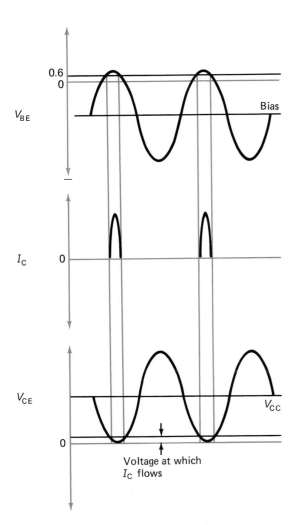

Fig. 8-28 Class C amplifier waveforms.

L network

Practical class C power amplifiers seldom use a separate bias supply for the base circuit. A better way to do it is to use signal bias. This is shown in Fig. 8-29. As the input signal goes positive, it forward-biases the base-emitter junction. Base current I_B flows as shown. The current charges C_1. Resistor R_1 discharges C_1 between positive peaks of the input signal. Resistor R_1 cannot completely discharge C_1, and the remaining voltage across C_1 acts as a bias supply. The polarity reverse-biases the base-emitter junction.

Figure 8-29 also shows a different type of tank circuit. It is known as an L network. It matches the impedance of the transistor to the load. Radio-frequency power transistors often have an output impedance of about 2 Ω. The standard load impedance in RF work is 50 Ω. Thus, the L network is necessary to match the transistor to 50 Ω.

Review Questions

Answer the following.

32. Refer to Fig. 8-24. The transistor is silicon, and $-V_{BB} = 4$ V. How positive will the input signal have to swing to turn on the transistor?

33. Refer to Fig. 8-24. Assume the input signal is a square wave. What signal can be expected across R_L, assuming a high-Q tank circuit?

34. Is class C more efficient than class B?

35. Does class C have the smallest conduction angle?

36. The input frequency to an RF tripler stage is 7 MHz. What is the output frequency?

not zero. The tank circuit will also cause some loss. A practical class C power amplifier might reach an efficiency of about 85 percent. This is still very good. Class C amplifiers are very popular at radio frequencies where tank circuits can restore the sine wave signal.

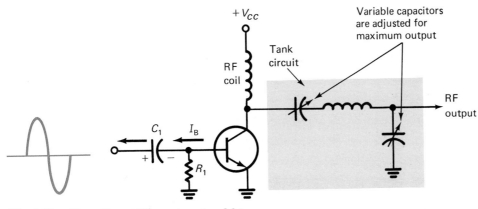

Fig. 8-29 Class C amplifier using signal bias.

131

37. A tank circuit uses a 4-μH coil and a 180-pF capacitor. What is the resonant frequency of the tank?

38. Refer to Fig. 8-29. Assume that the amplifier is being driven by a signal and the voltage at the base of the transistor is negative. What should happen to the base voltage if the drive signal increases?

Summary

1. All amplifiers are technically power amplifiers. Only those that handle large signals are called power amplifiers.

2. The power amplifier is usually the last stage in the signal processing system.

3. Power amplifiers should be efficient. Efficiency is a comparison of the signal power output to the dc power input.

4. Poor efficiency in a power amplifier means the power supply will have to be larger and more expensive.

5. Class A amplifiers operate at the center of the load line. They have low distortion and a conduction angle of 360°.

6. Class B operates at cutoff. The conduction angle is 180°.

7. Class B amplifiers do not present a fixed drain on the power supply. The drain is zero with zero signal.

8. Class B is more efficient than class A.

9. Bias controls the operating point and the class of operation for an amplifier.

10. The best efficiency possible for class A operation is 25 percent. With transformer coupling, it is 50 percent.

11. In a transformer-coupled amplifier, the impedance ratio is equal to the square of the turns ratio.

12. The fixed drain on the power supply is a major fault with class A circuits. Efficiency is very poor when signals are small.

13. A single class B transistor will amplify half the input signal.

14. Two class B transistors can be operated in push-pull.

15. The maximum theoretical efficiency for class B is 78.5 percent.

16. A class B amplifier draws less current from the supply for smaller signals.

17. For a given output power, class B transistors will require only one-fifth of the power rating needed for class A.

18. The biggest drawback to class B push-pull is crossover distortion.

19. Crossover distortion can be eliminated by providing some forward bias for the base-emitter junctions of the transistors.

20. Class AB amplifiers are forward-biased slightly above cutoff.

21. Class AB operation is the most popular for high-power audio work.

22. Push-pull operation can be obtained without center-tapped transformers by using a PNP-NPN pair.

23. An amplifier that uses a PNP-NPN pair for push-pull operation is called a complementary symmetry amplifier.

24. Complementary pairs have symmetrical characteristic curves.

25. Diodes may be used to stabilize the operating point in power amplifiers.

26. A bootstrap circuit may be used to raise the input impedance of an amplifier.

27. Class C amplifiers are biased beyond cutoff.

28. The conduction angle for class C is around 90°.

29. Class C amplifiers have too much distortion for audio work. They are useful at radio frequencies.

30. A tank circuit can be used to restore a sine wave signal in a class C amplifier.

31. The tank circuit should resonate at the signal frequency.

32. The class C amplifier has a maximum theoretical efficiency of 100 percent. In practice, it can reach about 85 percent.

Chapter Review Questions

Answer the following.

8-1. Refer to Fig. 8-1. In which of the three stages shown is efficiency the most important?

8-2. An amplifier delivers 80 W of signal power. Its power supply develops 28 V, and the current drain is 4 A. What is the efficiency of the amplifier?

8-3. An amplifier has an efficiency of 35 percent. It is rated at 5 W of output. How much current will it draw from a 12-V battery when delivering its rated output?

8-4. Which class of amplifier produces the least distortion?

8-5. What is the conduction angle of a class B amplifier?

8-6. The operating point for an amplifier is at the center of the load line. What class is the amplifier?

8-7. Refer to Fig. 8-7. What is the maximum theoretical efficiency for this circuit? At what signal level is this efficiency achieved?

8-8. What will happen to the efficiency of the amplifier in Fig. 8-7 as the signal level decreases?

8-9. Refer to Fig. 8-7. What turns ratio will be required to transform the 80-Ω load to a collector load of 720 Ω?

8-10. A class A transformer-coupled amplifier operates from an 18-V supply. What is the maximum peak-to-peak voltage swing at the collector?

8-11. Refer to Fig. 8-9. What would have to be done to V_{cc} so that the circuit could use PNP transistors?

8-12. Refer to Fig. 8-9. With zero signal level, how much current will be taken from the 16-V supply?

8-13. Refer to Fig. 8-9. What is the phase of the signal at the base of Q_1 as compared to the base of Q_2? What component causes this?

8-14. Refer to Fig. 8-9. Assume the peak-to-peak sine wave swing across the collectors is 32 V. The transformer is 100 percent efficient. Calculate
(A) V_{p-p} across the load (do not forget to use half the turns ratio)
(B) V_{rms} across the load (C) P_L (load power)

8-15. Calculate the minimum wattage rating for each transistor in a push-pull class B design rated at 50-W output.

8-16. Calculate the minimum wattage rating for a class A power transistor that is transformer-coupled and rated at 50-W output.

8-17. At what signal level is crossover distortion most noticeable?

8-18. Refer to Fig. 8-17. Which two components set the forward bias on the base-emitter junctions of Q_1 and Q_2?

8-19. Refer to Fig. 8-17. There is no input signal. Will the amplifier take any current from the power supply?

8-20. Refer to Fig. 8-17. What will happen to the current taken from the power supply as the signal level increases?

8-21. Refer to Fig. 8-18. What is the configuration of Q_1?

8-22. Refer to Fig. 8-18. What is the configuration of Q_2?

8-23. Refer to Fig. 8-21. When the input signal goes negative, what supplies the energy to the load?

8-24. The major reason for using class AB in a push-pull amplifier is to eliminate distortion. What name is given to this particular type of distortion?

8-25. Refer to Fig. 8-22. Assume that a signal drives the input positive. What happens to the current flow in Q_1 and Q_3?

8-26. Refer to Fig. 8-22. Assume that a signal drives the input positive. What happens at the top of R_L?

8-27. Refer to Fig. 8-22. Assume that Q_2 is turned on harder (conducts more). What should happen to Q_4?

8-28. Refer to Fig. 8-23. Which two transistors operate in complementary symmetry?

8-29. Which amplifier class has the best efficiency?

8-30. Refer to Fig. 8-24. What allows the output signal across the load resistor to be a sine wave?

8-31. Refer to Fig. 8-24. What makes the conduction angle of the amplifier so small?

8-32. Refer to Fig. 8-29. What does the charge on C_1 accomplish?

8-33. Refer to Fig. 8-29. What two things does the tank circuit accomplish?

Answers to Review Questions

1. F	16. F	remove forward bias and tend to make them run cooler.
2. F	17. T	
3. F	18. T	
4. F	19. 200 Ω, 200 Ω	28. Yes, because this would increase forward bias.
5. T	20. 20 W	
6. T	21. 2 W	29. Positive
7. 66.7 percent	22. 0.4 W	30. Q_2
8. 250 W	23. 1.02 W	31. 7, 7.6, and 6.4 V
9. 0.768 W	24. 39.2 percent (*Note:* This is half the efficiency achieved for driving the amplifier to its maximum swing.)	32. 4.6 V
10. 0.6 W		33. Sine wave
11. 320 Ω		34. Yes
12. 3 V peak-to-peak, 14.06 mW		35. Yes
		36. 21 MHz
13. F	25. Yes	37. 5.93 MHz
14. T	26. Class B	38. It should increase (go more negative).
15. F	27. No, because this would	

Operational Amplifiers

Integrated circuit technology has had quite an impact on modern electronic design. Integrated circuits have given designers high performance for low cost. In the linear field, perhaps the best example is the modern operational amplifier. These building blocks offer solutions to many circuit problems.

This chapter deals with the theory and characteristics of differential and operational amplifiers. Information is supplied on their advantages, their limitations, and many of the ways in which they are used.

9-1 THE DIFFERENTIAL AMPLIFIER

An amplifier can be designed to respond to the difference between two input signals. Such an amplifier has two inputs and is called a difference, or *differential*, amplifier. Figure 9-1 shows the basic arrangement of this amplifier. Notice that two power-supply polarities are shown ($+V_{CC}$ and $-V_{EE}$). This is common practice in differential amplifier work. Such supplies are called dual supplies. Two batteries can also be used to form a dual supply, as in Fig. 9-2. Figure 9-3 is a dual-supply rectifier circuit.

A differential amplifier can be driven at one of its inputs. This is shown in Fig. 9-4. An output signal will appear at both collectors.

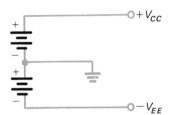

Fig. 9-2 A battery dual supply.

Assume that the input drives the base of Q_1 in a positive direction. The conduction in Q_1 will increase. More voltage will drop across Q_1's load resistor because of the increase in current. This will cause the collector of Q_1 to go less positive. Thus, an inverted output is available at the collector of Q_1.

What causes the signal at the collector of Q_2? As Q_1 is turned on harder by the positive-going input, the current through R_E will increase. This makes the drop across R_E increase. The emitters of both transistors will go in a positive direction. Making Q_2's emitter go positive has the same effect as making its base go negative. Transistor Q_2 responds by conducting less current. Less voltage drops across Q_2's load resistor, and its collector goes in a positive direction.

As can be seen in Fig. 9-4, both an inverted (out-of-phase) and a noninverted (in-phase) signal are available. Also a differential output is available from the collector of Q_1 to the collector of Q_2. The differential output has

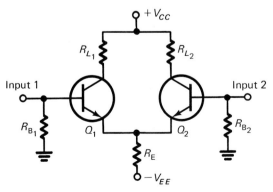

Fig. 9-1 A differential amplifier.

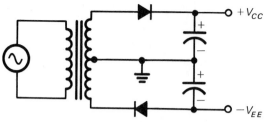

Fig. 9-3 A rectifier dual supply.

From page 135:
Differential
amplifier

Dual supply

On this page:
Hum and noise

Common-mode
signal

Common-mode
rejection ratio
(CMRR)

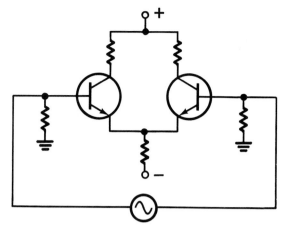

Fig. 9-5 Driving the amplifier differentially.

twice the swing. If, for example, Q_1's collector has gone 2 V negative and Q_2's collector has gone 2 V positive, the difference is $(+2) - (-2) = 4$ V. The differential output is 2 times the output of one collector signal referenced to ground.

The amplifier can also be driven differentially, as shown in Fig. 9-5. There is an advantage to this mode of operation. Hum and noise often can be reduced. If the hum is common to both inputs (same phase), it will not affect the differential output. Figure 9-6 shows how hum can affect a desired signal. This happens around power lines and other sources of electrical noise. The result is a noisy signal of poor quality. Sometimes the hum and noise can be so bad that the signal is not usable.

Refer to Fig. 9-7. A noisy differential signal is shown. Note that the phase of the hum is common. The hum goes positive to both inputs at the same time. Later, both inputs see a negative-going hum signal. This is called a *common-mode signal*. Common-mode signals are greatly reduced in differential amplifiers.

Refer to Fig. 9-5. Assume that a common-mode signal drives both inputs positive at the same time. This will increase the conduction in both transistors. The drop

across both load resistors will increase. *Both* collectors will go in a negative direction. Assuming a balanced circuit, the difference voltage from the collector of Q_1 to the collector of Q_2 will be zero. Figure 9-7 shows that the differential output signal does not contain the common-mode signal. This ability to reject hum and noise common to both inputs is one of the advantages of differential amplifiers.

In practice, a differential amplifier will not have perfect balance. For example, one transistor may show a little more gain than the other. This means that some common-mode signal will appear across the output. The ability to reject the common-mode signal is given by the common-mode rejection ratio (CMRR):

$$\text{CMRR} = \frac{A_{V,\text{diff}}}{A_{V,\text{com}}}$$

where $A_{V,\text{diff}}$ = voltage gain of amplifier
for differential signals
$A_{V,\text{com}}$ = voltage gain of amplifier for
common-mode signals

Assume the input signal to a differential amplifier is 1 V common mode. The common-mode output is measured at 0.05 V. The common-mode voltage gain is

$$A_{V,\text{com}} = \frac{\text{signal out}}{\text{signal in}} = \frac{0.05}{1} = 0.05$$

Also assume that the differential input signal is 0.1 V and produces an output of 10 V. The

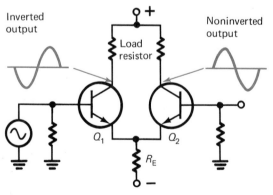

Fig. 9-4 Driving the differential amplifier at one input.

Inverted output

Load resistor

Noninverted output

Q_1 Q_2

R_E

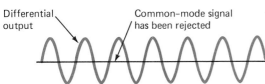

Fig. 9-6   Hum voltage can add to a signal.

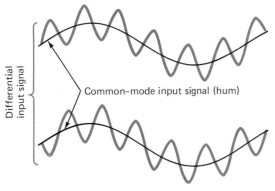

Fig. 9-7   The common-mode hum can be rejected.

differential voltage gain is then calculated by

$$A_{V,\,\text{diff}} = \frac{\text{signal out}}{\text{signal in}} = \frac{10}{0.1} = 100$$

The common-mode rejection ratio of the amplifier is

$$\text{CMRR} = \frac{100}{0.05} = 2000$$

The amplifier shows 2000 times as much gain for differential signals as it does for common-mode signals.   The CMRR is usually specified in decibels:

$$\text{CMRR (dB)} = 20 \times \log 2000 = 66 \text{ dB}$$

Practical differential amplifiers may give common-mode rejection ratios over 100 dB. They are quite effective in rejecting common-mode signals.

Review Questions

Answer the following.

1. Refer to Fig. 9-1.   What name is given to the energy source marked $+V_{CC}$ and $-V_{EE}$?

2. Refer to Fig. 9-1.   Assume that input 1 and input 2 are driven 1 V positive.   If the amplifier has perfect balance, what voltage difference should appear across the two collectors?

3. Refer to Fig. 9-1.   Assume that a signal appears at input 1 and drives it positive.   In what direction will the collector of Q_1 be driven?   The collector of Q_2?

4. When a signal drives input 2 in Fig. 9-1, what component couples the signal to Q_1?

5. Assume that in Fig. 9-1 a signal drives input 1 and produces an output at the collector of Q_1.   This output measures 5 V peak-to-peak with respect to ground.   What signal amplitude should appear across the two collectors?

6. Refer to Fig. 9-2.   Both batteries are 9 V. What is V_{CC} with respect to ground?   What is V_{EE} with respect to ground?   What is V_{CC} with respect to V_{EE}?

7. Refer to Fig. 9-4.   What reference point is used to establish the inverted output and the noninverted output?

8. Refer to Fig. 9-5.   Assume the signal source supplies 100 mV.   The differential output signal is 12 V.   Calculate the differential voltage gain of the amplifier.

9. Refer to Fig. 9-5.   The differential voltage gain is 80.   A common-mode hum voltage of 800 mV is applied to both inputs.   The differential hum output is 8 mV.   Calculate the CMRR for the amplifier.

9-2 OPERATIONAL AMPLIFIERS

There are no perfect amplifiers though modern operational amplifiers approach perfection.   They have characteristics that make them very useful in electronic circuits.   Some of their characteristics are as follows:

1. *Common-mode rejection*—this gives them the ability to reduce hum and noise.
2. *High input impedance*—they will not "load down" a high-impedance signal source.

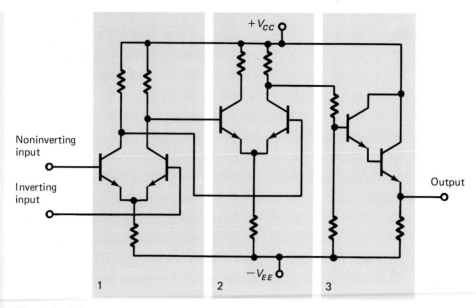

Fig. 9-8 The major sections of an operational amplifier.

3. *High gain*—they have "gain to burn" which can be reduced by using feedback.
4. *Low output impedance*—they are able to deliver a signal to a low-impedance load.

No single amplifier circuit can rate high in all the above characteristics. An operational amplifier is actually a combination of circuits. Refer to Fig. 9-8. The first section of this combination circuit is a differential amplifier. Differential amplifiers have common-mode

rejection and a high input impedance. Some operational amplifers may use field-effect transistors in this first section for an even better input impedance.

The second section of Fig. 9-8 is another differential amplifier. This allows the differential output of the first section to be used. Thus, common-mode rejection, maximum signal swing, and additional gain are all realized.

The third section of Fig. 9-8 is a common-

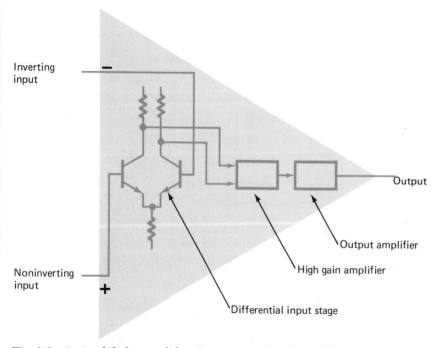

Fig. 9-9 A simplified way of showing an operational amplifier.

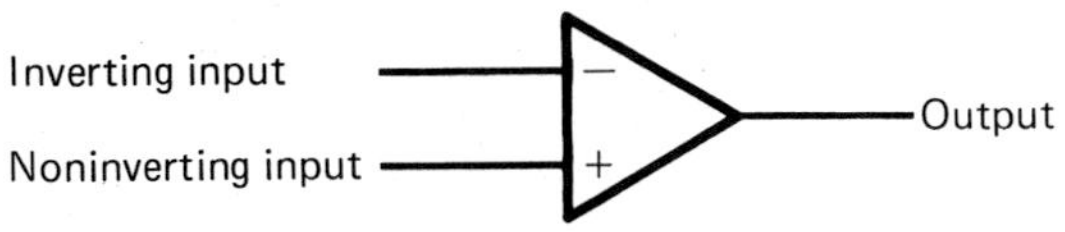

Inverting input ———

Noninverting input ———

——— Output

Fig. 9-10 The standard way of showing an operational amplifier.

collector, or emitter-follower, stage. This configuration is known for its low output impedance. Notice that the output is a single terminal. No differential output is possible. This is usually referred to as a *single-ended output*. Most electronic applications require only a single-ended output.

A single-ended output terminal can show only one phase with respect to ground. This is why in Fig. 9-8 one input is marked noninverting and the other is marked inverting. The noninverting input will be in phase with the output terminal. The inverting input will be 180° out of phase with the output terminal.

Figure 9-9 shows the amplifier in a simplified way. Notice the triangle. Often triangles are used on schematic diagrams to represent amplifiers. Also notice that the inverting input is marked with a minus (−) sign and that the noninverting input is marked with a plus (+) sign. This is standard practice in linear diagrams.

Many circuits today are integrated. Everything is in one little package. A technician cannot see inside the package or make any internal measurements. Therefore, it is seldom necessary to show the schematic details of the internal circuitry. Figure 9-10 is an example of the standard way of showing an operational amplifier.

Modern operational amplifiers approach perfection. They have the desirable characteristics listed earlier. Such amplifiers were originally used in analog computers. They performed mathematical operations. This is where they got the name "operational amplifiers," or "op-amps." Op-amps are very popular, mainly because they are available in compact, integrated form at very reasonable costs.

Figure 9-11 is the schematic diagram of an integrated circuit op-amp. It has a noninverting input, an inverting input, and a single-ended output. It also has two terminals marked "offset null." These terminals are used when the amplifier must be perfectly balanced. Any imbalance will cause a dc error or offset in the output. With no dc input, the dc output should be zero with respect to ground. Figure 9-12 shows how the terminals are used when the offset must be reduced.

The potentiometer in Fig. 9-12 can be adjusted to reduce offset in the amplifier. It is adjusted so the output terminal is at dc ground potential. The control has a limited range. It will overcome an offset in the millivolt range. If there is a large dc offset at the input

Single-ended output

Op-amp

Offset null

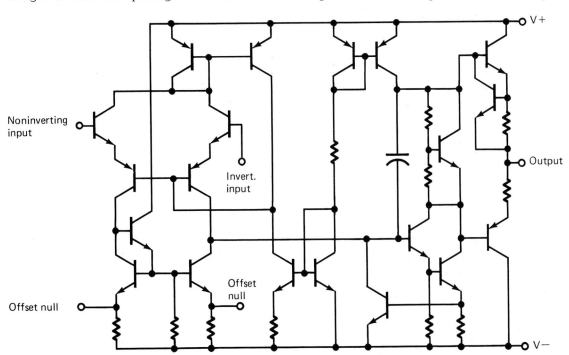

Fig. 9-11 A schematic of an operational amplifier.

Slew rate

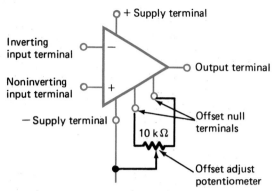

Fig. 9-12   Using the offset null terminals.

(measured across the minus and plus input terminals), the offset adjustment will not overcome it. In many applications, a small offset is not a problem. The offset null terminals are not used in these applications.

Quite a variety of integrated circuit operational amplifiers are available. Some use bipolar transistors, and some use field-effect transistors in combination with bipolars. Op-amps are available with good characteristics in areas such as input impedance and high-frequency performance. It is not possible to list all their characteristics here. The following list represents some general characteristics for a typical op-amp.

- Voltage gain: 200,000
- Output impedance: 75 Ω
- Input impedance: 1 MΩ
- CMRR: 90 dB
- Offset adjustment range: ± 15 mV
- Output voltage swing: ± 13 V
- Slew rate: 0.5 V/μs

The last characteristic in the list is the *slew rate*. This is the maximum output swing (slew) from one voltage to another in a given

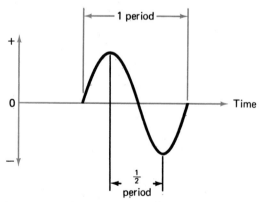

Fig. 9-13   Slew rate can limit the output swing.

period of time. The slew rate becomes a very important factor at high frequencies when the output must swing rapidly. For example, suppose the input signal to the op-amp is a sine wave at a frequency of 500,000 Hz. The maximum time available for the output to swing can be found by first determining the period of the input waveform. Period is the time required to complete one cycle and is found by

$$\text{Period} = \frac{1}{\text{frequency}} = \frac{1}{5 \times 10^5} = 2 \times 10^{-6} \text{ s}$$

Figure 9-13 shows that only half the period is available for the swing from the maximum positive voltage value to the maximum negative voltage value. Therefore, 1 μs is the maximum time available. The typical op-amp slews only 0.5 V in this time. This means that in order for the op-amp to accurately reproduce this signal, the output signal cannot change more than 0.5 V peak-to-peak in 1 μs. As shown in the list, the output is capable of swinging 26 V (± 13 V) peak-to-peak. At an input frequency of 500,000 Hz, the output has to be held to much less than this because of the slew rate of the op-amp. A general-purpose op-amp is not suited to high-frequency operation. Special op-amps with high slew rates are available.

Review Questions

Answer the following.

10. Refer to Fig. 9-8. Which section of the amplifier (1, 2, or 3) operates as an emitter follower to produce a low output impedance?

11. Refer to Fig. 9-8. A signal is applied to the inverting input terminal. What is the phase of the signal that appears at the output terminal as compared to the input signal?

12. Refer to Fig. 9-8. Is the output of the amplifier differential or single-ended?

13. Refer to Fig. 9-8. The output terminal measures 5 mV negative with respect to ground. Will it normally be possible to nullify this offset?

14. Refer to Fig. 9-8. The output terminal measures 3 V positive with respect to ground. Will it normally be possible to nullify this offset?

15. It is desired to have an output of 100 kHz at an amplitude of 10 V peak-to-peak. What slew rate does this signal represent? (Specify in volts per microsecond.)

16. An op-amp has a slew rate of 5 V/μs. What is the maximum amplitude that can be expected from this op-amp at a frequency of 1 MHz?

9-3 SETTING OP-AMP GAIN

The gain of modern integrated circuit op-amps is very high at low frequencies. With some op-amps, the gain may be as high as 1 million. This much gain is seldom needed. Usually it is much better to have less gain and have it over a broader range of frequencies. This is why op-amps are almost always operated at reduced gain by using negative feedback.

The gain listed on an op-amp specification sheet is the *open-loop gain*. "Open loop" means there is no feedback. Op-amps are seldom used in this way. They are used with negative feedback. Some of the output signal is fed back into the inverting input of the amplifier. Signals fed back to the noninverting input of the op-amp would produce positive feedback. Positive feedback is generally used only in digital circuit applications.

Figure 9-14 shows an op-amp with negative feedback. This negative feedback decreases the voltage gain and increases the bandwidth of the amplifier. If the open-loop gain is very high, the gain of Fig. 9-14 will be almost entirely set by two resistors. The gain will be

$$A_V = \frac{R_F}{R_1} = \frac{10 \text{ k}\Omega}{1 \text{ k}\Omega} = 10$$

The gain of Fig. 9-14 can be increased easily. For example, the feedback resistor R_F could be increased to 100,000 Ω. This would produce a gain of

$$A_V = \frac{R_F}{R_1} = \frac{100 \text{ k}\Omega}{1 \text{ k}\Omega} = 100$$

It is also easy to reduce the gain. A 1000-Ω feedback resistor would produce unity gain ($A_V = 1$). Setting the gain for such a circuit is not difficult. It is also accurate if the open-loop gain is much higher than the closed-loop (feedback) gain.

The feedback in Fig. 9-14 also affects the op-amp's input impedance. Notice that the

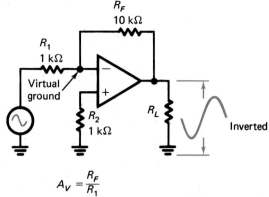

$$A_V = \frac{R_F}{R_1}$$

Fig. 9-14 Op-amps are operated by using negative feedback.

signal is applied to the inverting input. Since two signals are on this terminal, we may expect some effects. Suppose the source signal goes positive. This drives the inverting input positive. A negative-going signal is fed back. It will prevent the inverting input from going very far in a positive direction. In fact, the feedback keeps the inverting input very near ground potential. This is called a virtual ground. A virtual ground is a low impedance point. Therefore, the input impedance of Fig. 9-14 is equal to R_1.

Figure 9-15 shows another op-amp circuit. Notice that the signal is fed to the plus or noninverting terminal. This makes the circuit noninverting. The output signal is in phase with the input signal. The feedback is still applied to the inverting terminal. The voltage gain is slightly different:

$$A_V = 1 + \frac{R_F}{R_1} = 1 + \frac{100 \text{ k}\Omega}{10 \text{ k}\Omega} = 11$$

An advantage of the noninverting circuit is a higher input impedance. The noninverting

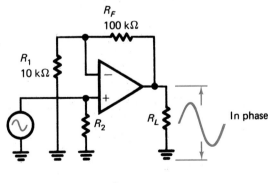

$$A_V = 1 + \frac{R_F}{R_1}$$

Fig. 9-15 This op-amp circuit is noninverting.

141

Bode plots

terminal is not a virtual ground. The input impedance of a differential amplifier is very high as long as there is no negative feedback to the input.

In Fig. 9-15, R_2 shunts the input to the amplifier. This resistor lowers the input impedance of the amplifier. One might be tempted to use a rather high value of resistance at R_2 to keep the input impedance high. This is not good practice. A small bias current flows through R_1 and R_2 in Fig. 9-15. If the two resistors are greatly different in value, the bias current will cause two different voltage drops to appear. This produces a dc offset across the input. The high gain of the op-amp will amplify this offset. The offset may be quite high at the output. It may prevent a good linear swing.

The graph of Fig. 9-16 shows gain versus frequency for a typical integrated circuit op-amp. Graphs of this type are known as *Bode plots*. They show how gain tends to drop off as frequency increases. Notice in Fig. 9-16 that the open-loop performance curve shows a sharp break or corner below a frequency of 10 Hz. This corner or break frequency is a very important point on a Bode plot. The gain will decrease at a uniform rate as frequency is increased beyond this point. The standard op-amp shows a gain decrease of 20 dB per decade above the break frequency.

What is the open-loop gain in Fig. 9-16 at 10 Hz? It is 100 dB. A *decade* increase in frequency means an increase of 10 times. Now, what is the gain at 100 Hz? It has dropped to

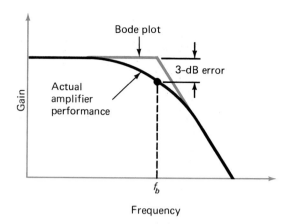

f_b = break frequency

Fig. 9-17 Bode plot error is greatest at the break frequency.

80 dB. The loss in gain is 20 dB. Beyond the break frequency, gain drops at 20 dB per decade.

Bode plots are only approximate. Figure 9-17 shows that the actual performance of an amplifier is 3 dB poorer at the break point. The error is much less as the frequency moves much above or below the break frequency. You should remember that a Bode plot is 3 dB in error at the break frequency. To find the true gain at f_b, subtract 3 dB.

The open-loop gain shown in Fig. 9-16 indicates a break frequency lower than 10 Hz. This is a Bode plot, so we know that the gain is already down 3 dB at this point. The gain must start dropping off at just a few hertz. Obviously, the typical op-amp is not exactly a wide-band amplifier.

Negative feedback really helps to increase the bandwidth of the op-amp. For example, the gain can be decreased to 20 dB. Now, the frequency response of the amplifier increases to over 10 kHz. This closed-loop performance is shown in Fig. 9-16.

Why does the gain of the typical op-amp drop so much as frequency increases? It is designed to do this. As frequency goes higher and higher, a phase error is created in the op-amp. The output signal starts lagging behind the input signal. If this happens to reach 180°, the amplifier may become unstable. The feedback is supposed to be negative in the op-amp circuit. It would become positive feedback if the phase delay reached 180°:

$$180° + 180° = 360°$$

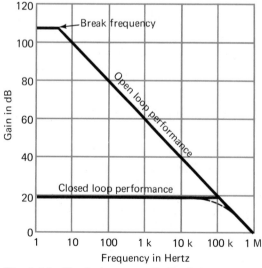

Fig. 9-16 Typical op-amp Bode plot.

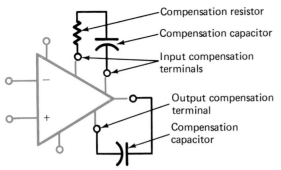

Fig. 9-18 An externally compensated op-amp.

At 360° the feedback becomes positive, and the amplifier could become unstable.

Some op-amps are not designed for a Bode response of 20 dB per decade. Their frequency response must be controlled by external components for stable operation. Figure 9-18 shows this type of op-amp. The amplifier has compensation terminals. Extra components are required to control the frequency response.

Most op-amps used today are internally compensated. They do not have compensation terminals. They are easier and less costly to use. This makes them more desirable for most applications.

There are op-amps available with very good high-frequency performance. They are more expensive than the typical operational amplifiers discussed to this point. The Bode plot for a high-performance op-amp might resemble that of Fig. 9-19. This particular device also has a much better slew rate. It is in excess of 50 V/μs. A high slew rate will allow the device to produce a good peak-to-peak output

over much of its operating range. In selecting an op-amp for high-frequency performance, both the Bode plot and the slew rate should be checked.

Review Questions

Answer the following.

17. Refer to Fig. 9-14. Is the amplifier operating in open loop or closed loop?

18. Refer to Fig. 9-14. Assume that $R_1 = 470\ \Omega$ and $R_F = 4.7\ k\Omega$. What is the low-frequency voltage gain of the amplifier? What is the input impedance of the amplifier?

19. Refer to Fig. 9-14. It is desired to use this circuit for an amplifier that has an input impedance of 3300 Ω and a voltage gain of 100. What value would you choose for R_1? For R_F?

20. Refer to Fig. 9-15. Assume that all three resistors equal 47,000 Ω. Calculate the voltage gain of the amplifier. What is the input impedance of the amplifier?

21. Refer to Fig. 9-16. Assume that the amplifier is to be used with negative feedback and the voltage gain will be set at 100 (40 dB). What gain can be expected at a frequency of 1 kHz? At 100 kHz?

22. What is the break frequency for the amplifier in question 21? What is the actual gain of the amplifier at this frequency?

23. The Bode plot for an op-amp shows a slope of 20 dB per decade and a break frequency of 100 Hz. If the open-loop gain of the amplifier is 80 dB, what is the greatest gain that can be expected at a frequency of 10 kHz?

24. Refer to Fig. 9-19. The amplifier is to be operated with a closed-loop gain of 40 dB. What is the maximum frequency that can be reached before the gain will drop below 37 dB? What other important op-amp specification may limit the performance of the circuit at this frequency?

9-4 OP-AMP APPLICATIONS

Operational amplifiers are used in many ways. Their good performance and low cost make them very desirable. They represent one of the areas where integrated circuit technology has provided many standard building blocks.

Op-amps make good low-frequency amplifi-

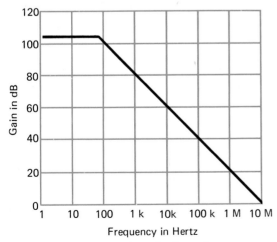

Fig. 9-19 Bode plot for a high-performance op-amp.

Summing mode

Mixer

Subtracting
mode

Passive filter

ers. They can be used in an inverting or a noninverting mode. The voltage gain is easy to set by using negative feedback. Generally, high-frequency performance (above 10 kHz) is not as good. High-performance op-amps are available when the high-frequency gain and slew rate must be better. The improved op-amps cost more.

Figure 9-20 shows an operational amplifier used in the *summing mode*. Two input signals V_1 and V_2 are applied to the inverting input. The output will be the inverted sum of the two input signals. Summing amplifiers can be used to add ac or dc signals. The output signal will equal

$$V_{out} = R_F \left(\frac{V_1}{R_1} + \frac{V_2}{R_2} \right)$$

Suppose, in Fig. 9-20, that all the resistors are 10 kΩ. Assume also that V_1 is 2 V and V_2 is 4 V. The output will be

$$V_{out} = 10 \text{ k}\Omega \left(\frac{2 \text{ V}}{10 \text{ k}\Omega} + \frac{4 \text{ V}}{10 \text{ k}\Omega} \right)$$

$$= \frac{2 \text{ V} \times 10 \text{ k}\Omega}{10 \text{ k}\Omega} + \frac{4 \text{ V} \times 10 \text{ k}\Omega}{10 \text{ k}\Omega}$$

$$= 2 \text{ V} + 4 \text{ V} = 6 \text{ V (negative)}$$

The output voltage is negative because the two inputs were summed at the inverting input.

The circuit of Fig. 9-20 can be changed to *scale* the inputs. For example, R_1 could be changed to 5 kΩ. The output voltage will now be

$$V_{out} = 10 \text{ k}\Omega \left(\frac{2 \text{ V}}{5 \text{ k}\Omega} + \frac{4 \text{ V}}{10 \text{ k}\Omega} \right) = 4 + 4$$

$$= 8 \text{ V (negative)}$$

The amplifier has scaled V_1 to 2 times its value and then added it to V_2.

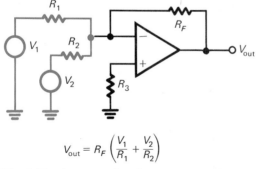

$$V_{out} = R_F \left(\frac{V_1}{R_1} + \frac{V_2}{R_2} \right)$$

Fig. 9-20 An operational summing amplifier.

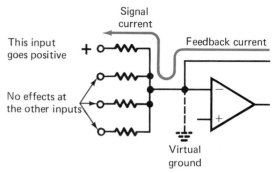

Fig. 9-21 The virtual ground isolates the inputs from one another.

Figure 9-20 can be expanded for more than two inputs. A third, fourth, and even a tenth input can easily be summed at the inverting input. Scaling of some or all the inputs is possible by selecting the input resistors and the feedback resistor.

Op-amp summing amplifiers are also called *mixers*. They could be used to mix the outputs from four microphones during a recording session, for example. They make good mixers since there is little interaction between inputs. The inverting input is a virtual ground. This prevents one input signal from appearing at the other inputs. Figure 9-21 shows how the virtual ground isolates the inputs.

Op-amps can be used in a *subtracting mode*. Figure 9-22 shows a circuit that can provide the difference between two inputs. With all resistors equal, the output is the nonscaled difference. If $V_1 = 2$ V and $V_2 = 5$ V, then

$$V_{out} = V_2 - V_1 = 5 - 2 = 3 \text{ V}$$

It is possible to have a negative output if the voltage to the inverting input is greater than the voltage to the noninverting input. If $V_1 = 6$ V and $V_2 = 5$ V,

$$V_{out} = 5 - 6 = -1 \text{ V}$$

Figure 9-22 can scale the inputs. Changing R_1 or R_2 would provide this type of operation.

Operational amplifiers can be used to filter signals. A *filter* is a circuit or device that allows some frequencies to pass through and rejects other frequencies. It rejects these frequencies by reducing, or attenuating, the strength of their signals. Filters using only resistors, capacitors, and inductors are called *passive filters*. Filter performance can be im-

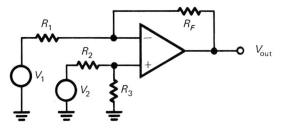

$$V_{out} = V_2 - V_1 \quad \text{for } R_F = R_1 = R_2 = R_3$$

Fig. 9-22 An operational subtracting amplifier.

proved by adding active devices such as transistors or op-amps. These devices are called *active filters*. Low-cost integrated op-amps have helped to make active filters very popular.

Figure 9-23(a) is a low-pass filter circuit. The graph in Fig. 9-23(b) shows that the gain is good at low frequencies. As frequency increases, the gain starts to drop. Thus the filter rejects, or attenuates, the high frequencies and passes the low frequencies.

Active filters have a big advantage at the lower frequencies. A passive low-pass filter

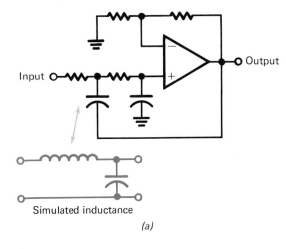

Simulated inductance

(a)

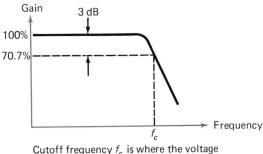

Cutoff frequency f_c is where the voltage gain drops to 70.7% of its maximum value.

(b)

Fig. 9-23 An active low-pass filter.

Active filter

Band-pass filter

designed with a cutoff point around 800 Hz would need a high-value inductor to do a good job. The inductor would be physically large and expensive. Inductance can be simulated by using feedback of the correct phase. An inductor makes the circuit current lag the voltage by 90°. The feedback through the capacitor in Fig. 9-23 produces this same phase effect at the cutoff frequency. The active filter produces a response very similar to an inductor-capacitor low-pass filter.

It is possible to cascade active filters for a sharper cutoff. Figure 9-24(a) shows two active low-pass filters in cascade. The improved cutoff can be seen in the graph [Fig. 9-24(b)]. The active filter may seem more complicated than the simulated inductor-capacitor filter. It is, but at low frequencies it will be smaller and cost less to build.

Figure 9-25(a) is an active high-pass filter. The graph [Fig. 9-25(b)] indicates attenuation of the low frequencies. This action is opposite to that of the low-pass circuit. Notice that the resistors and capacitors have apparently changed places. You should compare Fig. 9-25 with Fig. 9-23.

A band-pass filter is shown in Fig. 9-26(a). The graph [Fig. 9-26(b)] shows maximum gain at a single frequency. Higher or lower frequencies receive less gain. Filters of this type are useful for separating signals. For example, a radio receiver may pick up several stations at the same time. Good reception is difficult when this happens. It is sometimes possible to select the one station of interest by using a band-pass filter.

Figure 9-27 shows the application of an active band-pass filter in a radio receiver. This receiver is designed for voice reception and Morse code reception. A very narrow bandpass is helpful when receiving Morse code. Two sections of the integrated circuit, IC_{203A} and IC_{203C}, form a cascade band-pass filter. Notice how their feedback circuits compare to Fig. 9-26. The narrow-band audio output from this filter is used for code reception.

In Fig. 9-27 the operational amplifier is energized from a single supply voltage. This is made possible by R_{219} and R_{214}. They form a voltage divider which provides 6 V for the noninverting inputs of the op-amps. This is about half the supply voltage. The op-amp outputs are also at 6 V. Thus, it is possible to float an op-amp at the center of a single supply voltage.

145

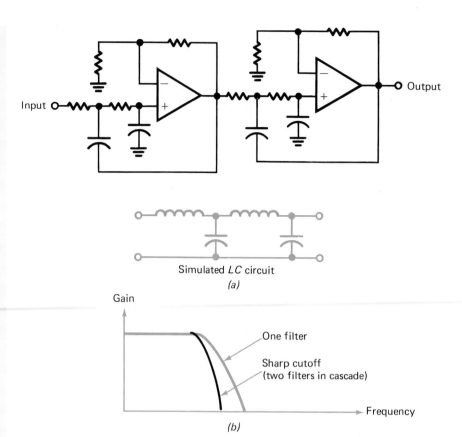

Input

Output

Simulated *LC* circuit

(a)

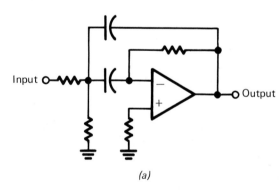

Gain

One filter

Sharp cutoff
(two filters in cascade)

Frequency

(b)

Fig. 9-24   Cascade filters give a sharper cutoff.

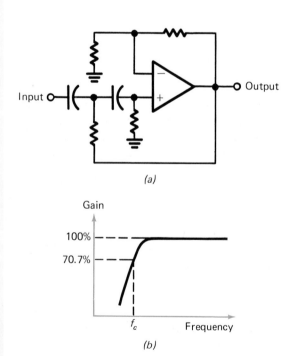

Input

Output

(a)

Gain

100%

70.7%

f_c

Frequency

(b)

Fig. 9-25   An active high-pass filter.

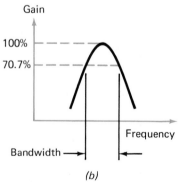

Input

Output

(a)

Gain

100%

70.7%

Frequency

Bandwidth

(b)

Fig. 9-26   An active band-pass filter.

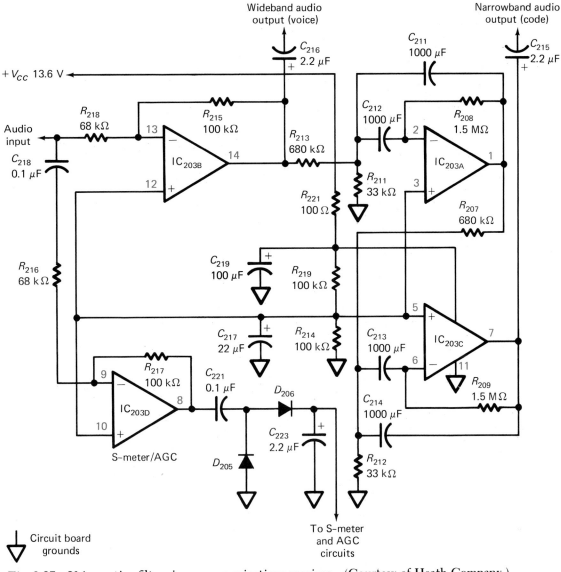

Band-stop filter

Notch filter

Trap

Integrator circuit

Fig. 9-27 Using active filters in a communications receiver. (Courtesy of Heath Company.)

Figure 9-28(*a*) shows a band-stop filter. It may also be called a *notch filter*, or a *trap*. It gives maximum attenuation (minimum gain) at a single frequency [Fig. 9-28(*b*)]. Higher and lower frequencies receive more gain through the filter. Filter circuits of this type are useful when a signal at one particular frequency is causing problems.

Another way the operational amplifier is used is in *integrator circuits*. Integration is usually thought of as a mathematical operation. It is a process of continuous addition. Integrators were very useful in analog computers. As we will see, there are other uses for integrators.

An op-amp integrator is shown in Fig. 9-29. Notice the capacitor in the feedback circuit.

Suppose a positive-going signal is applied to the input. The output must go negative because the inverting input is used. The feedback keeps the inverting input at virtual ground. The current through resistor R is supplied by charging the feedback capacitor as shown.

If the input signal in Fig. 9-29 is at some constant positive value, the feedback current will also be constant. We can assume that the capacitor is being charged by a constant current. When a capacitor is charged by a constant current, the voltage across the capacitor increases in a *linear* fashion. Figure 9-29 shows that the output of the integrator is ramping negative, and the ramp is linear.

Now look at Fig. 9-30. This circuit is a

147

Voltage-to-frequency converter

Comparator

Speed-up capacitor

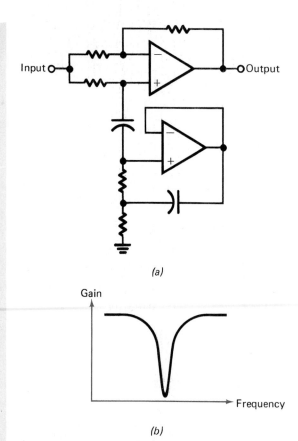

(a)

(b)

Fig. 9-28 An active band-stop filter.

voltage-to-frequency converter. It is a very useful circuit. It uses an op-amp integrator to convert positive voltages to a frequency. If the frequency is sent to a digital counter, a digital voltmeter is the result. If the voltage V_{in} represents a temperature, a digital thermometer is the result. Voltage-to-frequency converters form the basis for many of the digital instruments now in use.

What happens when a dc voltage is applied in the circuit of Fig. 9-30? If the voltage is positive, we know that the integrator will ramp in a negative direction. Note that the output of the integrator goes to a second op-amp used as a *comparator*. It compares two inputs.

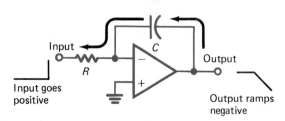

Fig. 9-29 An op-amp integrator.

One input is a fixed -7.5 V. It comes from the voltage divider formed by the two 1000-Ω resistors. They divide -15 V in half.

The integrator output in Fig. 9-30 will continue to ramp negative until its level exceeds -7.5 V. At this time, the comparator sees a greater voltage at its inverting input. The voltage is negative. The output of the comparator switches positive. This positive comparator output turns on Q_1. Since the emitter of Q_1 is negative, the input of the integrator is now quickly driven in a negative direction. This makes the integrator ramp quickly in a positive direction. Finally, the comparator again sees a greater negative voltage at its noninverting input. The comparator output goes negative and turns off Q_1.

The waveforms of Fig. 9-30 explain the voltage-to-frequency conversion process. With a constant positive dc voltage applied to the input, a series of negative ramps appears at the integrator output. When a ramp exceeds -7.5 V, Q_1 is switched on. The current through the transistor causes a voltage pulse across the emitter resistors. The transistor is on for a very short time. The output is a series of very narrow pulses.

With good op-amps, the circuits of Fig. 9-30 can be very linear. For example, if the dc input voltage is exactly doubled, the output frequency will double. This means the output frequency is a linear function of the input voltage. This is very desirable in a voltage-to-frequency converter.

Why does the frequency output double when the input voltage doubles? The input voltage causes a current to flow in the 12-kΩ resistor of Fig. 9-30. If the input voltage increases, so will the input current. Recall that this current is supplied by charging the integrator capacitor. We can assume that the charging current is now twice what it was. This means that the voltage across the capacitor will increase twice as fast. It will take half the time to reach -7.5 V. This doubles the output frequency. Figure 9-31 shows the graph for output frequency versus input voltage for this circuit.

Two of the components in Fig. 9-30 should be mentioned. The 50-pF capacitor in the base circuit of the transistor helps the transistor turn on fast. It is called a *speed-up capacitor*. Speed-up capacitors are often used to improve the switching time of transistors. The other component worth noting is the

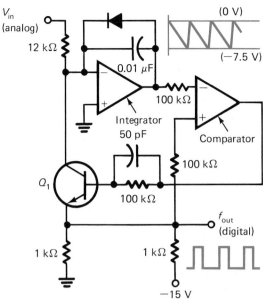

Fig. 9-30 A voltage-to-frequency converter.

diode across the integrating capacitor. It improves the response time of the integrator by reducing delay and overshoot. The circuit will work without these two components, but not nearly as well.

The last op-amp application to be discussed is a *voltage regulator*. Voltage regulators are used to hold an output voltage constant under varying loads or a fluctuating input voltage. Power-supply circuits often use output regulators to keep the voltage constant.

Op-amps make good voltage regulators.

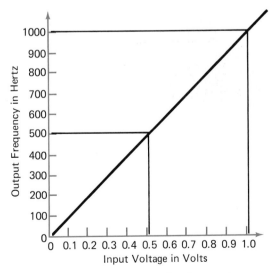

Fig. 9-31 Performance of the voltage-to-frequency converter.

Their differential input makes them ideal for comparing two voltages. If one of the two voltages changes, the output of the op-amp will change.

Figure 9-32 shows how an operational amplifier may be used to regulate voltage. A Zener diode D_1 is used to provide a constant voltage input to the op-amp. The other op-amp input is a portion of the load voltage. Resistors R_2 and R_3 form a voltage divider across the load circuit. Suppose the load current increases. As in any power supply, this tends to make the output voltage drop. Resistors R_2 and R_3 will now be dividing a smaller voltage. The inverting input of the op-amp goes in a negative direction. This makes the output of the op-amp go in a positive direction. Transistor Q_1 is now turned on harder because its base has been made more positive. This tends to increase the voltage across the load.

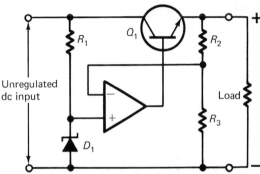

Fig. 9-32 An op-amp voltage regulator.

Transistor Q_1 in Fig. 9-32 is called a *pass transistor*. All load current must pass through this transistor. The transistor is in its active region. It will drop more or less voltage from collector to emitter depending on how much base current is flowing. The base current is supplied by the op-amp. The op-amp compares a fixed voltage with a sample of the output voltage. Any change in output voltage will cause the op-amp to supply more or less base current to the pass transistor. The pass transistor is adjusted by the op-amp to keep the output voltage constant.

We have not covered all the applications for operational amplifiers. We have covered enough to show how widely they are used in modern electronic applications. Their high performance and low cost resulting from integrated-circuit techniques have made them very popular.

149

Review Questions

Answer the following.

25. Refer to Fig. 9-20. All resistors are the same value. If $V_1 = +1$ V and $V_2 = +3$ V, what will the output voltage be (value and polarity)?

26. Refer to Fig. 9-20. All resistors are the same value. If $V_1 = -2$ V and $V_2 = -3$ V, what will the output voltage be?

27. Refer to Fig. 9-20. All resistors are the same value. If $V_1 = +2$ V and $V_2 = -3$ V, what will the output voltage be?

28. Refer to Fig. 9-20. $R_F = 20$ kΩ, $R_1 = 10$ kΩ, and $R_2 = 5$ kΩ. If $V_1 = 2$ V and $V_2 = 1$ V, what will the output voltage be?

29. Refer to Fig. 9-21. What circuit feature prevents a signal at one of the inputs from appearing at the other inputs?

30. Refer to Fig. 9-22. All the resistors are the same value. If $V_1 = 3$ V and $V_2 = 4$ V, what will the output voltage be?

31. Refer to Fig. 9-22. All resistors are the same value. If $V_1 = 5$ V and $V_2 = 3$ V, what will the output voltage be?

32. Refer to Fig. 9-22. All resistors are the same value. If $V_1 = -2$ V and $V_2 = 1$ V, what will the output voltage be?

33. Refer to Fig. 9-23. This circuit is checked with a variable-frequency signal generator and an oscilloscope. The following data are collected:

$V_{out} = 10$ V peak-to-peak at 100 Hz
$V_{out} = 10$ V peak-to-peak at 1 kHz
$V_{out} = 7$ V peak-to-peak at 10 kHz
$V_{out} = 1$ V peak-to-peak at 20 kHz

What is the cutoff frequency f_c of the filter?

34. Refer to Fig. 9-25. What can you expect to happen to the circuit gain as the signal frequency drops below f_c?

35. Refer to Fig. 9-26. Assume the gain to be maximum at 2500 Hz. Also assume that the gain drops 3 dB at 2800 Hz and 3 dB at 2200 Hz. What is the filter bandwidth?

36. Refer to Fig. 9-27. What is the voltage gain of IC_{203B}? What is its input impedance? Is it inverting or noninverting?

37. Refer to Fig. 9-30. Assume the converter is perfectly linear. If $f_{out} = 400$ Hz when $V_{in} = 0.4$ V, what should f_{out} be at $V_{in} = 0.8$ V?

38. Refer to Fig. 9-32. Assume the dc input voltage increases because of a power line change.
    a. What tends to happen at the inverting input of the op-amp?
    b. What happens at the output of the op-amp?
    c. What happens to the pass transistor?
    d. Is the output corrected?

Summary

1. A differential amplifier responds to the difference between two input signals.
2. A dual supply develops both positive and negative voltages with respect to ground.
3. A differential amplifier can be driven at one of its inputs.
4. It is possible to use a differential amplifier as an inverting or a noninverting amplifier.
5. A differential amplifier rejects common-mode signals.
6. The common-mode rejection ratio (CMRR) is the ratio of differential gain to common-mode gain.
7. Most op-amps have a single-ended output (one output terminal).
8. Op-amps have two inputs. One is the inverting input, and the other is the noninverting input. The inverting input is marked with a minus (−) sign, and the noninverting input is marked with a plus (+) sign.
9. An op-amp's offset null terminals can be used to reduce dc error in the output. With no input, the output terminal is adjusted to 0 V with respect to ground.
10. An op-amp's slew rate limits its output swing at high frequencies.
11. General-purpose op-amps are low-frequency devices. They perform well from 0 Hz (dc) to around 10 kHz.
12. The open-loop (no feedback) gain of op-amps is very high at 0 Hz (dc). It drops off rapidly as frequency increases.

13. The feedback is usually from the output to the inverting input. This is negative feedback.

14. Negative feedback decreases the voltage gain and increases the bandwidth of the amplifier.

15. Op-amp gain is set by the ratio of feedback resistance to input resistance.

16. Negative feedback makes the impedance of the inverting input very low. The terminal is called a virtual ground.

17. The feedback is not applied to the noninverting terminal. Its impedance is very high.

18. The Bode plot for a standard op-amp shows the gain decreasing at 20 dB per decade beyond the break frequency.

19. The actual gain at the break frequency is 3 dB lower than shown on the Bode plot.

20. Internally compensated op-amps are easier to use and are desirable for most applications.

21. The high-frequency performance of an op-amp is limited by both its Bode plot and its slewing rate.

22. Op-amps can be used as summing amplifiers.

23. By adjusting input resistors, a summing amplifier can scale some of or all the inputs.

24. Summing amplifiers may be called mixers. A mixer can sum several audio inputs.

25. Op-amps can be used as subtracting amplifiers. The signal at the inverting input is subtracted from the signal at the noninverting input.

26. Op-amps are used in active filter circuits.

27. Active filters can be cascaded (connected in series) for sharper cutoff.

28. Op-amp integrators use capacitive feedback.

29. Op-amp integrators can be used as analog to digital converters.

30. An op-amp integrator and an op-amp comparator can be combined to form a voltage-to-frequency converter. This is one type of analog-to-digital conversion.

31. Op-amps make good voltage regulators. They compare a fixed voltage with the output voltage.

32. Integrated circuit op-amps are very versatile. Their low cost and high performance have made them popular.

Chapter Review Questions

Answer the following.

9-1. What name is given to an amplifier that responds to the difference between two input signals?

9-2. A dual supply provides how many polarities with respect to ground?

9-3. Refer to Fig. 9-4. Assume that the input signal drives the base of Q_1 in a positive direction. What effect does this have on the emitter of Q_2? On the collector of Q_2?

9-4. Refer to Fig. 9-4. If a single-ended output is taken from the amplifier, will common-mode signals be rejected?

9-5. Refer to Fig. 9-4. If the output is taken from the collector of Q_1 and the collector of Q_2, what is the output called? Will this output connection reduce common-mode signals?

9-6. Refer to Fig. 9-4. If the single-ended output signal is 3.4 V peak-to-peak, what will the differential output be?

9-7. The differential input of an amplifier is 300 mV, and the output is 9 V. What is the differential gain of the amplifier?

9-8. In using the same amplifier as in question 9-7, it is noted that a 2-V common-mode signal is reduced to 50 mV in the output. What is the CMRR?

9-9. Refer to Fig. 9-8. Does the operational amplifier provide a single-ended or differential output?

151

9-10. Refer to Fig. 9-8. Is common-mode rejection possible with the output shown?

9-11. What geometric shape is often used on schematic diagrams to represent an amplifier?

9-12. What polarity sign will be used to mark the noninverting input of an operational amplifier?

9-13. What op-amp terminals can be used to correct for any slight dc imbalance across the inputs?

9-14. An op-amp has a slew rate of 1 V/μs. What is the maximum peak-to-peak output swing that can be expected at a frequency of 500 kHz?

9-15. What is the gain of an op-amp called if there is no feedback?

9-16. What does negative feedback do to the open-loop gain of an op-amp?

9-17. What does negative feedback do to the bandwidth of an op-amp?

9-18. Refer to Fig. 9-14. To what value will R_F have to be changed in order to produce a voltage gain of 22?

9-19. Refer to Fig. 9-14. Change R_1 to 220 Ω. What is the voltage gain? What is the input impedance?

9-20. Refer to Fig. 9-15. What component sets the input impedance of this amplifier?

9-21. Refer to Fig. 9-15. What can happen to the op-amp if R_1 and R_2 are very different in value?

9-22. Refer to Fig. 9-15. Resistors R_1 and R_2 are 47,000 Ω, and R_F is 220 kΩ. What is the voltage gain of the amplifier? What is the input impedance?

9-23. Refer to Fig. 9-16. Where does the maximum error occur in a Bode plot? What is the magnitude of this error?

9-24. Refer to Fig. 9-16. The gain of the op-amp is to be set at 80 dB by using negative feedback. Where will the break frequency be?

9-25. Refer to Fig. 9-16. Is it possible to use this op-amp to obtain a 30-dB gain at 100 Hz?

9-26. Refer to Fig. 9-16. Is it possible to use this op-amp to obtain a 50-dB gain at 1 kHz?

9-27. Refer to Fig. 9-20. Assume that R_1 and R_2 are 4700 Ω. What impedance does source V_1 see? Source V_2?

9-28. Refer to Fig. 9-20. Resistors R_1 and R_2 are both 10 kΩ, and R_F is 68 kΩ. If $V_1 = 0.3$ V and $V_2 = 0.4$ V, what will V_{out} be?

9-29. Refer to Fig. 9-22. All the resistors are 1000 Ω. If $V_1 = 2$ V and $V_2 = 2$ V, what will the output voltage be?

9-30. The cutoff frequency of a filter can be defined as the frequency at which the output drops to 70.7 percent of its maximum value. What loss does this represent in decibels?

9-31. Refer to Fig. 9-27. Which two op-amps act as inverting amplifiers?

9-32. Refer to Fig. 9-27. Which two op-amps act as band-pass filters?

9-33. Is it possible to use op-amps from a single power supply?

9-34. Refer to Fig. 9-30. What component is used to discharge the integrator?

9-35. Refer to Fig. 9-30.   Which op-amp is used to turn on Q_1?

9-36. Refer to Fig. 9-31.   Is the relationship between input voltage and output frequency linear?

9-37. Refer to Fig. 9-32.   Assume that Q_1 fails and shorts from emitter to collector.   Will the output voltage stay the same, go up, or go down?

Answers to Review Questions

1. Dual supply
2. 0 V
3. Negative; positive
4. R_E
5. 10 V peak-to-peak
6. $+9$ V; -9 V; $+18$ V
7. Ground
8. 120
9. 8000 (78 dB)
10. Section 3
11. $180°$ out of phase
12. Single-ended
13. Yes
14. No
15. 2 V/μs

16. 2.5 V peak-to-peak
17. Closed loop
18. 10; 470 Ω
19. 3300 Ω; 330 kΩ
20. 2; 47 kΩ
21. 100 (40 dB); 10 (20 dB)
22. 10 kHz; 37 dB
23. 40 dB
24. 100 kHz; slew rate
25. -4 V
26. $+5$ V
27. $+1$ V
28. -8 V
29. Virtual ground
30. $+1$ V

31. -2 V
32. $+3$ V
33. 10 kHz
34. It will decrease.
35. 600 Hz
36. 1.47; 68 kΩ; inverting
37. 800 Hz
38. *a.* It goes in a positive direction.
    b. It goes in a negative direction.
    c. It receives less forward bias and increases its resistance.
    d. Yes.

Amplifier Troubleshooting

- The amplifier is the basic circuit of linear electronics. Very few types of electronic equipment exist that do not have at least one stage of gain. It is very important for a technician to be skilled in amplifier troubleshooting and repair.

    There are many types of amplifiers. Some of the names used to describe them are dc amplifiers, audio amplifiers, video amplifiers, RF amplifiers, operational amplifiers, and power amplifiers. Most of the skills and techniques used in amplifier troubleshooting are the same for all amplifier types. It has been found that a technician can often troubleshoot an unfamiliar circuit by using basic skills developed for other circuits.

10-1 PRELIMINARY CHECKS

When troubleshooting, remember the word "GOAL."

Good troubleshooting always involves

1. Observing the symptoms
2. Analyzing the possible causes
3. Limiting the possibilities

Observing the symptoms must include a preliminary check of all control settings, proper connections, and power-supply operation. Preliminary checks are easy, they do not take too much time, and they can save precious time. Many television technicians have entered a customer's home to find the set unplugged and no other difficulty. Always check the obvious.

Figure 10-1 shows a modern stereo amplifier. Several front panel features have been emphasized to show how important preliminary checks are. The power-on light will indicate if the unit is plugged in and turned on. Generally, if the fuse has blown, the light will not come on. Thus, simply looking at the light will verify several important conditions. Notice the source selector control in Fig. 10-1. It is used to select various amplifier inputs: phonograph, tape deck, tuner, microphone, or auxiliary. This control must be set prop-

erly for the amplifier to receive an input signal.

Figure 10-1 also shows a speaker selector. This can be set for main speakers, remote speakers, and auxiliary speakers. The speaker selector must be set properly for the output signal to reach the load. This amplifier also has a tape monitor switch. When it is switched on, the signal is interrupted and does not reach the speakers. The headphone jack also interrupts the signal to the speakers when a plug is inserted.

Preliminary checks must include the rear panel of the equipment. Are the plugs connected to the right jacks? Are the speaker leads connected to the correct terminals? Is

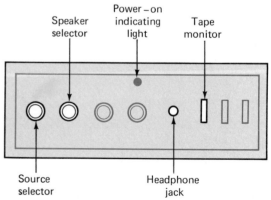

Fig. 10-1 Stereo amplifier.

the circuit breaker or fuse open? Checking all these items takes only a minute or two. It usually takes longer to remove the chassis from the cabinet. Do not waste time. <u>Check the obvious first.</u>

If all the external preliminary checks have been made and the unit is still not working, an internal inspection must be made. <u>Do not attempt to remove the unit from its cabinet until you have disconnected it from the ac line.</u> A slip of the tool or finger can cause damage or a severe shock when the power is on. Charged filter capacitors can do the same even after the unit is unplugged. <u>It is a good idea to turn on the unit after it has been unplugged.</u> A five-minute wait will help to ensure complete discharge of the filter capacitors.

Follow the manufacturer's procedures in taking apart equipment. Often service literature will show exactly how to do it. Many technicians overlook this and just start removing parts and fasteners. This often causes internal components to fall apart. Damage and long delays in reassembly may result. It saves time in the long run to work carefully and use the service literature.

Use the proper tools. The wrong wrench or screwdriver may slip and damage fasteners or other parts. <u>A scratched front panel usually cannot be repaired.</u> It may take weeks to get one and several hours to replace. It may not be possible to obtain a new one. The old saying "haste makes waste" fits perfectly in electronics.

Sort and save all fasteners and other parts. There is nothing more disturbing to a customer or a supervisor than to find an expensive piece of equipment with missing screws, shields, and other parts. The manufacturer includes all those pieces for a very good reason: they are necessary for proper and safe operation of the equipment.

The next part of the preliminary check is a visual inspection of the interior of the equipment. Look for the following:

1. Burned and discolored components
2. Broken wires and components
3. Cracked or burned circuit boards
4. Foreign objects (paper clips, etc.)
5. Bent transistor leads that may be touching (this includes other noninsulated leads as well)
6. Leaking components (especially electrolytic capacitors and batteries)

Obvious damage should be repaired at this point. However, do not energize the unit immediately. For example, suppose a resistor is burned black. In many cases, the new one will quickly do the same. Inspect the schematic to see what the resistor does in the circuit. Try to determine what kinds of problems could have caused the overload.

Refer to Fig. 10-2. Suppose the visual inspection revealed that R_1 was badly burned. What kinds of other problems are likely? There are several:

1. Capacitor C_1 could be shorted.
2. Zener diode D_1 could be shorted.
3. There is a short somewhere in the regulated output circuit.
4. The unregulated input voltage is too high (this is not too likely, but it is a possibility).

When the preliminary visual inspection is complete, a preliminary electrical check should be made. Be sure to use an isolation transformer. Never troubleshoot equipment operating directly from the ac line. There may be a chance for a dangerous electrical shock. The equipment and the test instruments could be damaged. Figure 10-3 shows the proper use of an isolation transformer.

The first part of the preliminary electrical check involves any signs of overheating. Your nose may give important information. Hot electronic components often give off a distinct odor. Many technicians use a finger to test for excessive heat. This must be done with extreme caution. It must *never* be done if there is any chance of high voltage being present. Electronic temperature probes are available for accurately measuring heat.

The next part of the electrical check is to verify the power-supply voltages. Power-supply problems can produce an entire range of symptoms. This is why it can save a lot of time to check here first. Consult the manufacturer's specifications. The proper voltages are usually indicated on the schematic diagram. Some error is usually allowed. A 20

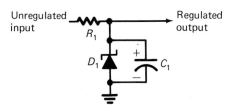

Fig. 10-2 Zener shunt regulator.

From page 154:
GOAL

Preliminary checks

155

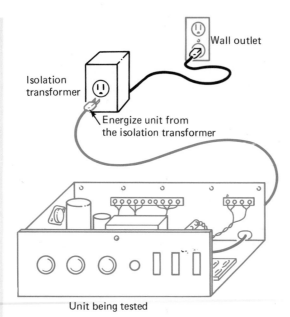

Isolation transformer

Wall outlet

Energize unit from the isolation transformer

Unit being tested

Fig. 10-3   Using an isolation transformer.

percent variation is not unusual in many circuits. Of course, if a precision voltage regulator is in use, this much error is not acceptable. Be sure to check all the supply voltages. Remember, it only takes one incorrect voltage to keep the system from working.

Review Questions

Choose the letter that best completes each statement.

1. Good troubleshooting is
   A. A matter of luck
   B. A logical and orderly procedure
   C. Begun by measuring all voltages
   D. Always done with the power off

2. The first step in troubleshooting should involve
   A. A check of all control settings and connections
   B. Verifying all resistances
   C. Removing the chassis from the cabinet
   D. Measuring line current

3. Refer to Fig. 10-1. The power light will not come on, and the amplifier is completely dead. The *least* likely cause of this is
   A. The wall outlet is off
   B. The fuse is blown
   C. The power light is burned out
   D. The line cord is defective

4. Refer to Fig. 10-1. The power light is on, and the amplifier is completely dead. The *least* likely cause of this is

   A. The speaker selector is set wrong
   B. The tape monitor is set wrong
   C. The headphone jack is defective
   D. The fuse is blown

5. Refer to Fig. 10-2. A visual check shows C_1 is bulging. This may be a sign of excessive voltage. This could have been caused by
   A. A short in D_1
   B. An open in R_1
   C. The output being shorted to ground
   D. An open in D_1

6. After a piece of equipment has been removed from its cabinet and inspected visually, the next step should be
   A. To check supply voltages
   B. To check the transistors
   C. To check the integrated circuits
   D. To check the electrolytic capacitors

10-2 NO OUTPUT

This section deals with the symptom of no output. The material and procedures presented here assume that all the preliminary checks described in the previous section have been made.

There are several causes for no output in an amplifier. Perhaps the most obvious is no input. It is worth the effort to check this early. You may find that a wire has been pulled loose. With no input signal, there can be no output signal.

The output device may be defective. For example, in an audio amplifier, the output is sent to a loudspeaker or perhaps headphones. These devices fail from time to time. You can check them easily. An ordinary flashlight cell can be used to make the test. Figure 10-4 shows how. A good speaker will make a click when the test leads touch the speaker terminals. Most ohmmeters, on the $R \times 1$ range, will make the same click when connected across a speaker. Either technique tells you the speaker is capable of changing electricity

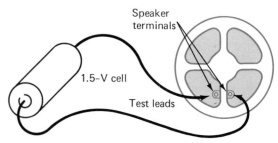

Speaker terminals

1.5-V cell

Test leads

Fig. 10-4   Checking a loudspeaker.

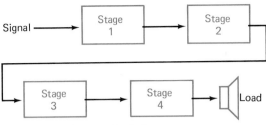

Fig. 10-5   The signal chain.

into sound. This test is a simple one and cannot be used to check the quality of a speaker.

If there is nothing wrong with the speaker, there is probably a break in the signal processing system. Figure 10-5 illustrates this. The system can be thought of as a signal chain. The signal must travel the chain, stage by stage, to reach the load. A break at any point in the chain will usually cause the no-output symptom.

A four-stage amplifier contains many parts. There are dozens of measurements to be taken. Therefore, the efficient way to troubleshoot is to isolate the problem to one stage. One way to do this is to use *signal injection*. Figure 10-6 shows what needs to be done. A signal generator is used to provide a test signal. The test signal is injected at the input to the last stage. If an output signal is heard, then the last stage is good. The test signal is then moved to the input of the next-to-last stage. When the signal is injected to the input of the broken stage, no output will be heard. This eliminates the other stages, and you can zero in on the defective circuit.

Signal injection is not without its problems. One danger is the possibility of overdriving an amplifier and damaging a part. More than one technician has ruined a loudspeaker by feeding too large a signal into the amplifier.

A high-power amplifier must be treated with respect!

Another danger in signal injection is improper connection. A schematic diagram is a must. The common ground is generally used for connecting the ground lead from the generator. Assuming that the chassis is common will not always work out. If the common connection is made in error, a large hum voltage may be injected into the system. Damage can result.

Most amplifiers are tested with ac signals. The signal must be capacitively coupled to avoid upsetting the bias on a transistor or integrated circuit. If the generator is dc-coupled, a capacitor must be used in series with the hot lead. This capacitor will block the dc component yet allow the ac signal to be injected. A 0.1-μF capacitor is usually good for audio work, and a 0.001-μF capacitor can be used for radio frequencies.

The test frequency varies depending on the amplifier being tested. A 1000-Hz signal is standard for audio work. A radio-frequency amplifier should be tested at its design frequency. This is especially important in band-pass amplifiers. Some are so narrow that an error of a few kilohertz will cause the signal to be blocked. It may be necessary to vary the generator frequency slowly while watching for output.

Signal injection can be performed without a signal generator in many amplifiers. A resistor can be used to inject a click into the chain. This click is really a signal pulse caused by upsetting the bias on a transistor. A resistor is connected momentarily from the collector lead to the base lead as in Fig. 10-7. This will cause a sudden increase in transistor current. The collector voltage will drop suddenly, and the click travels down the chain and reaches the output device. When the

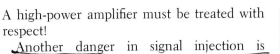

Fig. 10-6   Using signal injection.

157

Signal tracing

Voltage analysis

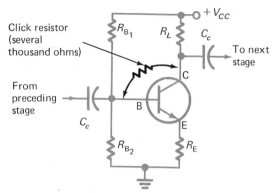

Fig. 10-7 The click test.

stage is reached where the click cannot reach the output, the problem has been isolated. As with other types of signal injection, start at the last stage and work toward the first stage.

The click test must be used carefully. Use only a resistor of several thousand ohms. Never use a screwdriver or a jumper wire. This can cause severe damage to the equipment. Never use a click test in high-voltage equipment. It is not safe for you or for the equipment. Always be careful when probing in circuits. If you slip and short two leads, damage often results.

Signal tracing is another way to isolate the defective stage. This technique may utilize a meter, an oscilloscope, a signal tracer, or some related instrument. Signal tracing starts at the input to the first stage of the amplifier chain. Then the tracing instrument is moved to the input of the second stage, and so on. Suppose that a signal is found at the input to the third stage but not at the input to the fourth stage. This would mean that the

signal is being lost in the third stage. The third stage is probably defective.

The important thing to remember in signal tracing is the gain and frequency response of the instrument being used. For example, do not expect to see a low-level audio signal on an ordinary ac voltmeter. Also, do not expect to see a 10.7-MHz radio signal on an oscilloscope rated to 500 kHz. Even if the oscilloscope is rated to 10 MHz, the signal must be in the millivolt range to be detectable. Some radio signals are in the microvolt range. Not knowing the limitations of your test equipment will cause you to reach false conclusions!

Once the fault has been localized to a particular stage, it is time to determine which part has failed. Of course, it is possible that more than one part is defective. More often than not, one component will be found to be defective.

Most technicians use *voltage analysis* to determine which part is defective. They also use their knowledge of circuit laws. The voltage readings and the knowledge will usually point the experienced technician in the right direction. Study Fig. 10-8. Suppose the collector of Q_2 measures 21 V. The manufacturer's schematic shows that the collector of Q_2 should be 12 V with respect to ground. What could cause this large an error? It is likely that Q_2 is in cutoff. A 21-V reading at the collector tells us that the voltage is the same on both ends of R_6. Ohm's law tells us that no voltage drop means no current flow. Transistor Q_2 must be cut off.

Now, what are some possible causes for Q_2 to be cut off? First, the transistor could be defective. Second, R_7 could be open. Re-

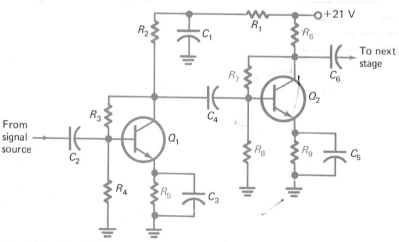

Fig. 10-8 Troubleshooting with voltage analysis.

sistor R_7 supplies the base current for Q_2. If it opens, no base current will flow. This cuts off the transistor. This can be checked by measuring the base voltage of Q_2. With R_7 open, the base voltage will be zero. Third, R_9 could be open. If it opens, there will be no emitter current. This cuts off the transistor. A voltage check at the emitter of Q_2 will show 21 V. The full voltage will always drop across the open component. Fourth, R_8 could be shorted. This seldom happens, but a troubleshooter soon learns that all things are possible. With R_8 shorted, no base current can flow and the transistor is cut off. The base voltage will measure zero.

Let us try another symptom. Suppose the collector voltage at Q_1 measures 0 V. A check on the manufacturer's service notes shows that it is supposed to be 11 V. What could be wrong? First, C_1 could be shorted. The combination of R_1 and C_1 acts as a low-pass filter to prevent any hum or other unwanted ac signal from reaching Q_1. If C_1 shorts, R_1 will drop the entire 21-V supply. This can be checked by measuring the voltage at the junction of R_2 and C_1. With C_1 shorted, it will be 0 V. Second, R_2 could be open. This can be checked by measuring the voltage at the junction of R_2 and C_1. A normal voltage here indicates R_2 must be open. Could Q_1 be shorted? The answer is no. Resistor R_5 would drop at least a small voltage, and the collector would be above 0 V.

Sometimes it helps to ask yourself what might happen to the circuit given a specific component failure. This thoughtful question-and-answer game is used by most technicians. Again, refer to Fig. 10-8. What would happen if C_4 shorts? This short circuit would apply the dc collector potential of Q_1 to the base of Q_2. Chances are that this would greatly increase the base voltage and drive Q_2 into saturation. The collector voltage at Q_2 will drop to some low value.

What if C_2 in Fig. 10-8 shorts? Transistor Q_1 could be driven to cutoff or to saturation. If the signal source has a ground or negative dc potential, the transistor will be cut off. If the signal source has a positive dc potential, the transistor will be driven toward saturation.

The beauty of voltage analysis is that it is easy to make the measurements. Often, the expected voltages are indicated on the schematic diagram. A small error is usually not a sign of trouble. Many schematics will indi-cate that all voltages are to be within a ±10 percent range. *Current analysis* is not easy. Circuits must be broken to measure current. Often, a technician can find a known resistance in the circuit where current is to be measured. A voltage reading can be converted to current by Ohm's law.

Resistance analysis can also be used to isolate defective components. This technique can be tricky, however. Multiple paths may produce confusing readings. Refer to Fig. 10-9. An ohmmeter check is being made to verify the value of the 3.3-kΩ resistor. The reading is incorrect because the internal supply in the ohmmeter forward-biases the base-emitter junction. Current flows in both the resistor and the transistor. This makes the reading too low. The ohmmeter polarity can be reversed. This will reverse-bias the transistor and give much greater accuracy. Some ohmmeters use a very low supply voltage to overcome this problem. The voltage is not enough to turn on a semiconductor junction.

Even if a junction is not turned on by the ohmmeter, in many cases it is still impossible to obtain useful readings. There will be other components in the circuit to draw current from the ohmmeter. Any time that you are using resistance analysis, remember that a low reading could be caused by multiple paths.

It is usually poor practice to unsolder parts for resistance analysis unless you are reasonably sure the part is defective. Unsoldering can cause damage to circuit boards and to the parts. It is also time-consuming.

Current analysis

Resistance analysis

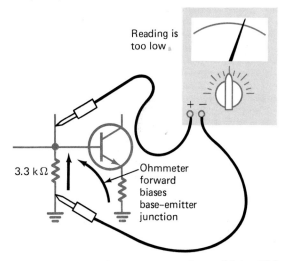

Reading is too low

3.3 kΩ

Ohmmeter forward biases base–emitter junction

Fig. 10-9 An ohmmeter may forward-bias PN junctions.

As mentioned before, most technicians use voltage analysis to locate defective parts. This is valid and effective since most circuit faults will change at least one dc voltage. There is the possibility of an ac fault that breaks the signal chain without changing any of the dc readings. Here are some of the possibilities:

1. An open coupling capacitor
2. A defective coupling coil or transformer
3. A break in the printed circuit board
4. A dirty or bent connector (plug-in modules often suffer this fault)
5. An open switch or control such as a relay

To find this type of fault, signal tracing or signal injection can be used. You will find different conditions at either end of the break in the chain. Some technicians use a coupling capacitor to bypass the signal around the suspected part. The value of the capacitor can usually be 0.1 μF for audio work and 0.001 μF for radio circuits. Do not use this approach in high-voltage circuits. Never use a jumper wire. Severe circuit damage may result from jumping the wrong two points.

Review Questions

Choose the letter that best answers each question.

7. Refer to Fig. 10-5. A signal generator is applied to the input of stage 4, then stage 3, and then stage 2. When the input of stage 2 is reached, it is noticed that nothing reaches the speaker. The defective stage is most likely number
   A. 1
   B. 2
   C. 3
   D. 4

8. The procedure used in question 7 is called
   A. Signal tracing
   B. Signal injection
   C. Current analysis
   D. Voltage analysis

9. Refer to Fig. 10-5. A signal generator is first applied to the input of stage 1. It is noticed that nothing reaches the speaker. The difficulty is in
   A. Stage 1
   B. Stage 2
   C. Stage 3
   D. Any of the stages

10. A good test frequency for audio troubleshooting is
    A. 20 Hz   C. 20 kHz
    B. 1 kHz   D. 100 kHz

11. It is a good idea to use a coupling capacitor in signal injection to
    A. Improve the frequency response
    B. Block the ac signal
    C. Provide an impedance match
    D. Prevent any dc shift or loading effect

12. Refer to Fig. 10-7. When the click resistor is added, the collector voltage should
    A. Not change
    B. Go in a positive direction
    C. Go in a negative direction
    D. Change for a moment and then settle back to normal

13. Refer to Fig. 10-7. When the click resistor is added, the emitter voltage should
    A. Not change
    B. Go in a positive direction
    C. Go in a negative direction
    D. Change for a moment and then settle back to normal

14. Refer to Fig. 10-8. Assume that C_4 is open. It is most likely that
    A. The collector voltage of Q_1 will read high
    B. The collector voltage of Q_1 will read low
    C. The base of Q_2 will be 0 V
    D. All the dc voltages will be correct

15. Refer to Fig. 10-8. Resistor R_9 is open. It is most likely that the
    A. Collector voltage of Q_2 will be high (near 21 V)
    B. Collector voltage of Q_2 will be low (near 1 V)
    C. Emitter of Q_2 will be at 0 V
    D. Transistor will go into saturation

16. Refer to Fig. 10-8. Suppose it is necessary to know the collector current of Q_1. The easiest technique is to
    A. Break the circuit and measure it
    B. Measure the voltage drop across R_2 and use Ohm's law
    C. Measure the collector voltage
    D. Measure the emitter voltage

17. The technique used in question 16 is not 100 percent accurate because
    A. Resistor R_2 may be off in resistance value
    B. Resistors R_3 and R_4 add to the current in R_2
    C. The base current flows through R_2
    D. All the above

18. Refer to Fig. 10-8. It is desired to check the value of R_8. The power is turned off; the negative lead of the ohmmeter is applied at the top of R_8, and the positive lead is applied at the bottom. This will prevent the ohmmeter from forward-biasing the base-emitter junction. The ohmmeter reading is still going to be less than the actual value of R_8 because

    A. Capacitor C_4 is in the circuit
    B. Transistor Q_1 is in the circuit
    C. Resistors R_7 and R_6 and the power supply provide a path to ground
    D. None of the above

10-3 REDUCED OUTPUT

Low output from an amplifier tells us there is lack of gain in the system. In an audio amplifier, for example, normal volume cannot be reached at the maximum setting of the volume control. Do not troubleshoot for low output until you have made the preliminary checks described in Sec. 10-1.

Low output from an amplifier often is caused by low input to the amplifier. The signal source is weak for some reason. A microphone may deteriorate with time and rough treatment. The same is true of a phonograph cartridge. The check is simple: Try a new signal source or substitute a signal generator.

Another possible cause for low output is reduced performance in the output device. The loudspeaker circuit may have developed a high resistance. This will prevent normal loudness. In a video system, there may be a difficulty in the cathode-ray tube (picture tube), which causes poor contrast. This can be checked by substituting for the output device or replacing the device with a known load and measuring the output.

In Fig. 10-10 a loudspeaker has been replaced with an 8-Ω resistor in order to measure the output power of an audio amplifier. This resistor must match the output impedance requirement of the amplifier. It must also be rated to safely dissipate the output power of the amplifier. The signal generator is usually adjusted for a sinusoidal output of 1 kHz. The signal level is set carefully so as not to overdrive the amplifier being tested.

Suppose you need to check an audio amplifier for rated output power with an oscilloscope and signal generator. The specifications for the amplifier rate it at 100 W of continuous sine wave power output. How could you be sure the amplifier meets its specification and does not suffer from low output? The power formula shows the relationship between output voltage and the output resistance:

$$P = \frac{V^2}{R}$$

In this case, P is known from the specifications, and R is the substitute resistor. What you must determine is the expected output voltage:

$$V^2 = PR \qquad \text{or} \qquad V = \sqrt{PR}$$

From the known data

$$V = \sqrt{100 \times 8} = 28.28 \text{ V}$$

The 100-W amplifier can be expected to develop 28.28 V across the 8-Ω load resistor.

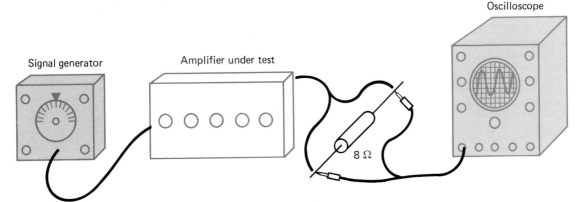

Fig. 10-10   Replacing the speaker with a resistor.

Dummy load

However, the oscilloscope is usually used for making peak-to-peak measurements. Thus, it would be a good idea to convert 28.28 V to its peak-to-peak value:

$$V_{p-p} = V_{rms} \times 1.414 \times 2 = 80 \text{ V}$$

To test the 100-W amplifier, the gain control would be advanced until the oscilloscope showed an output sine wave of 80 V peak-to-peak. There should be no sign of clipping on the peaks of the waveform. If the amplifier passes this test, you know it is within specifications.

Figure 10-11 shows another method of testing amplifiers that is often used in the radio communications industry. Normally, the two-way radio delivers its RF output power to an antenna. For test purposes the antenna has been replaced by a dummy load. This load is a noninductive resistor of usually 50 Ω. This provides a way of testing without producing interference and ensures the proper load for the transmitter. The RF wattmeter will allow the technician to determine if the RF amplifier has normal output.

If the input signal and output device are both normal, the problem must be in the amplifier itself. One or more stages are giving less than normal gain. You can expect the problem to be limited to one stage in most cases. It is more difficult to isolate a low-gain stage than it is to find a total break in the signal chain. Signal tracing and signal injection can both give misleading results.

Suppose you are troubleshooting the three-stage amplifier shown in Fig. 10-12. Your oscilloscope shows the input to stage 1 is

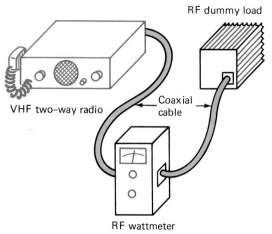

Fig. 10-11   Measuring transmitter power output.

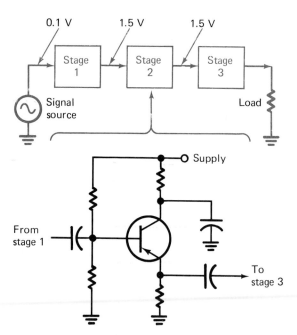

Fig. 10-12   Troubleshooting a three-stage amplifier.

0.1 V and the output is 1.5 V. A quick calculation gives a gain of 15:

$$\text{Gain} = \frac{1.5}{0.1} = 15$$

This seems acceptable, so you move the probe to the output of stage 2. The voltage here also measures 1.5 V. This seems strange. Stage 2 is not giving any gain. However, a close inspection of the schematic shows that stage 2 is an emitter follower. You should remember that emitter followers do not produce any voltage gain. Perhaps the problem is really in stage 1. The normal gain for this stage may be 150 rather than 15. Knowledge of the circuit is required for troubleshooting the low-output symptom.

Beginners in electronics often feel that a dead circuit is going to be more difficult to fix than a weak one. Just the opposite is generally true. It is usually easier to find the broken link in the signal chain because the symptoms are more definite.

If a technician is experienced with a given circuit, the low-output problem is usually not too difficult, because the technician knows approximately what to expect from each stage. In addition to experience, service notes and schematics can be a tremendous help. Often they include pictures of the expected waveforms at various circuit points. If

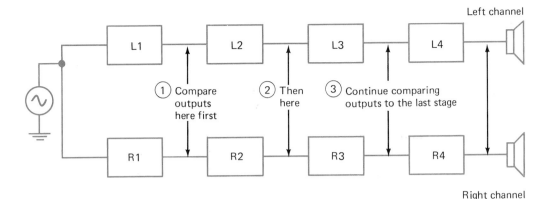

Fig. 10-13   A stereo amplifier.

the oscilloscope shows a low output at one stage, then the input can be checked. If the input signal is normal, it is a safe to assume that the low-gain stage has been found.

Sometimes the needed information is found in the equipment itself. A good example is a stereo amplifier (Fig. 10-13). Assume that the left channel is weak. By checking back and forth between the right channel and the left channel, the low-gain stage can be isolated. Remember, when you are signal-tracing, to work toward the output.

Once the weak stage is located, the fault can be narrowed down to one component. Several components can be defective, but usually only one is at fault. Some possible causes for low gain are

1. Low supply voltage
2. Open bypass capacitor
3. Partially open coupling capacitor
4. Improper transistor bias
5. Defective transistor
6. Defective coupling transformer
7. Misaligned or defective tuned circuit

Supply voltages are supposed to be verified in the preliminary checks. However, it is still possible that a stage will not receive the proper voltage. Note the resistor and capacitor in the supply line in Fig. 10-14. This *RC* decoupling network may be defective. Such a defect may drastically lower the collector voltage, and this may decrease gain.

Figure 10-14 also shows that the emitter bypass capacitor may be open. This can lower the voltage gain from over 100 to less than 4. The coupling capacitors may have lost capacity, causing loss of signal. The voltage checks at the transistor terminals shown

in Fig. 10-14 will determine whether the bias is correct.

In the dual-gate MOSFET RF amplifier of Fig. 10-15, the input signal is applied to gate 1 of the transistor. Gate 2 is connected to the supply through a resistor and to a separate AGC circuit. The letters AGC stand for *automatic gain control*. An AGC fault will often reduce the gain of an amplifier. The gain reduction can be more than 20 dB. Thus, if an amplifier is controlled by AGC, this control voltage must be measured to determine if it is normal.

Figure 10-15 also shows that the drain load is a tuned circuit. This circuit is adjusted for the correct resonant frequency by moving a tuning slug in the transformer. The possibility exists that someone turned the slug. This can produce a severe loss of gain at the operating frequency of the amplifier. In such cases, refer to the service notes for the proper adjustment procedure.

Troubleshooting for loss of gain in amplifiers can be difficult. Many things can give this

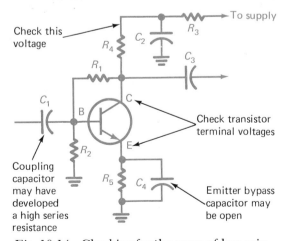

Fig. 10-14   Checking for the cause of low gain.

163

Electronics:
Principles and
Applications
CHAPTER 10

Distortion

Noise

Motorboating

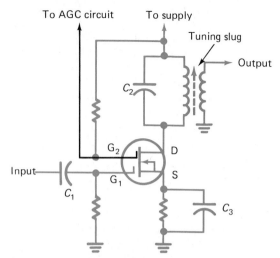

Fig. 10-15   A MOSFET RF amplifier.

symptom. Voltage analysis will locate some of them. Others must be found by substitution. For example, a good capacitor can be temporarily placed in parallel with one that is suspected of being open. If gain is restored, the technician's suspicion that the original capacitor was defective is correct.

Review Questions

Choose the letter that best answers each question.

19. Refer to Fig. 10-10. The amplifier is rated at 40-W power output. The oscilloscope should show at least
    A. 17.9 V peak-to-peak before clipping
    B. 35.8 V peak-to-peak before clipping
    C. 50.6 V peak-to-peak before clipping
    D. 98.7 V peak-to-peak before clipping

20. The normal voltage gain for an emitter-follower amplifier is
    A. 250
    B. 150
    C. 50
    D. Less than 1

21. Refer to Fig. 10-14. The power supply has been checked, and it is normal; yet the collector of the transistor is quite low in voltage. It is unlikely that the cause is
    A. An open in R_1
    B. A leaky capacitor C_2
    C. A short in C_3
    D. A short in C_2

22. Refer to Fig. 10-14. The stage is supposed to have a gain of 50, but a test shows that it is much less. It is unlikely that the cause is

A. A defective transistor
B. A short in C_4
C. An open in C_4
D. A leaky capacitor C_2

23. Refer to Fig. 10-15. The stage has very low gain. It is unlikely that the cause is
    A. An incorrect AGC voltage
    B. A misadjusted tuning slug
    C. A short in C_2
    D. A short in C_3

10-4 DISTORTION AND NOISE

Distortion and noise in an amplifier mean that output signal contains different information from the input signal. A linear amplifier is not supposed to change the quality of the signal. The amplifier is used to increase the amplitude of the signal.

Noise can produce a variety of symptoms. Some noise problems that may be found in an audio amplifier are

1. Constant frying or hissing noise
2. Popping or scratching sound
3. Hum
4. Motorboating (a "putt putt" sound)

Noise problems can often be traced to the power supply. In troubleshooting for this symptom, it is a very good idea to use an oscilloscope to check the various supply lines in the equipment. The preliminary checks detailed in Sec. 10-1 rely on meter readings to check the supply. An oscilloscope will show things that a meter cannot. For example, Fig. 10-16 shows a power-supply waveform with excess ac ripple. The average dc value of the waveform is correct. This means the meter reading will be acceptable.

It is possible to obtain some idea of the ac content in a power-supply line without an oscilloscope. Most volt-ohm-milliammeters (VOMs) have a separate jack or function

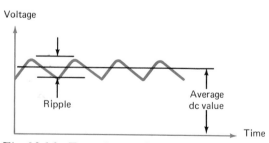

Fig. 10-16   Excessive ripple.

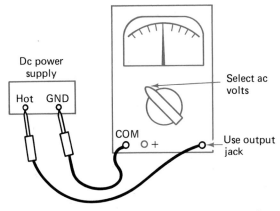

Dc power
supply

Hot   GND

Select ac
volts

COM

Use output
jack

Fig. 10-17 Blocking the dc component of a supply voltage.

marked "output." This places a dc blocking capacitor in series with the test leads and allows a measurement of just the ac content on the supply line. Figure 10-17 shows the conections from the output of the dc supply to the VOM. The oscilloscope test is preferred since it gives more information and more accurate information.

The most common noise problem is hum. *Hum* refers to the introduction of a 60-Hz signal into the amplifier. It can also refer to 120-Hz interference. If the hum is coming from the power supply, it will be 60 Hz for half-wave supplies and 120 Hz for full-wave supplies. Hum can also get into the amplifier because of a broken ground connection. High-gain amplifiers often use shielded cables in areas where the ac line frequency can induce signals into the circuitry. Shielded cable is used in a stereo amplifier system for connections between the turntable and amplifier (Fig. 10-18). Be sure to check all shielded

cables when hum is a problem. The braid may be open, which allows hum to get into the amplifier.

Some high-gain amplifiers also use shields to prevent noise from entering the circuits. These shields must be in the proper position and fastened securely.

Another cause for hum is poor grounding of circuit boards. In some equipment, the fasteners do double duty. They mechanically hold the board and provide an electric contact to the chassis. Check the fasteners to make sure they are secure.

Sometimes amplifier noise can be limited to general sections of the circuit by checking the effect of the various controls. Figure 10-19 is a block diagram of a four-stage amplifier. The gain control is located between stage 2 and stage 3. It is a good idea to operate this control to see whether it affects the noise reaching the output. If it does, then the noise is most likely originating in stage 1 or stage 2. Of course, if the control has no effect on the noise level, then it is probably originating in stage 3 or stage 4.

Another good reason for checking the controls is that they are often the source of the noise. Scratchy noises and popping sounds can often be traced to variable resistors. Special cleaner sprays are available for reducing or eliminating noise in controls. However, the noise often returns. The best approach is to replace noisy controls.

A constant frying or hissing noise usually indicates a defective transistor or integrated circuit. Signal tracing is effective in finding out in which stage the noise is originating. Resistors can also become somewhat noisy.

Hum

Shielded cable

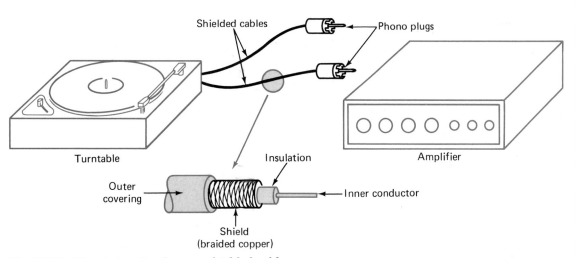

Shielded cables

Phono plugs

Turntable

Amplifier

Outer
covering

Insulation

Inner conductor

Shield
(braided copper)

Fig. 10-18 Signal circuits often use shielded cable.

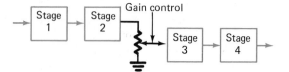

Fig. 10-19 A four-stage amplifier.

Motorboating

Distortion

The problem is generally limited to early stages in the signal chain. Because of the high gain, it does not take a large noise signal to cause problems at the output.

It is worth mentioning that the noise may be coming from the signal source itself. It may be necessary to substitute another source or disconnect the signal. If the noise disappears, the problem has been found.

Motorboating is a problem that usually indicates an open filter capacitor, an open bypass capacitor, or a defect in the feedback circuit of the amplifier. An amplifier can become an oscillator (make its own signal) under certain conditions. This topic is covered in detail in Sec. 11-6.

Amplifier distortion may be caused by bias error in one of the stages. You should remember that bias sets the operating point for an amplifier. Incorrect bias can shift the operating point to a nonlinear region, and distortion will result. Of course, the transistor itself can be defective and produce severe distortion.

It may help to determine whether the distortion is present at all times or just on large signals. A large-signal distortion may indicate a defect in the power (large-signal) stage. This is not conclusive, but it may be a good clue.

Another way to isolate the stage causing distortion is to feed a test signal into the amplifier and "walk" through the circuit with an oscilloscope. Many technicians prefer using a triangle waveform [Fig. 10-20(a)] for making this test. The distorted triangle waveform is seen

in Fig. 10-20(b). The sharp peaks of the triangle make it very easy to spot any clipping or compression. The linear ramp from positive to negative makes it very easy to see any crossover distortion. In using such a test, it is a good idea to try different signal levels. Some problems show up at low levels and some at high levels. For example,

1. Crossover distortion in a push-pull amplifier is most noticeable at low levels.
2. Operating point error in an early class A stage is most noticeable at high levels.

Review Questions

Choose the letter that best answers each question.

24. An audio amplifier has severe hum in the output only when the selector is switched to PHONO. Which of the following is *least* likely to be wrong?
    A. A defective phono jack
    B. A defective shielded cable to the turntable
    C. A bad filter capacitor in the power supply
    D. An open ground in the turntable

25. An audio amplifier has severe hum all the time. Which of the following is *most* likely at fault?
    A. The volume control
    B. The power cord
    C. The filter in the power supply
    D. The output transistor

26. Refer to Fig. 10-19. The amplifier has a loud, hissing sound only when the volume control is turned up. Which of the following conclusions is the best?
    A. The problem is in stage 3
    B. The problem is in stage 4
    C. The speaker is defective
    D. The problem is in stage 1 or 2

27. An audio amplifier has bad distortion when played at low volume. At high volume, it is noticed that the distortion is only slight. Which of the following is most likely to be the cause?
    A. A bias error in the push-pull output stage
    B. A defective volume control
    C. A defective tone control
    D. Supply voltage is too low

28. An amplifier makes a putt-putt sound at high volume levels (motorboating).

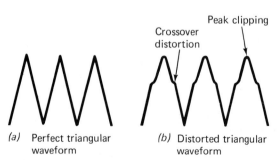

(a) Perfect triangular waveform

(b) Distorted triangular waveform

Fig. 10-20 Triangle waveform used for distortion analysis.

Which of the following is most likely to be the cause?
A. A defective volume control
B. A defective transistor
C. An open filter or bypass capacitor
D. A defective speaker

29. An amplifier has bad distortion when played at high volume. Which of the following is most likely to be the cause?
A. A cracked circuit board
B. A bias error in one of the amplifiers
C. A defective volume control
D. A defective tone control

10-5 INTERMITTENTS

An electronic device is *intermittent* when it will work only some of the time. It may become defective after being on for a few minutes. It may come and go with vibration. The source of these kinds of problems can be very difficult to locate. Technicians generally agree that intermittents are the most difficult to troubleshoot.

There are two basic ways to find the cause of an intermittent problem. One way is to run the equipment until the problem appears and then use ordinary troubleshooting practice to isolate it. The second way is to use various procedures to force the problem to show up. Some of these are

1. Heat various parts of the circuit.
2. Cool various parts of the circuit.
3. Change the supply voltage.
4. Vibrate various parts of the circuit.

The actual technique used will depend on the symptoms and how much time is available to service the equipment. Some intermittents will not show up in a week of continuous operation. In such a case, it is probably best to try to make the problem occur.

Many intermittents are *thermal*. That is, they appear at one temperature extreme or another. If the problem shows up only at a high temperature, it may be very difficult to find with the cabinet removed. With the cabinet removed the circuits usually run much cooler, and a thermal intermittent will not show. In such a case, it may be necessary to use a little heat to find the problem.

Figure 10-21 shows some of the ways that electronic equipment and components can be safely heated to check for thermal intermittents. The bench lamp is useful for heating many components at one time. By placing a 100-W lamp near the equipment, the circuits will become quite warm after a few minutes. Be careful not to overheat the circuits. Certain plastic materials can be easily damaged. A vacuum desoldering tool makes a good heat source for small areas. Squeezing the bulb will direct a stream of hot air where needed. Be careful not to spray solder onto the circuit. A heat gun is useful for heating larger components and several parts at one time. Finally, an ordinary soldering pencil may be used by touching the tip to a component lead or to a metal case.

Spray coolers are available for tracing thermal intermittents. They use a chemical such as Freon to rapidly cool circuit compo-

Intermittent

Thermal intermittent

Spray cooler

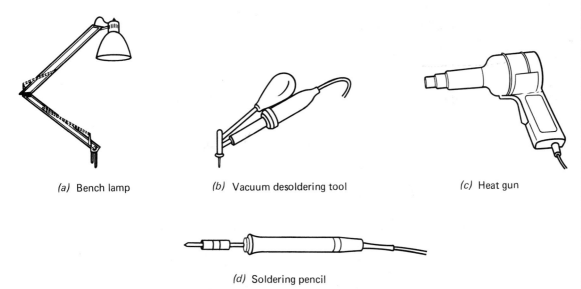

(a) Bench lamp   (b) Vacuum desoldering tool   (c) Heat gun

(d) Soldering pencil

Fig. 10-21 Methods for heating components and circuits.

167

Voltage-sensitive
intermittent

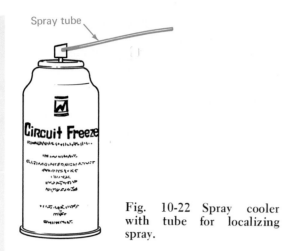

Fig. 10-22 Spray cooler with tube for localizing spray.

Fig. 10-24 Insulated tools are the safest to use when applying vibrations to components.

nents. A spray tube is included to control the application closely (Fig. 10-22). Thus it is easy to check one component at a time.

Some intermittents are voltage-sensitive. The ac line voltage is nominally rated at 115 V. However, it can and does fluctuate. It may go below 110 V, and it may go above 125 V. Most electronic equipment is designed to work over this range. In some cases, a circuit or a component can become critical and voltage-sensitive. This type of situation may show up as an intermittent. Figure 10-23 shows the test arrangement used in many shops. The equipment is connected to a variable ac transformer. This allows the voltage-sensitive problem to appear. An isolation transformer may still be required for safety. Most variable ac transformers are autotransformers and do not give any isolation.

Intermittents are often sensitive to vibration. This may be caused by a bad solder joint, a bad connector, or a defective component. The only way to trace this kind of a problem is to use vibration. Careful tapping with an *insulated tool* (Fig. 10-24) may allow you to isolate the defect. In addition to tap-

ping, try flexing the circuit boards and the connectors. These tests are made with the power on, so be very careful.

You may find it impossible to localize the intermittent to a single point in the circuit. Turn off the power and use some fresh solder to reflow every joint in the suspected area. Joints often fail electrically but look perfect. Resoldering is the only way to be sure.

Do not overlook sockets. Try plugging and unplugging several times to clean the sliding contacts. The power must be off. Never unplug a transistor or any device with the power on. Severe damage may result.

Circuit board connectors may require cleaning. An ordinary pencil eraser does a good job on the board contacts. Use just enough pressure to brighten the contacts. Clear away any debris left by the eraser before reconnecting the board.

Intermittents can be tough to work on, but they are not impossible. Use every clue and test possible to localize the problem. It is far easier to check a few things than to check every joint, contact, and component in the system.

Review Questions

Choose the letter that best completes each statement.

30. A circuit works intermittently as the chassis is tapped with a screwdriver. The problem may be
    A. Thermal
    B. An open filter capacitor
    C. A cold solder joint
    D. Low supply voltage

31. A problem appears as one part of a large circuit board is flexed. The proper procedure is to
    A. Replace the components in that part of the board
    B. Reflow the solder joints in that part of the board
    C. Heat that part of the board
    D. Cool that part of the board

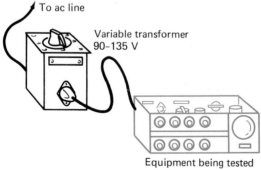

To ac line

Variable transformer
90–135 V

Equipment being tested

Fig. 10-23 Checking for a voltage-sensitive intermittent.

32. A circuit always works fine when first turned on but then fails after about 20 minutes of operation. Out of its cabinet, it works fine indefinitely. The problem is
    A. Thermal
    B. An open ground
    C. High line voltage
    D. The on-off switch

33. The correct procedure to isolate the defect in question 32 is to
    A. Run it at reduced line voltage
    B. Remove the cabinet and cool various parts
    C. Remove the cabinet and heat various parts
    D. Resolder the entire circuit

Summary

1. When troubleshooting, remember to check all connections and control settings.
2. If the unit is ac-operated, disconnect it from the line before taking it apart.
3. Use service literature and the proper tools.
4. Sort and save all fasteners, knobs, and other small parts.
5. Make a thorough visual inspection of the interior of the equipment.
6. Try to determine why a component failed before turning on the power.
7. Check for overheating.
8. Verify all power-supply voltages.
9. Lack of amplifier output may not be in the amplifier itself. There could be a defective output device or no input signal.
10. A multistage amplifier can be viewed as a signal chain.
11. Signal injection begins at the load end of the chain.
12. Signal tracing begins at the input end of the chain.
13. Voltage analysis is generally used to limit the possibilities to one defective component.
14. Some circuit defects cannot be found by dc voltage analysis. These defects are usually the result of an open device or coupling component.
15. Low output from an amplifier may be due to low input.
16. A dummy load resistor is often substituted for the output device when amplifier performance is measured.
17. Both signal tracing and signal injection can give misleading results when troubleshooting for the low-gain stage.
18. Voltage analysis will lead to some causes of low gain.
19. A capacitor suspected of being open can be checked by bridging it with a new one.
20. A linear amplifier is not supposed to change anything but the amplitude of the signal.
21. Noise may be originating in the power supply.
22. Hum refers to a 60-Hz or a 120-Hz signal in the amplifier.
23. Hum may be caused by a defective power supply, an open shield, or a poor ground.
24. Operate all controls to see if the noise occurs before or after the control.
25. Motorboating noise means the amplifier is oscillating.
26. Distortion can be caused by bias error or defective transistors.
27. Thermal intermittents may show up after the equipment is turned on for some time.
28. Use heat or cold to localize thermal problems.
29. Vibration intermittents can be isolated by careful tapping with an insulated tool.

Chapter Review Questions

Choose the letter that best answers each question.

10-1. When troubleshooting, which of the following questions is not part of a preliminary check?
(A) Are all cables plugged in? (B) Are all controls set properly?
(C) Are all transistors good? (D) Is the power supply on?

10-2. What is the quickest way to check a speaker for operation (not for quality)?
(A) A click test using a dry cell (B) Substitution with a good speaker (C) Connecting an ammeter in series with the speaker (D) Connecting an oscilloscope across the speaker

10-3. Refer to Fig. 10-2. The regulated output is zero. The unregulated input is normal. Which of the following could be the cause of the problem?
(A) C_1 is open (B) D_1 is open (C) R_1 is open (D) R_1 is shorted

10-4. Refer to Fig. 10-2. The regulated output is low. The unregulated input is normal. Which of the following could be the cause of the problem?
(A) D_1 is open (B) D_1 is shorted (C) C_1 is open (D Excessive current at the output

10-5. Refer to Fig. 10-3. What is the purpose of the isolation transformer?
(A) To boost the line voltage (B) To drop the line voltage (C) To prevent shock and ground loops (D) To uncover any intermittents

10-6. Refer to Fig. 10-5. The amplifier is dead. The speaker is known to be good. Where should the signal be injected first?
(A) At the input of stage 1 (B) At the input of stage 2 (C) At the input of stage 3 (D) At the input of stage 4

10-7. Refer to Fig. 10-5. The amplifier is dead. A known good signal has been connected to the input of stage 1. Signal tracing should begin at
(A) The output of stage 1 (B) The output of stage 2 (C) The output of stage 3 (D) The output of stage 4

10-8. Refer to Fig. 10-8. The collector of Q_1 measures 21 V, and it should be 12 V. Which of the following is *least* likely to be wrong?
(A) Q_1 is open (B) Q_1 is shorted (C) R_3 is open (D) C_2 is shorted

10-9. Refer to Fig. 10-8. The collector of Q_1 measures 2 V, and it is supposed to be 12 V. Which of the following could be the problem?
(A) R_1 is shorted (B) C_1 is open (C) R_2 is open (D) C_4 is shorted

10-10. Refer to Fig. 10-8. Resistor R_1 is open. Which of the following statements is correct?
(A) The collector of Q_1 will be at 0 V (B) Q_2 will go into saturation (C) Q_2 will go into cutoff (D) Q_1 will go into saturation

10-11. Refer to Fig. 10-8. Resistor R_9 is open. Which of the following statements is correct?
(A) The collector of Q_2 will be at 0 V (B) Q_2 will go into saturation (C) Q_2 will go into cutoff (D) Q_1 will go into saturation

10-12. Refer to Fig. 10-14. Resistor R_1 is open. Which of the following is correct?
(A) The collector voltage will be very low (B) The collector voltage will be very high (C) The transistor will be in saturation (D) The emitter voltage will be very high

10-13. Refer to Fig. 10-15. Capacitor C_3 is open. Which of the following is correct?
(A) The dc voltages will all be wrong (B) The AGC voltage will drop (may even go negative) (C) Extreme distortion will result (D) The gain will be low

10-14. Refer to Fig. 10-19. A scratching sound is heard as the volume control is rotated. Where is the problem likely to be?

(A) Stage 1 or 2 (B) The volume control (C) Stage 3 or 4 (D) The speaker

10-15. Refer to Fig. 10-19. There is severe hum in the output, but turning down the volume control makes it stop completely. Where is the problem likely to be?
(A) Third stage or fourth stage (B) Power-supply filter (C) The volume control (D) Input cable (broken ground)

10-16. An amplifier is capacitively coupled. What is the best way to find an open coupling capacitor?
(A) Look for transistors in cutoff (B) Look for transistors in saturation (C) Look for dc bias errors on the bases (D) Look for a break in the signal chain

10-17. An amplifier has a push-pull output stage. Bad distortion is noted at high volume levels only. The problem could be which of the following?
(A) Bias error in an earlier stage (B) A shorted output transistor (C) Crossover distortion (D) A defective volume control

10-18. What is probably the slowest way to find an intermittent problem?
(A) Try to make it show by using vibration (B) Use heat (C) Use cold (D) Wait until it shows up by itself

10-19. An automobile radio works fine except while traveling over a bumpy road. What is likely to be the cause of the problem?
(A) Thermal (B) An open capacitor (C) A low battery (D) A loose antenna connection

Answers to Review Questions

1. *B*	12. *C*	23. *D*
2. *A*	13. *B*	24. *C*
3. *C*	14. *D*	25. *C*
4. *D*	15. *A*	26. *D*
5. *D*	16. *B*	27. *A*
6. *A*	17. *D*	28. *C*
7. *B*	18. *C*	29. *B*
8. *B*	19. *C*	30. *C*
9. *D*	20. *D*	31. *B*
10. *B*	21. *A*	32. *A*
11. *D*	22. *B*	33. *C*

Oscillators

- Oscillators are basic signal sources in electronics. They can be used to generate almost any wave shape at almost any frequency. This is a rather large order, and many different kinds of oscillators exist to fill it.

  Oscillators can be designed in several basic ways. This chapter covers the most popular circuits. This chapter also discusses undesired oscillations. You will see that almost any amplifier can become an oscillator. When this happens, the amplifier is useless.

11-1 OSCILLATOR CHARACTERISTICS

An *oscillator* is an electronic circuit that converts dc to ac. Figure 11-1 is an oscillator in block diagram form. The output is an ac sine wave. Not all oscillator circuits produce sine waves. Oscillators are usually divided into two types: sinusoidal and nonsinusoidal.

Sinusoidal oscillators usually are based on an amplifier circuit. A feedback circuit is added to the amplifier so that it will oscillate (produce ac). The amplifiers that you have studied so far depend on an ac or a dc signal arriving at their input terminals. An oscillator is an amplifier that provides its own input signal. Figure 11-2 shows how the feedback circuit is used to make an amplifier provide its own input signal.

Most amplifiers will oscillate if the conditions are correct. For example, you probably know what happens when someone turns the volume control too high on a public address system. The squeals and howls that come out

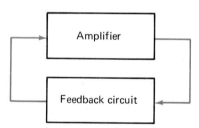

Fig. 11-2 An amplifier with feedback.

of the loudspeaker are a result of oscillations. The feedback in this case is in the form of sound waves from the loudspeakers reentering the system through the microphone (Fig. 11-3).

Thus far, a very important idea has been discussed—feedback. The feedback in most oscillator circuits is electrical. It is accomplished by resistive, capacitive, or inductive coupling between the output and the input of the amplifier. This feedback allows the amplifier to develop and maintain its own input signal.

Feedback alone will not guarantee oscillations. Look at Fig. 11-3 again. You know that turning down the volume control will stop the public address system from oscillating. The feedback is still present. But now there is not enough gain to overcome the loss in the feedback path. This is one of the basic criteria for an oscillator circuit: The gain must be greater than the loss.

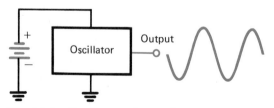

Fig. 11-1 Oscillators change dc to ac.

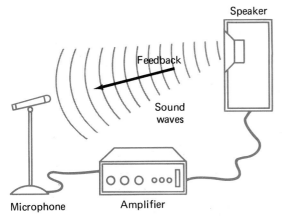

Fig. 11-3 Feedback can make an amplifier unintentionally oscillate.

There is one other idea that is also important—phase. The signal fed back to the input of the amplifier must be in phase with the input signal. In-phase feedback is also referred to as *positive* feedback, or *regenerative* feedback. When the amplifier input and output are normally out of phase (such as in the common-emitter amplifier), the feedback circuit will have to achieve a phase reversal.

Oscillators can be made to work over quite a range of frequencies. For example, some oscillators work in the audio-frequency range, and others work in the radio-frequency range. An electronic organ uses oscillators to develop the audio signals that you hear. A radio transmitter uses an oscillator to develop the high-frequency radio signal.

Different oscillators have different requirements. In addition to frequency, there is the question of *stability*. A stable oscillator will produce a signal of constant amplitude and frequency. Another requirement for some oscillators is the capacity for tuning. Tuning will change the output frequency. These oscillators are often called VFOs. The letters VFO stand for *variable-frequency oscillator*.

Review Questions

Choose the letter that best answers each question.

1. What conditions are required for an amplifier to oscillate?
   A. There must be feedback.
   B. The feedback must be in phase.
   C. The gain must be greater than the loss.
   D. All the above are required.

2. Which of the following statements is *not* true?
   A. An oscillator is a circuit that converts dc to ac.
   B. An oscillator is an amplifier that supplies its own input signal.
   C. All oscillators generate sine waves.
   D. In-phase feedback is called positive feedback.

3. Refer to Fig. 11-3. The system oscillates. Which of the following is most likely to correct the problem?
   A. Increase the gain of the amplifier.
   B. Use a microphone with higher output.
   C. Move the microphone closer to the speaker.
   D. Decrease the feedback by using acoustical (sound-absorbing) materials, changing the position of the speaker, and so on.

11-2 RC CIRCUITS

It is possible to control the output frequency of an oscillator by using *RC* (resistive and capacitive) components. The *RC* circuit often used in oscillators is shown in Fig. 11-4(*a*). This circuit will have a maximum output at only one frequency. This frequency is called the *resonant frequency* f_r. It can be found with this equation:

$$f_r = \frac{1}{6.28RC}$$

The frequency response curve of Fig. 11-4(*b*) shows that the *RC* network will show less output at frequencies above and below f_r. Low frequencies find the series capacitor a

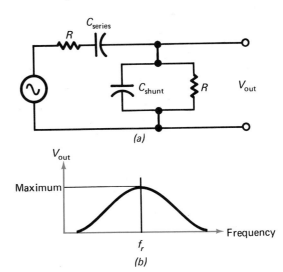

Fig. 11-4 An *RC* lead-lag network.

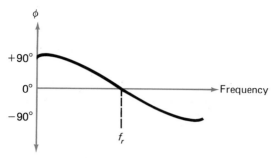

Fig. 11-5   Phase response of the lead-lag network.

Lead-lag
network

Wien-bridge
oscillator

high reactance. This high series reactance causes signal loss, and V_{out} drops. High frequencies find the shunt capacitor a low reactance. This produces a high current, resulting in a high voltage drop across the series resistor. Again, V_{out} drops. You should review the formula for capacitive reactance to verify this action:

$$X_C = \frac{1}{6.28fC}$$

In this formula, as f (frequency) becomes a small number, X_C becomes large. As f becomes a large number, X_C becomes small.

The circuit of Fig. 11-4 is called a *lead-lag network*. This is because of its phase response (Fig. 11-5). You can see that for frequencies lower than f_r the phase angle is positive and approaches 90°. Frequencies higher than f_r give a negative phase angle that approaches $-90°$. The phase response of the network can *lead* or *lag*, depending on the input frequency and f_r. The lead-lag network produces a 0° phase angle when the input frequency is equal to f_r.

Figure 11-6 shows a lead-lag network used to control the frequency of an oscillator. Note that the feedback is applied through the lead-lag network to the noninverting input of the op-amp. Feedback applied to the noninverting input is *positive* feedback. However, only one frequency will arrive at the noninverting input exactly in phase. That frequency is the resonant frequency f_r of the network. All other frequencies will lead or lag. This means that the oscillator will operate at a single frequency.

The oscillator circuit of Fig. 11-6 is called a *Wien-bridge* oscillator. The lead-lag network forms one leg of the bridge, and the resistors marked R' and $2R'$ form the other leg of the bridge. The operational amplifier is connected across the legs. Resistor R' is a device with a large positive temperature coefficient. A good choice for R' is a tungsten lamp. The filaments of such lamps have the desired temperature response.

The Wien-bridge oscillator will start with R' low in resistance. This is because there has been no feedback signal to heat it. As the amplitude of the oscillations builds up, the tungsten filament will become warmer. This will make its resistance increase. Resistors R' and $2R'$ form a voltage divider. As R' increases, the voltage applied to the inverting input of the operational amplifier will increase. As we learned earlier, negative feedback tends to decrease the gain of an op-amp. This stabilizes the Wien-bridge oscillator.

You may be wondering how the oscillator starts. When the circuit is first turned on, no oscillations are present. However, the amplifier begins generating wide-band noise. All

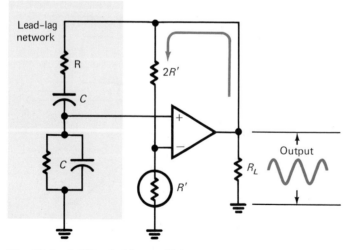

Fig. 11-6   A Wien-bridge oscillator.

frequencies are present at the output of the amplifier. Of course, only one frequency arrives at the noninverting input exactly in phase. This is the resonant frequency of the lead-lag network. This frequency is amplified and appears at the output of the amplifier. Again it is fed back and appears at the noninverting input. In a very short time, strong oscillations can be found at the output.

The Wien-bridge circuit satisfies the basic demands of all oscillator circuits: (1) the gain is adequate to overcome the loss in the feedback circuit, and (2) the feedback is in phase. The gain of the circuit is very high at the beginning. This ensures rapid starting of the oscillator. After that, the gain decreases because of the heating of R'. This keeps the output stable and eliminates amplifier clipping and distortion. Wien-bridge oscillators are noted for their low-distortion sinusoidal output.

A variable-frequency Wien-bridge oscillator can be made (Fig. 11-7). The variable capacitors shown are ganged. One control shaft operates both capacitors. What will the frequency range be? It will be necessary to use the resonant frequency formula twice to answer this question:

$$f_r = \frac{1}{6.28RC}$$

$$= \frac{1}{6.28 \times 47 \times 10^3 \times 100 \times 10^{-12}}$$

$$= 33,880 \text{ Hz}$$

$$f_r = \frac{1}{6.28 \times 47 \times 10^3 \times 500 \times 10^{-12}}$$

$$= 6776 \text{ Hz}$$

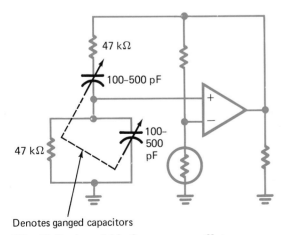

Denotes ganged capacitors

Fig. 11-7 A variable-frequency oscillator.

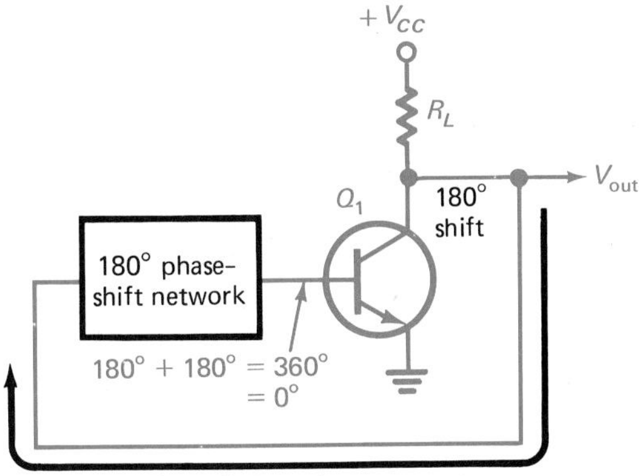

Phase-shift oscillator

+V_{CC}

R_L

Q_1

180° shift

V_{out}

180° phase-shift network

180° + 180° = 360° = 0°

Feedback

Fig. 11-8 A phase-shift network in a common-emitter oscillator.

The range is from 6776 to 33,880 Hz. A 5:1 capacitor range produces a 5:1 frequency range:

$$6776 \text{ Hz} \times 5 = 33,880 \text{ Hz}$$

There is another way to use RC networks to control the frequency of an oscillator. They can be used to produce a 180° phase shift at the desired frequency. This is useful when the common-emitter configuration is used. Figure 11-8 shows what will be needed for a common-emitter oscillator. The signal at the collector is 180° out of phase with the signal at the base. By including a network which gives an additional 180° phase shift, the base receives in-phase feedback. This is because 180° + 180° = 360° and 360° is the same as 0° in circular measurement.

In Fig. 11-9 the phase shift network is divided into three separate sections. Each section is designed to produce a 60° phase shift, and the total phase shift will be 3 × 60°, or 180°. The frequency of oscillations can be predicted by

$$f = \frac{1}{10.87RC}$$

Figure 11-10 is the schematic of a phase-shift oscillator circuit with component values. Each of the three phase-shift networks has been designed to produce a 60° response. Note, however, that the value of R_B is 100 times higher than the values of the other two resistors in the network. This may seem to be an error since all three networks should be the same. Actually, R_B does appear to be much lower in value as far as the ac signal is con-

175

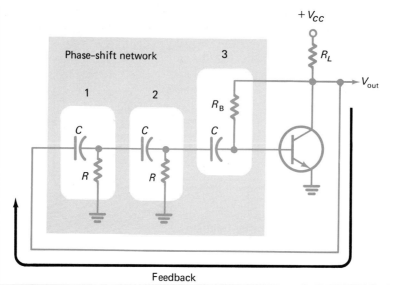

Fig. 11-9 A phase-shift oscillator.

cerned. This is because it is connected to the collector of the transistor. There is an ac signal present at the collector when the oscillator is running which is 180° out of phase with the base signal. This makes the voltage difference across R_B much higher than would be produced by the base signal alone. Thus, more signal current flows through R_B. Resistor R_B produces an ac loading effect at the base that is set by the voltage gain of the amplifier and the value of R_B:

$$R_B' = \frac{R_B}{A_V}$$

This equation tells us that the actual ac loading effect R_B' is equal to R_B divided by the voltage gain of the amplifier. If we assume that the gain of the amplifier is 100, then

$$R_B' = \frac{920 \text{ k}\Omega}{100} = 9.2 \text{ k}\Omega$$

We may conclude that all three phase-shift networks are the same. The frequency of oscillations will be

$$f = \frac{1}{10.87 RC}$$

$$= \frac{1}{10.87 \times 9.2 \times 10^3 \times 0.02 \times 10^{-6}}$$

$$= 500 \text{ Hz}$$

The circuit of Fig. 11-10 may not oscillate at exactly 500 Hz. The formula deals with only the values of the RC network. It ignores some effects caused by the transistor. The formula is good for most applications.

Figure 11-11 shows that the formula is correct as far as the RC network is concerned. Note that the equivalent circuit of one RC section is shown. The phasor diagram indicates that the values in the RC network do

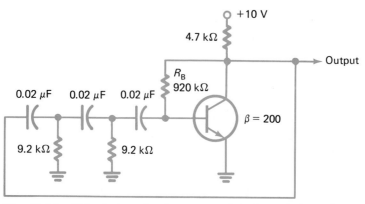

Fig. 11-10 Calculating the frequency of a phase-shift oscillator.

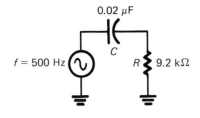

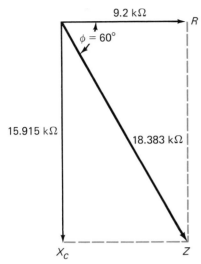

Fig. 11-11 Phase-shift phasor diagram.

produce a 60° phase angle for a frequency of 500 Hz.

Phase-shift oscillators must have good voltage gain to operate. This is due to large losses in the phase-shift network. Remember, an oscillator must have more gain than loss in the feedback circuit. The phase requirement is met by the phase-shift network.

Phase-shift oscillators must be designed carefully for low-distortion sinusoidal output. They are subject to clipping and other nonlinear modes. The Wien-bridge circuit is usually more desirable when the output sine waves must be nearly perfect.

Review Questions

Choose the letter that best answers each question.

4. Refer to Fig. 11-4 where $R = 4700 \, \Omega$ and $C = 0.01 \, \mu F$. At what frequency will V_{out} be in phase with the signal source?
   A. 486 Hz
   B. 1200 Hz
   C. 3386 Hz
   D. 9834 Hz

5. Refer to Fig. 11-6 where $R = 6800 \, \Omega$ and $C = 0.001 \, \mu F$. What will the frequency of the output signal be?
   A. 23.41 kHz
   B. 46.79 kHz
   C. 78.90 kHz
   D. 98.94 kHz

6. Refer to Fig. 11-6. What is the phase relationship of the output signal and the signal at the noninverting (+) input of the amplifier?
   A. 180°
   B. 0°
   C. 90°
   D. 270°

7. Refer to Fig. 11-8. What is the configuration of Q_1?
   A. Common emitter
   B. Common collector
   C. Common base
   D. Emitter follower

8. Refer to Fig. 11-9 where $R_B = 1 \, M\Omega$ and the voltage gain of the circuit is 150. What is the actual loading effect of R_B as far as the phase-shift network is concerned?
   A. 1 MΩ
   B. 500 kΩ
   C. 6667 Ω
   D. 384 Ω

9. Refer to Fig. 11-10. The capacitors are all changed to 0.05 μF. What is the frequency of oscillation?
   A. 60 Hz
   B. 200 Hz
   C. 1.84 kHz
   D. 0.95 MHz

10. Refer to Fig. 11-10. What is the phase relationship of the signal arriving at the base to the output signal?
    A. 0°
    B. 90°
    C. 180°
    D. 270°

11. What do phase-shift oscillators and Wien-bridge oscillators have in common?
    A. They use RC frequency control.
    B. They have a sinusoidal output.
    C. They use amplifier gain to overcome feedback loss.
    D. All the above.

11-3 *LC* CIRCUITS

An RC circuit is limited to frequencies below 1 MHz. Higher frequencies will require a different approach to oscillator design. An

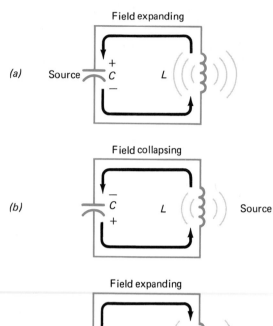

Field expanding

(a) Source ⎓ C   L

Field collapsing

(b)   C   L   Source

Field expanding

(c) Source ⎓ C   L

Field collapsing

(d)   C   L   Source

Fig. 11-12   Tank circuit action.

inductive-capacitive (LC) circuit can be used to carry oscillator performance into hundreds of megahertz. Such circuits are called *tank circuits*, or *flywheel circuits*.

Figure 11-12 shows how a tank circuit can be used to develop sinusoidal oscillations. Figure 11-12(a) assumes that the capacitor is charged. As the capacitor discharges through the inductor, a field expands about the turns of the inductor. After the capacitor has been discharged, the field collapses and current continues to flow. This is shown in Fig. 11-12(b). Note that the capacitor is now being charged in the opposite polarity. After the field collapses, the capacitor again acts as the source. Now, the current is flowing in the opposite direction. Figure 11-12(c) shows the second capacitor discharge. Finally, Fig. 11-12(d) shows the inductor acting as the source and charging the capacitor back to the

original polarity shown in Fig. 11-12(a). The cycle will repeat over and over.

Inductors and capacitors are both energy storage devices. In a tank circuit, they exchange energy back and forth at a rate fixed by the values of inductance and capacitance. The frequency of oscillations will be

$$f_r = \frac{1}{6.28\sqrt{LC}}$$

You should recognize this formula. It is the resonance equation for an inductor and a capacitor. It is based on the resonant frequency, where the inductive reactance and the capacitive reactance are equal. An LC tank circuit will oscillate at its resonant frequency.

Actual circuits have resistance in addition to inductance and capacitance. This resistance will cause the tank circuit oscillations to decay with time (Fig. 11-13). Note that the amplitude of the sine wave gradually decreases. This is known as a *damped sine wave*.

To build a practical LC oscillator, an amplifier must be added. The gain of the amplifier will overcome resistive losses, and a sine wave of constant amplitude can be generated.

One way to combine an amplifier with an LC tank circuit to provide constant-amplitude sine waves is shown in Fig. 11-14. The circuit is called a *Hartley oscillator*. Note that the inductor is tapped. The tap position is important since the ratio of L_A to L_B determines the feedback ratio for the circuit. In practice, the feedback ratio is selected for reliable operation. This ensures that the oscillator will start every time the power is turned on. Too much feedback will distort the output waveform.

The transistor amplifier of Fig. 11-14 is in the common-emitter configuration. This means that a 180° phase shift will be required

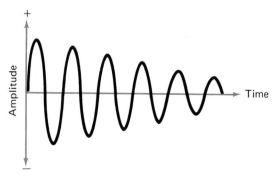

Fig. 11-13   A damped sine wave.

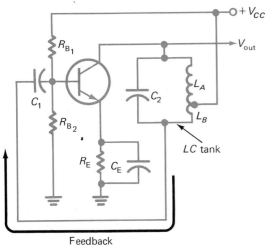

Fig. 11-14   The Hartley oscillator.

somewhere in the feedback path. The tank circuit gives this. A parallel resonant circuit produces a 180° phase shift at its resonant frequency. Thus, the collector signal arrives in phase at the base. Knowing the total inductance and the capacitance of the tank circuit will allow a solution for the resonant frequency. For example, if the total inductance $L_A + L_B$ is 20 μH and the capacitance C_2 is 400 pF, then

$$f_r = \frac{1}{6.28 \sqrt{20 \times 10^{-6} \times 400 \times 10^{-12}}}$$
$$= 1.78 \times 10^6 \text{ Hz}$$

Another way to control the feedback in an LC oscillator is to tap the capacitive leg of the tank circuit. When this is done, the circuit is called a *Colpitts oscillator* (Fig. 11-15).

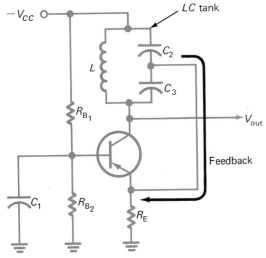

Fig. 11-15   The Colpitts oscillator.

Capacitor C_1 grounds the base of the transistor for ac signals. The amplifier is operating in the common-base configuration. The feedback path is not across the tank circuit since the 180° phase reversal is not desired.

Capacitors C_2 and C_3 in Fig. 11-15 act in series as far as the tank circuit is concerned. Assume that $C_2 = 1000$ pF and $C_3 = 100$ pF. Let us use the series capacitor formula to determine the effect of the series connection:

$$C_T = \frac{C_2 \times C_3}{C_2 + C_3} = \frac{1000 \times 100}{1000 + 100} = 90.91 \text{ pF}$$

This means that 90.91 pF, along with the value of L, would be used to predict the frequency of oscillation.

Figure 11-16 is a variable-frequency oscillator followed by a buffer amplifier. Both stages are operating in the common-drain configuration and use insulated gate field-effect transistors. This circuit represents a design that might be used when maximum frequency stability is needed.

Transistor Q_1 of Fig. 11-16 provides the needed gain to sustain the oscillations. Transistor Q_2 serves as a buffer amplifier. This protects the oscillator circuit from loading effects. Changing the load on an oscillator tends to change both the amplitude and the frequency of the output. For best stability, the oscillator circuit should be isolated from the stages that follow. Transistor Q_2 has a very high input impedance and a low output impedance. This helps the buffer amplifier to isolate the oscillator from loading effects.

The tank circuit of Fig. 11-16 is made up of L, C_1, C_2, and C_3. This arrangement is known as a *series-tuned Colpitts*, or *Clapp*, circuit. It is one of the most stable of all LC oscillators. Assume that C_1 varies from 10 to 100 pF and that C_2 and C_3 are both 1000 pF. We will have to use the series capacitor formula to determine the capacitive range of the tank circuit. When $C_1 = 10$ pF,

$$C_T = \frac{1}{1/C_1 + 1/C_2 + 1/C_3}$$
$$= \frac{1}{1/10 + 1/1000 + 1/1000} = 9.8 \text{ pF}$$

When $C_1 = 100$ pF,

$$C_T = \frac{1}{1/100 + 1/1000 + 1/1000} = 83.3 \text{ pF}$$

Colpitts oscillator

Clapp circuit

179

Quartz crystal

Piezoelectric
material

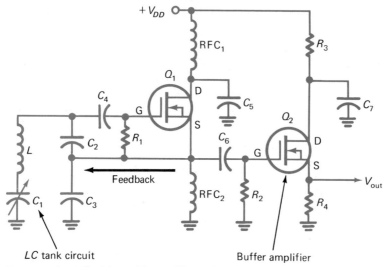

Fig. 11-16   A highly stable oscillator design.

The calculations show that the effective value C_T of the capacitors is determined mainly by C_1. The stray and shunt capacities of Fig. 11-16 appear in parallel with C_2 and C_3. These stray and shunt capacities can change and cause frequency drift in LC oscillator circuits. The Clapp design minimizes these effects by making the series-tuned capacitor have the major effect on the tank circuit.

Review Questions

Choose the letter that best answers each question.

12. Refer to Fig. 11-14. What is the configuration of the amplifier?
    A. Common emitter
    B. Common base
    C. Common collector
    D. Emitter follower

13. Refer to Fig. 11-14. Where is the feedback signal shifted 180°?
    A. Across C_1
    B. Across R_{B_2}
    C. Across R_E
    D. Across the tank circuit

14. Refer to Fig. 11-14 where $C_2 = 270$ pF and $L_A + L_B = 1.8$ μH. Calculate the frequency of the output signal.
    A. 484 kHz
    B. 1.85 MHz
    C. 5.58 MHz
    D. 7.22 MHz

15. Refer to Fig. 11-14. What is the waveform of V_{out}?
    A. Sawtooth wave

B. Sine wave
C. Square wave
D. Triangle wave

16. Refer to Fig. 11-15. The amplifier is in what configuration?
    A. Common emitter
    B. Common base
    C. Common collector
    D. Emitter follower

17. Refer to Fig. 11-15 where $C_2 = 470$ pF, $C_3 = 68$ pF, and $L = 0.8$ μH. What is the frequency of oscillation?
    A. 1.85 MHz
    B. 9.44 MHz
    C. 23.1 MHz
    D. 27.55 MHz

18. Refer to Fig. 11-16. Transistor Q_1 is operating as what type of oscillator?
    A. Clapp oscillator
    B. Hartley oscillator
    C. Phase-shift oscillator
    D. Buffer oscillator

19. Refer to Fig. 11-16. What is the major function of Q_2?
    A. It provides voltage gain.
    B. It provides the feedback signal.
    C. It isolates the oscillator from loading effects.
    D. It provides a phase shift.

11-4 CRYSTAL CIRCUITS

Another way to control the frequency of an oscillator is to use a quartz crystal. Quartz is a piezoelectric material. Such materials can change electric energy into mechanical energy. They can also change mechanical en-

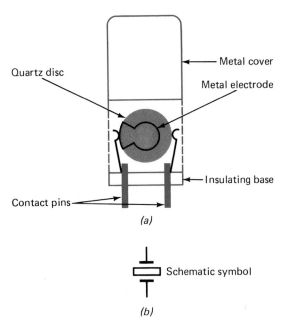

Quartz disc

Metal cover

Metal electrode

Insulating base

Contact pins

(a)

Schematic symbol

(b)

Fig. 11-17 A crystal used for frequency control.

ergy into electric energy. A quartz crystal will tend to vibrate at its resonant frequency. The resonant frequency is determined by the physical characteristics of the crystal. Crystal thickness is the major determining factor for the resonant point.

Figure 11-17(*a*) shows the construction of a quartz crystal. The quartz disk is usually thin, especially for high-frequency operation. A metal electrode is fused to each side of the disk. When an ac signal is applied across the electrodes, the crystal vibrates. The vibration will be at the resonant frequency of the crystal. The crystal will develop a voltage with a frequency equal to the resonant frequency. When a crystal is excited by a frequency equal to its resonant point, a large voltage appears across the electrodes. The schematic symbol for a crystal is shown in Fig. 11-17(*b*).

Crystals can become the frequency-determining components in high-frequency oscillator circuits. They can replace *LC* tank circuits. Crystals have the advantage of producing very stable output frequencies. A crystal oscillator can have a stability better than 1 part in 10^6 per day. This is equal to an accuracy of 0.0001 percent. By carefully controlling temperature, the stability of a crystal oscillator can be held to better than 1 part in 10^{10} per day. Precision ovens are used to achieve this stability.

An *LC* oscillator circuit is subject to frequency variations. Some things that can

cause a change in oscillator output frequency are:

1. Temperature
2. Supply voltage
3. Mechanical stress and vibration
4. Component drift
5. Movement of metal parts near the oscillator circuit

Crystal-controlled circuits can greatly reduce all these effects.

A quartz crystal can be represented by an equivalent circuit (Fig. 11-18). The *L* and *C* represent the resonant action of the crystal. Their values set the series resonant point. Notice that there is another capacitor in the circuit. This is the capacitance of the electrodes and the holder assembly. The electrode capacitance causes the crystal to show a parallel resonant point. Since the capacitors act in series, the net capacitance is lower. This makes the parallel resonant frequency slightly higher than the series resonant frequency.

The equivalent circuit of a crystal predicts that oscillations can occur in two modes: parallel and series. In practice, the parallel mode is from 2 to 15 kHz higher. Oscillator circuits may be designed to use either mode. When a crystal is replaced, it is very important to obtain the correct type. For example, if a series-mode crystal is substituted in a parallel-mode circuit, the oscillator will run high in frequency.

Refer again to Fig. 11-18. Note that the quartz equivalent circuit also contains resistance *R*. This represents losses in the quartz. Most crystals have small losses. In fact, the

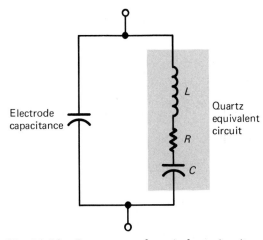

Electrode capacitance

L

R

C

Quartz equivalent circuit

Fig. 11-18 Quartz crystal equivalent circuit.

181

Circuit Q

Overtone crystal

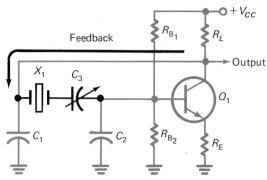

Fig. 11-19 A crystal-controlled oscillator.

losses are small enough to give the crystal a very high Q. Circuit Q is very important in an oscillator circuit. High Q gives frequency stability. Crystal Qs can be in excess of 3000. By comparison, *LC* tank circuit Qs seldom exceed 200. This is why a crystal oscillator is so much more stable than an *LC* oscillator.

Figure 11-19 shows the schematic diagram of a crystal oscillator. The amplifier configuration is common emitter. This means that the feedback will require a 180° phase shift for oscillations to occur. This phase shift is produced by the crystal X_1. The crystal is operating in its parallel mode, and a parallel tuned circuit produces a 180° phase shift. Capacitor C_3 is a trimmer capacitor. It is used to precisely set the frequency of oscillations. Capacitors C_1 and C_2 form a voltage divider to control the amount of feedback. Excess feedback causes distortion and drift. Too little feedback causes unreliable operation: for example, the circuit may not start every time it is turned on. The remaining components in Fig. 11-19 are standard for the common-emitter configuration.

Very-high-frequency crystals present problems. The thickness of the quartz must decrease as frequency goes up. Above 15 MHz, the quartz becomes so thin that it is too fragile. Higher frequencies require the use of overtone crystals which utilize harmonics of the fundamental frequency. This technique extends the range of crystal oscillators to around 150 MHz.

Oscillator circuits designed to use overtone crystals must include an *LC* tuned circuit. This *LC* tuned circuit must be tuned to the correct harmonic. This more or less forces the crystal to operate in the proper mode. Otherwise, it would tend to oscillate at a lower frequency.

Figure 11-20 is an overtone oscillator circuit. Capacitor C_1 grounds the base for ac signals. This makes the amplifier operate in the common-base configuration. In this case, feedback can be direct. A 180° phase reversal will not be needed. Capacitors C_3 and C_4 form a divider to set the amount of feedback from collector to emitter. Crystal X_1 is in the feedback path. It is operating in the series mode. No phase reversal occurs across a series resonant circuit. All overtone crystals operate in the series mode.

Inductor L_1 of Fig. 11-20 is part of the tuned circuit used to select the proper overtone. It resonates with C_3, C_4, and C_5 to form a tank circuit. Coil L_1 is adjusted to the correct overtone frequency. Capacitor C_2 is a trimmer capacitor used to set the crystal frequency. In practice, L_1 is adjusted first until the oscillator starts and works reliably. Then, C_2 is adjusted for the exact frequency required.

There are quite a few crystal oscillator cir-

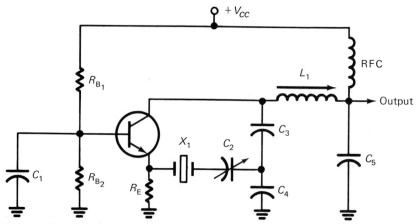

Fig. 11-20 An overtone crystal oscillator.

cuits. Some are simple, and others are complex. The major disadvantage of them all is cost. This can become quite a problem in equipment such as a multichannel transmitter. A separate crystal will be required for every channel. The cost soon reaches the point where another solution must be found. This solution is a *frequency synthesizer*. These are digital circuits that can synthesize many frequencies from one or more crystals.

Review Questions

Choose the letter that best completes each statement.

20. The quartz crystals used in oscillators show a
    A. Piezoelectric effect
    B. Semiconductor effect
    C. Diode action
    D. Transistor action

21. An oscillator may use a quartz crystal in place of
    A. A transistor
    B. An *LC* tank circuit
    C. A VFO
    D. None of the above

22. A 10-MHz crystal oscillator has a stability of 1 part in 10^6. The largest frequency error expected of this crystal is
    A. 1 Hz
    B. 10 Hz
    C. 100 Hz
    D. 1000 Hz

23. A series-mode crystal is marked 10.000 MHz. It is used in a circuit that operates the crystal in its parallel mode. The circuit can be expected to
    A. Run below 10 MHz
    B. Run at 10 MHz
    C. Run above 10 MHz
    D. Not oscillate

24. Refer to Fig. 11-19. The phase relationship of the signal at the collector of Q_1 to the base signal is
    A. 0°
    B. 90°
    C. 180°
    D. 360°

25. Refer to Fig. 11-19. The required phase shift is produced across
    A. C_1
    B. X_1
    C. C_2
    D. C_3

26. Refer to Fig. 11-20. The configuration of the amplifier is
    A. Common emitter
    B. Common collector
    C. Common base
    D. Emitter follower

27. Refer to Fig. 11-20. The function of C_2 is to
    A. Act as an emitter bypass
    B. Adjust the tank circuit to the crystal harmonic
    C. Produce the required phase shift
    D. Set the exact frequency of oscillation

Oscillators
CHAPTER 11

Frequency
synthesizer

Relaxation
oscillator

11-5 RELAXATION OSCILLATORS

All the oscillator circuits discussed so far have been the sinusoidal type. There is another major class of oscillators that do not produce sine waves. They are known as *relaxation oscillators*. The output waveforms can be sawtooths, pulses, or square waves.

Figure 11-21 is one type of relaxation oscillator. A unijunction transistor is the key component. You recall that a UJT is a negative-resistance device. Specifically, the resistance from the emitter terminal to the B_1 terminal will drop when the transistor fires.

When the power is applied to the oscillator circuit of Fig. 11-21, the capacitor begins charging through R_1. As the capacitor voltage increases, the emitter voltage of the transistor also increases. Eventually, the emitter voltage will reach the firing point, and the emitter diode begins to conduct. At this time, the resistance of the UJT suddenly decreases. This quickly discharges the capacitor. With the capacitor discharged, the UJT switches back to its high resistance state, and the

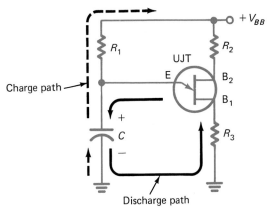

Fig. 11-21 A UJT relaxation oscillator.

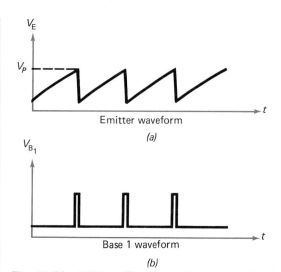

(a)

(b)

Fig. 11-22 UJT oscillator waveforms. (*a*) Sawtooth and (*b*) pulse.

Sawtooth wave

Pulse waveform

Intrinsic standoff ratio

Programmable unijunction transistor

Astable multivibrator

Free-running flip-flop

capacitor begins charging again. This cycle will repeat over and over.

Figure 11-22 shows two of the waveforms that can be expected from the UJT oscillator. Notice that a sawtooth wave appears at the emitter of the transistor. This wave shows the gradual increase of capacitor voltage. When the firing voltage V_P is reached, the capacitor is rapidly discharged. The discharge current flows through the base 1 resistor, causing a voltage drop. Therefore the base 1 waveform shows narrow voltage pulses that correspond to the falling edge of the sawtooth. The pulses are narrow since the capacitor is discharged rapidly.

In practice, either the sawtooth or the pulse or both waveforms may become the output of the circuit. Such circuits are very useful in timing and control applications. The frequency of oscillations may be roughly predicted by

$$f = \frac{1}{RC}$$

Assume that $R_1 = 10,000\ \Omega$ and $C = 10\ \mu\text{F}$ in Fig. 11-21. The approximate frequency of oscillations is given by

$$f = \frac{1}{10 \times 10^3 \times 10 \times 10^{-6}} = 10\ \text{Hz}$$

The UJT itself will also have an effect on the frequency of oscillation. The most important UJT parameter is the *intrinsic standoff ratio*. This ratio is a measure of how the transistor will internally divide the supply voltage and thus bias the emitter terminal. The

ratio will set the firing voltage V_P. Standard UJTs are found to have intrinsic standoff ratios from around 0.4 to 0.85. If the intrinsic standoff ratio is near 0.63, then the formula given previously will be accurate. If the ratio is near 0.4, the oscillator circuit discussed previously will actually run at 20 Hz.

The variation in intrinsic standoff ratio can be overcome with a device called a *programmable unijunction transistor*. In this device, the ratio is programmed with external resistors. This makes it possible to build a circuit and have it oscillate very near the design frequency. Figure 11-23 shows a programmable unijunction transistor oscillator. Resistor R_1 and Capacitor C will set the frequency of oscillation along with R_3 and R_4 which determine the intrinsic standoff ratio.

Figure 11-24 is another type of relaxation oscillator, the *astable multivibrator*. The circuit has no stable states. The circuit voltages switch constantly as it oscillates. This is in contrast to the *monostable* version that has one stable state and the bistable circuit with two stable states. The monostable and bistable circuits will not be discussed since this chapter is limited to oscillators.

Astable multivibrators are also called *free-running flip-flops*. This name is more descriptive of how the circuit behaves. Notice in Fig. 11-24 that two transistors are used. If Q_1 is on (conducting), Q_2 will be off. After a period of time, the circuit flips and Q_1 goes off while Q_2 comes on. After a second period, the circuit flops, turning on Q_1 again and turning off Q_2. The flip-flop action continues as long as the power is applied.

Study the waveforms shown in Fig. 11-25. They are for transistor Q_1 of Fig. 11-24.

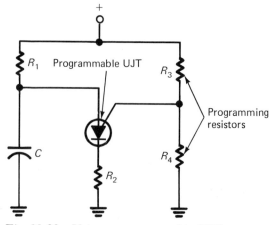

Fig. 11-23 Using a programmable UJT.

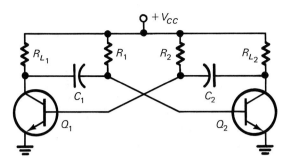

Fig. 11-24 The astable multivibrator.

gives us a way to estimate the time that each transistor will be held in the off state:

$$t = 0.69RC$$

Assume that R_1 and R_2 are both 47,000-Ω resistors and C_1 and C_2 are both 0.05-μF capacitors. Each transistor should be held off for

$$t = 0.69 \times 47 \times 10^3 \times 0.05 \times 10^{-6}$$
$$= 1.62 \times 10^{-3} \text{ s}$$

The period will be

$$T = (1.62 \times 10^{-3}) + (1.62 \times 10^{-3})$$
$$= 3.24 \times 10^{-3} \text{ s}$$

It will take 3.24 ms for the oscillator to produce one square wave. Now that the period is known, it will be easy to calculate the frequency of oscillation:

$$f = \frac{1}{T} = \frac{1}{3.24 \times 10^{-3}} = 309 \text{ Hz}$$

Square wave

With $R_1 = R_2$ and $C_1 = C_2$, the oscillator can be expected to give symmetrical square waves. Connecting an oscilloscope to either collector will show the positive-going part of the signal equal in time to the negative-going part.

What happens when the timing components are not equal? Assume in Fig. 11-24 that R_1 and R_2 are both 10,000 Ω, $C_1 = 0.01$ μF, and $C_2 = 0.1$ μF. What waveform can be expected at the collector of Q_2? Computing both time constants will answer the questions:

$$t_1 = 0.69 \times 10 \times 10^3 \times 0.1 \times 10^{-6}$$
$$= 0.69 \times 10^{-3} \text{ s}$$
$$t_2 = 0.69 \times 10 \times 10^3 \times 0.01 \times 10^{-6}$$
$$= 0.069 \times 10^{-3} \text{ s}$$

Transistor Q_1 will be held in the off mode 10 times longer than Q_2. Figure 11-26 shows the expected collector waveform for Q_1. Such a circuit is *nonsymmetrical*.

What is the frequency of the nonsymmetrical square wave? First, the period must be determined:

$$T = (0.69 \times 10^{-3}) + (0.069 \times 10^{-3})$$
$$= 0.759 \times 10^{-3} \text{ s}$$

Transistor Q_2's waveforms will look the same, but they will be inverted. Suppose that Q_2 has just turned on, making its collector less positive. This means the collector of Q_2 is going in a negative direction. This negative signal is coupled by C_2 to the base of Q_1. This turns off Q_1. Capacitor C_2 will hold off Q_1 until R_2 can allow the capacitor to charge sufficiently positive to allow Q_1 to come on. The circuit works on RC time constants. Transistor Q_1 is being held in the "off" state by the time constant of R_2 and C_2.

As Q_1 is turning on, its collector will be going less positive. This negative-going signal is coupled by C_1 to the base of Q_2, and Q_2 is turned off. It will stay off for a period determined by the RC time constant of R_1 and C_1.

Again, refer to Fig. 11-25. One square wave will be produced during one period. The period has two parts; thus it is equal to

$$T = t_1 + t_2$$

It takes 0.69 time constants for the RC network to reach the base turn-on voltage. This

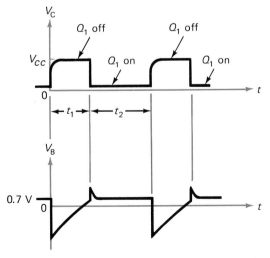

Fig. 11-25 Multivibrator waveforms.

185

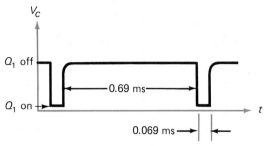

Fig. 11-26 Waveform for a nonsymmetrical multivibrator.

The frequency will be given by

$$f = \frac{1}{0.759 \times 10^{-3}} = 1318 \text{ Hz}$$

Review Questions

Choose the letter that best answers each question.

28. Refer to Fig. 11-21. What waveform should appear across the capacitor?
    A. Sawtooth
    B. Pulse
    C. Sinusoid
    D. Square

29. Refer to Fig. 11-21. What waveform should appear across R_3?
    A. Sawtooth
    B. Pulse
    C. Sinusoid
    D. Square

30. Refer to Fig. 11-21 where $R_1 = 10,000 \text{ }\Omega$ and $C = 0.05 \text{ }\mu\text{F}$. What is the approximate frequency of operation?
    A. 200 Hz
    B. 1000 Hz
    C. 2000 Hz
    D. 20 kHz

31. Refer to Fig. 11-23. What is the purpose of resistors R_3 and R_4?
    A. To set the desired intrinsic standoff ratio
    B. To set the exact frequency of oscillation
    C. Both of the above
    D. None of the above

32. Refer to Fig. 11-24. What waveform can be expected at the collector of Q_1?
    A. Sawtooth
    B. Pulse
    C. Sinusoid
    D. Square

33. Refer to Fig. 11-24. What waveform can be expected at the collector of Q_2?
    A. Sawtooth
    B. Pulse
    C. Sinusoid
    D. Square

34. Refer to Fig. 11-24. What is the phase relationship of the signal at the collector of Q_2 to the signal at the collector of Q_1?
    A. 0°
    B. 90°
    C. 180°
    D. 360°

35. Refer to Fig. 11-24 where $C_1 = C_2 = 1 \text{ }\mu\text{F}$ and $R_1 = R_2 = 22,000 \text{ }\Omega$. What is the frequency of oscillation?
    A. 16 Hz
    B. 33 Hz
    C. 66 Hz
    D. 99 Hz

11-6 UNDESIRED OSCILLATIONS

It was mentioned earlier that a public address system can oscillate if the gain is too high. Such oscillations are undesired. Many kinds of amplifiers may become oscillators. This is a serious problem. Amplifiers are not supposed to oscillate. Now that you have studied oscillators, it will be easier to understand how amplifiers can oscillate and what can be done to prevent it.

Negative feedback often is used in amplifiers to decrease distortion and improve frequency response. A three-stage amplifier is shown in simplified form in Fig. 11-27. Each stage uses the common-emitter configuration, and each will produce a 180° phase shift. This makes the feedback from stage 3 to stage 1 negative. Positive feedback is required for oscillation; therefore the amplifier should be stable. But, at very high or at very low frequencies, the feedback can become positive.

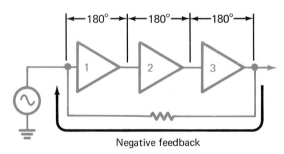

Fig. 11-27 A three-stage amplifier with negative feedback.

The *RC* coupling networks within the stages act as lead-lag networks. Each *RC* network causes some individual phase shift, and the total phase shift can reach 180°.

An amplifier, such as the one shown in Fig. 11-27, can become an oscillator at the frequency where the internal phase shifts sum to 180°. If amplifier gain is high enough at that frequency, it will oscillate. Such an amplifier is unstable and useless. *Frequency compensation* can be used to make such an amplifier stable. A compensated amplifier has networks added which decrease gain at the frequency extremes. Thus, by the time the frequency is reached where the phase shifts total 180°, the gain is too low for oscillations to occur. A good example of this technique is modern operational amplifiers. They are internally compensated for gain reductions of 20 dB per decade. At the higher frequencies where feedback might become positive, the gain is too low and stability is certain.

Another way that amplifiers can become unstable is that feedback paths can occur which do not show on the schematic diagram. For example, a good power supply is expected to have a very low internal impedance. This will make it very difficult for ac signals to appear across it. However, a power supply might have a high impedance. This could be caused by a defective filter. An old battery supply may develop a high internal impedance because it is drying out. The impedance of the power supply can provide a common load where signals are developed.

In the simplified three-stage amplifier of Fig. 11-28, Z_p represents the internal imped-

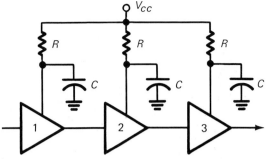

Fig. 11-29 Preventing supply feedback.

ance of the power supply. Suppose that stage 3 is drawing varying amounts of current because it is amplifying an ac signal. The varying current will produce a signal across Z_P. This signal will obviously affect stage 1 and stage 2. It is a form of feedback, and the amplifier may become unstable.

Figure 11-29 shows a solution for the feedback problem. An *RC* network has been added in the power-supply lines to each amplifier. These networks act as low-pass filters. The capacitors are chosen to have a low reactance at the signal frequency. This effectively shorts any ac signal appearing on the supply lines to ground. Sometimes the capacitors are used alone. They are called bypass capacitors. They can be very effective in preventing unwanted feedback paths.

Ground impedances can also produce feedback paths that do not appear on schematics. Heavy currents flowing through printed circuit foils or the metal chassis can cause voltage drops. The voltage drop from one amplifier may be fed back to another amplifier. Refer to Fig. 11-28. The impedance of the ground path is Z_G. As before, signal currents from stage 3 could produce a voltage across Z_G that will be fed back to the other stages. Ground currents cannot be eliminated, but proper layout can prevent them from producing feedback. The idea is to prevent later stages from sharing ground paths with earlier stages.

High-frequency amplifiers such as those used in radio receivers and transmitters can be very troublesome. These circuits can be coupled by stray capacitive and magnetic paths. When such circuits can see each other in the electrical sense, oscillations are likely to occur. These circuits must be shielded. Metal partitions and covers are used to keep the circuits isolated and prevent feedback.

Another feedback path often found in

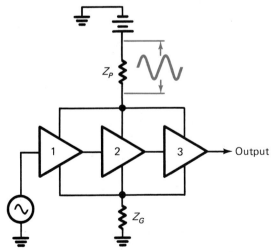

Fig. 11-28 The effect of supply and ground impedances.

187

Neutralization

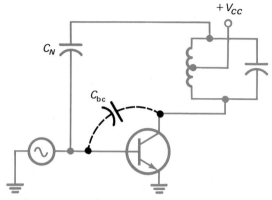

Fig. 11-30 Feedback inside the transistor.

high-frequency amplifiers lies within the transistor itself. This path can also produce oscillations and make the amplifier useless. In Fig. 11-30 C_{bc} represents the capacitance from the collector to the base of the transistor in a tuned high-frequency amplifier. This capacitance will feed some signal back. The feedback can become positive at a frequency where enough internal phase shift is produced.

Nothing can be done about the feedback inside a transistor. However, it is possible to create a second path external to the transistor. If the phase of the external feedback is correct, it can cancel the internal feedback. This is called *neutralization*. Figure 11-30 is a neutralized amplifier. Capacitor C_N feeds back from the collector circuit to the base of the transistor. The phase of the signal fed

back by C_N is opposite to the phase fed back by C_{bc}. This stabilizes the amplifier. Notice that the phase reversal is produced across the tuned circuit. Another possibility is to use a separate winding coupled to the tuned circuit.

Figure 11-31 is an actual radio-frequency amplifier used in an FM tuner. You will note that several of the techniques discussed in this section have been employed. Remember, the objective is to make the amplifier as stable as possible.

Review Questions

Choose the letter that best answers each question.

36. Examine Fig. 11-27. Assume at some frequency extreme that the actual phase shift in each stage is 240°. What happens to the feedback at that frequency?
    A. It does not exist.
    B. It becomes positive.
    C. It decreases the gain for that frequency.
    D. None of the above.

37. Refer to question 36. Assume that the amplifier has more gain at that frequency than it has loss in the feedback path. What happens to the amplifier?
    A. It burns out.
    B. It shorts out the signal source.
    C. It becomes unstable (oscillates).
    D. It can no longer deliver an output signal.

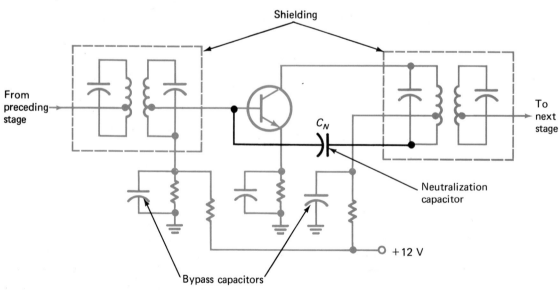

Fig. 11-31 A neutralized RF amplifier.

38. Why are most operational amplifiers internally compensated for gain reductions of 20 dB per decade?
    A. To prevent them from becoming unstable
    B. To prevent any phase error at any frequency
    C. To give them a pleasant tone
    D. To increase their gain at high frequencies

39. Refer to Fig. 11-28 where $Z_p = 10 \, \Omega$ and stage 3 is taking a current from the supply that fluctuates 100 mA peak-to-peak. What signal is developed across Z_p?
    A. 100 mV peak-to-peak
    B. 1 V peak-to-peak
    C. 10 V peak-to-peak
    D. None of the above

40. Refer to Fig. 11-28. Stage 3 draws current from the supply, and a signal is produced across Z_p. This signal
    A. Is delivered to the output
    B. Is canceled in stage 1
    C. Is dissipated in Z_G
    D. Becomes feedback to stage 1 and stage 2

41. Refer to Fig. 11-29. The *RC* networks shown are often called *decoupling* networks. This is because they
    A. Prevent unwanted signal coupling
    B. Bypass any dc to ground
    C. Act as high-pass filters
    D. Disconnect each stage from V_{CC}

42. Refer to Fig. 11-30. The function of C_N is to
    A. Bypass the base of the transistor
    B. Filter V_{CC}
    C. Tune the tank circuit
    D. Cancel the effect of C_{bc}

43. Refer to Fig. 11-31. How many techniques are shown for ensuring the stability of the amplifier?
    A. One
    B. Two
    C. Three
    D. Four

11-7 OSCILLATOR TROUBLESHOOTING

Oscillator troubleshooting uses the same skills as amplifier troubleshooting. Since most oscillators are amplifiers with positive feedback added, many of the faults are the same. When troubleshooting any electronic circuit, remember the word "GOAL." Good troubleshooting involves:

1. Observing the symptoms
2. Analyzing the possible causes
3. Limiting the possibilities

It is possible to observe the following symptoms when troubleshooting oscillators:

1. No output
2. Reduced amplitude
3. Unstable frequency
4. Frequency error

It is also possible that two symptoms may be observed at the same time. For example, an oscillator circuit may show reduced amplitude and frequency error.

Certain instruments are very useful for proper symptom identification. A frequency meter (especially a digital type) is very valuable. An oscilloscope is also a good instrument for oscillator troubleshooting. As always, a voltmeter is needed for supply and bias voltage checks. When using instruments in and around oscillator circuits, always remember this: oscillators can be subject to loading effects. More than one technician has been misled because the instrument pulled the oscillator off frequency or reduced the amplitude. It is even possible for the instrument to prevent the circuit from oscillating at all.

Loading effects can be reduced by using high-impedance instruments. It is also possible to reduce loading effects by taking readings at the proper point. If an oscillator is followed by a buffer stage, readings should be taken here. The buffer will help prevent the instrument from loading the oscillator.

Do not forget to check the effect of any and all controls when troubleshooting. If the circuit is a VFO, it is a good idea to tune it over its entire range. You may find that the trouble appears and disappears as the oscillator is tuned. Variable capacitors can short over a portion of their range.

The power supply can have several effects on oscillator performance. Frequency and amplitude are both sensitive to the supply voltage. It is worth knowing if the supply is correct and if it is stable. Power-supply checks should be made early in the troubleshooting process. They are easy to make and can save a lot of time.

It is important to review the theory of the circuit when troubleshooting. This will help you analyze possible causes. Determine what

Table 11-1 Troubleshooting oscillators

Problem	Possible Cause
No output	Supply voltage. Defective transistor. Shorted component (check tuning capacitor in VFO). Open component. Severe load (check buffer amplifier). Defective crystal. Defective joint (check printed circuit board).
Reduced amplitude	Supply voltage low. Transistor bias (check resistors). Circuit loaded down (check buffer amplifier). Defective transistor.
Frequency unstable	Supply voltage changes. Defective connection (vibration test). Temperature sensitive (check with heat and/or cold spray). Tank circuit fault. Defect in *RC* network. Defective crystal. Load change (check buffer amplifier). Defective transistor.
Frequency error	Wrong supply voltage. Loading error (check buffer amplifier). Tank circuit fault (check trimmers and/or variable inductors). Defect in *RC* network. Defective crystal. Transistor bias (check resistors).

controls the operating frequency. Is it a lead-lag network, an *RC* network, a tank circuit, or a crystal? Remember that loading effects can pull an oscillator off frequency. The problem could be in the next stage which is fed by the oscillator circuit.

Unstable oscillators can be quite a challenge. Technicians often resort to tapping components and circuit boards with an insulated tool to localize the difficulty. If this fails, they can use heat and cold to isolate a sensitive component. Desoldering pencils make excellent heat sources. A squeeze on the bulb will direct a stream of hot air just where it is needed. Chemical "cool sprays" are available for selective cooling of components.

Table 11-1 is a summary of cause and effects to help you troubleshoot oscillators.

Review Questions

Choose the letter that best answers each question.

44. What can loading do to an oscillator?
    A. Cause a frequency error
    B. Reduce the amplitude of the output
    C. Kill the oscillations completely
    D. All the above

45. An astable multivibrator is a little off frequency. Which of the following is least likely to be the cause?
    A. The supply voltage is wrong.
    B. A resistor has changed value.
    C. A capacitor has changed value.
    D. The transistors are defective.

46. A technician notes that a tool or a finger brought near a high-frequency oscillator circuit causes the output frequency to change.
    A. This is to be expected.
    B. It is a sign that the power supply is unstable.
    C. The tank circuit is defective.
    D. There is a bad transistor in the circuit.

47. A technician replaces the UJT in a relaxation oscillator. The circuit works fine, but the frequency is off a little. What is wrong?
    A. The new transistor is defective.
    B. The intrinsic standoff ratio is different.
    C. The resistors are burned out.
    D. The new transistor is in wrong.

Summary

1. Oscillators convert dc to ac.
2. Sinusoidal oscillators are amplifiers that produce their own input signal.
3. The gain of the amplifier must be greater than the loss in the feedback circuit to produce oscillation.

4. The feedback must be in phase to produce oscillation.
5. It is possible to choose the frequency of an oscillator by using the appropriate *RC* network.
6. The resonant frequency of a lead-lag net-

work produces maximum output voltage and a 0° phase angle.

7. The Wien-bridge oscillator uses a lead-lag network for frequency control.

8. It is possible to make the lead-lag network tunable by using variable capacitors or variable resistors.

9. Phase-shift oscillators use three RC networks, each giving a 60° phase angle.

10. An LC tank circuit can be used in very-high-frequency oscillator circuits.

11. The combination of an amplifier and a tank circuit will give constant-amplitude sinusoidal oscillations.

12. A Hartley oscillator uses a tapped inductor in the tank circuit.

13. The Colpitts oscillator uses a tapped capacitive leg in the tank circuit.

14. A buffer amplifier will improve the frequency stability of an oscillator.

15. The series tuned Colpitts, or Clapp, circuit is noted for good frequency stability.

16. A quartz crystal can be used to control the frequency of an oscillator.

17. Crystal oscillators are more stable in frequency than LC oscillators.

18. Crystals can operate in a series mode or a parallel mode.

19. The parallel frequency of a crystal is a little above the series frequency.

20. Crystals have a very high Q.

21. Relaxation oscillators produce nonsinusoidal outputs.

22. Relaxation oscillators can be based on negative-resistance devices such as the UJT.

23. Relaxation oscillator frequency can be predicted by RC time constants.

24. The intrinsic standoff ratio of a UJT will affect the frequency of oscillation.

25. The intrinsic standoff ratio of a programmable UJT can be set by the use of external resistors.

26. The astable multivibrator produces square waves.

27. A nonsymmetrical multivibrator is produced by using different RC time constants for each base circuit.

28. Feedback amplifiers use frequency compensation to achieve stability.

29. Feedback signals can develop across the internal impedance of the power supply.

30. An RC network or a bypass capacitor is used to prevent feedback on supply lines.

31. High-frequency circuits often must be shielded to prevent feedback.

32. Oscillator symptoms include no output, reduced amplitude, instability, and frequency error.

33. Test instruments can load an oscillator circuit and cause errors.

34. Unstable circuits can be checked with vibration, heat, or cold.

Chapter Review Questions

Choose the letter that best answers each question.

11-1. An amplifier will oscillate if
(A) There is feedback from output to input (B) The feedback is in phase (positive) (C) The gain is greater than the loss (D) All the above are true

11-2. It is desired to build a common-emitter oscillator that operates at frequency f. The feedback circuit will be required to provide
(A) 180° phase shift at f (B) 0° phase shift at f (C) Severe loss at f (D) Band-stop action for f

11-3. In Fig. 11-4, $R = 3300\ \Omega$ and $C = 1\ \mu F$. What is f_r?
(A) 48 Hz (B) 120 Hz (C) 383 Hz (D) 914 Hz

11-4. Examine Fig. 11-4. Assume the signal source develops a frequency above f_r. What is the phase relationship of V_{out} to the source?
(A) Positive (leading) (B) Negative (lagging) (C) In phase (0°) (D) None of the above

11-5. In Fig. 11-6, $R = 8200\ \Omega$ and $C = 0.1\ \mu F$. What is the frequency of oscillation?
(A) 39 Hz (B) 60 Hz (C) 194 Hz (D) 382 Hz

11-6. Refer to Fig. 11-6. What is the function of R'?
(A) It provides the required phase shift (B) It prevents clipping and distortion (C) It controls the frequency of oscillation (D) None of the above

11-7. In Fig. 11-9, $R_B = 1$ MΩ and the voltage gain of the amplifier is 120. What is the actual loading effect of R_B to a signal arriving at the base?
(A) 0 Ω (B) 8333 Ω (C) 1 MΩ (D) Infinite

11-8. Refer to Fig. 11-10. Assume the phase-shift capacitors are changed to 0.1 μF. What is the frequency of oscillation?
(A) 10 Hz (B) 40 Hz (C) 75 Hz (D) 100 Hz

11-9. Refer to Fig. 11-10. How many frequencies will produce an exact 60° phase shift in each RC section of the feedback network?
(A) One (B) Two (C) Three (D) An infinite number

11-10. Refer to Fig. 11-14. What is the major effect of C_E?
(A) It increases the frequency of oscillation (B) It decreases the frequency of oscillation (C) It makes the transistor operate common base (D) It filters V_{out}

11-11. Refer to Fig. 11-14. What would happen if C_2 were increased in capacity?
(A) The frequency of oscillation would increase (B) The frequency of oscillation would decrease (C) The inductance of L_A and L_B would change (D) Not possible to determine

11-12. In Fig. 11-15. $L = 0.5$ μH, $C_2 = 330$ pF, and $C_3 = 33$ pF. What is the frequency of oscillation?
(A) 11 MHz (B) 22 MHz (C) 33 MHz (D) 41 MHz

11-13. Refer to Fig. 11-15. What is the purpose of C_1?
(A) It bypasses power-supply noise to ground (B) It determines the frequency of oscillation (C) It makes the amplifier operate common-base (D) It filters V_{out}

11-14. Refer to Fig. 11-16. What is the configuration of Q_1?
(A) Common source (B) Common gate (C) Common drain (D) Drain follower

11-15. Crystal-controlled oscillators, as compared to LC-controlled oscillators, are generally
(A) Less expensive (B) Capable of a better output power (C) Superior for VFO designs (D) Superior for frequency stability

11-16. Why can the circuit of Fig. 11-19 not be used for overtone operation?
(A) The common-emitter configuration is used (B) Trimmer C_3 makes it impossible (C) The feedback is wrong (D) There is no LC circuit to select the overtone

11-17. The Q of a crystal, as compared to the Q of an LC tuned circuit, will be
(A) Much higher (B) About the same (C) Lower (D) Impossible to determine

11-18. In Fig. 11-21, $R_1 = 470,000$ Ω and $C = 10$ μF. What is the frequency of oscillation?
(A) 0.21 Hz (B) 9.45 Hz (C) 200 Hz (D) 382 Hz

11-19. In Fig. 11-24, $R_1 = R_2 = 10,000$ Ω, $C_1 = 0.5$ μF, and $C_2 = 0.05$ μF. What is the frequency of oscillation?
(A) 112 Hz (B) 264 Hz (C) 312 Hz (D) 989 Hz

11-20. In question 11-19, what will the square wave show?
(A) Symmetry (B) Nonsymmetry (C) Poor rise time (D) None of the above

11-21. Refer to Fig. 11-27. How may the stability of such a circuit be ensured?
(A) Operate each stage at maximum gain (B) Decrease losses in the feedback circuit (C) Use more stages (D) Compensate the circuit so the gain is low for those frequencies that give a critical phase error

11-22. Refer to Fig. 11-28. How may signal coupling across Z_G be reduced?
(A) Do not allow stages to share a ground path (B) By careful circuit layout (C) By using low-loss grounds (D) All the above

11-23. What is the purpose of neutralization?
(A) To ensure oscillations (B) To stabilize an amplifier (C) To decrease amplifier output (D) To prevent amplifier overload

Answers to Review Questions

1. *D*	17. *C*	33. *D*
2. *C*	18. *A*	34. *C*
3. *D*	19. *C*	35. *B*
4. *C*	20. *A*	36. *B*
5. *A*	21. *B*	37. *C*
6. *B*	22. *B*	38. *A*
7. *A*	23. *C*	39. *B*
8. *C*	24. *C*	40. *D*
9. *B*	25. *B*	41. *A*
10. *C*	26. *C*	42. *D*
11. *D*	27. *D*	43. *C*
12. *A*	28. *A*	44. *D*
13. *D*	29. *B*	45. *D*
14. *D*	30. *C*	46. *A*
15. *B*	31. *C*	47. *B*
16. *B*	32. *D*	

Radio Receivers

- Communications represents a large part of the electronics industry. This chapter introduces the basic ideas used in electronic communication. Once these basics are understood, it is easier to move into special areas such as television and two-way radio.

  Modulation is the fundamental process of electronic communication. It allows voice, pictures, and other information to be transferred from one point to another at the speed of light. The modulation process is reversed at the receiver. This chapter covers the circuits and the theory used in various types of receivers.

12-1 MODULATION AND DEMODULATION

Any high-frequency oscillator can be used to produce a radio wave. Figure 12-1 shows an oscillator that feeds its output energy to an antenna. The antenna converts the high-frequency ac to a radio wave. Such a radio wave is not useful for sending information to some distant point. Some form of *modulation* is required to place information on the signal.

A radio wave can be modulated by using a key switch to interrupt the antenna current (Fig. 12-2). This is the basic scheme of radio telegraphy. A code, such as the Morse code, can be used to form patterns that represent numbers, letters, and punctuation. This basic modulation form is known as interrupted

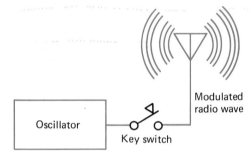

Fig. 12-2   A CW transmitter.

continuous wave, or CW. Continuous wave modulation is very simple, but it has many disadvantages. It is difficult to learn a code, such as the Morse code, and it cannot be used for music, pictures, and other kinds of information.

The block diagram of Fig. 12-3 shows *amplitude modulation*, or AM. In this modulation system, the intelligence or information is used to control the amplitude of the high-frequency signal. Amplitude modulation overcomes the disadvantages of CW modulation. It can be used to transmit voice, music, data, or even picture information (video). In amplitude modulation, the information controls the amplitude of the high-frequency oscillator signal.

In the amplitude modulator of Fig. 12-4 the audio is coupled by T_1 to the collector circuit

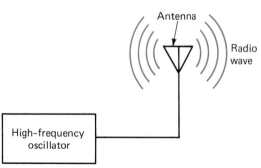

Fig. 12-1   A basic radio transmitter.

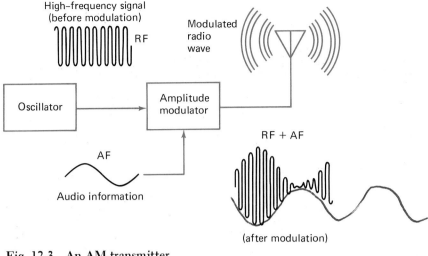

Fig. 12-3   An AM transmitter.

of the transistor. The audio voltage induced across the secondary of T_1 will aid V_{CC} or oppose V_{CC} depending on the phase at the time. This means that the collector supply for the transistor is not constant. It is varying with the audio input. This is how the amplitude control is achieved.

Transformer T_2 and capacitor C_2 in Fig. 12-4 form a resonant tank circuit. The resonant frequency will match the RF input. Capacitor C_1 and resistor R_1 form the input circuit for the transistor. Reverse bias is developed by the base-emitter junction, and the amplifier operates in class C.

Amplitude modulation does more than produce a change in the amplitude of the oscillator signal. In addition, new frequencies are created. Suppose a 500-kHz oscillator is amplitude-modulated by a 3-kHz audio tone. It can be shown that three frequencies are present in the output of the modulator. The original RF oscillator signal is called the *carrier*, and it is shown at 500 kHz on the frequency axis in Fig. 12-5. Also note that an upper sideband appears at 503 kHz and a lower sideband appears at 497 kHz. Thus, a total of three frequencies is present in the output.

Figure 12-5 is the type of display shown on a *spectrum analyzer*. Ordinary oscilloscopes will display AM as shown in Fig. 12-6. This is because they run in the *time domain*. The spectrum analyzer uses the *frequency domain*. Both displays have their uses in electronics.

Thus we find the AM signal to contain sidebands. This also means the AM signal has bandwidth. It will occupy so many kilohertz of the spectrum available. The sidebands appear above and below the carrier according to the frequency of the modulating information. If you whistle into the microphone at 1 kHz,

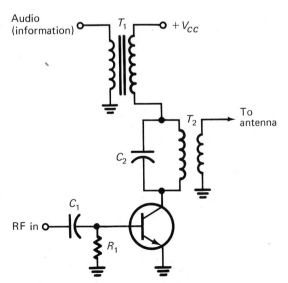

Fig. 12-4   An amplitude modulator.

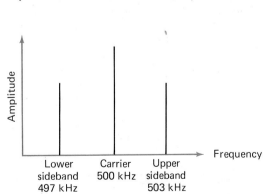

Fig. 12-5   Amplitude modulation produces sidebands.

195

Demodulation

Detection

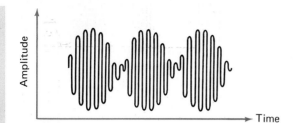

Fig. 12-6 An AM waveform as it appears on an oscilloscope.

the upper sideband will appear 1 kHz above the carrier frequency. The lower sideband will appear 1 kHz below the carrier frequency. We may conclude that the bandwidth of an AM signal is twice the modulating frequency. Your 1 kHz whistle produced a total signal width of 2 kHz.

Bandwidth is important because it limits the number of stations that can use a band of frequencies without interference. The AM broadcast band ranges from 540 to 1600 kHz. If each station modulates with audio frequencies up to 7.5 kHz, what is the maximum number of stations that can use the band? First, we must determine how much bandwidth is needed for each station:

Bandwidth = 2 × highest audio frequency
= 2 × 7.5 kHz = 15 kHz

Next, we must determine the total bandwidth available:

Total bandwidth = 1600 − 540 = 1060 kHz

Finally, we can divide and find that

No. of possible AM stations in one area

$$= \frac{\text{total bandwidth}}{\text{bandwidth}} = \frac{1060}{15} = 70$$

The above calculation is too simplified. It is based on the assumption of perfect receivers. Actually, it is not easy to separate stations so close in frequency. For this reason, you will not find 70 stations operating in any given area. The frequencies are assigned by geographic area to make it easier for the receiver circuits to select just one station at a time.

While modulation occurs in the transmitter, *demodulation* occurs in the receiver. It is the process of recovering the original information placed on the carrier wave. Demodulation is usually called *detection*.

The most common AM detector is a diode (Fig. 12-7). The modulated signal is applied across the primary of T_1. Transformer T_1 is tuned by capacitor C_1 to the carrier frequency. The passband of the tuned circuit is wide enough to pass the carrier and both sidebands. The diode detects the signal and recovers the original information used to modulate the carrier at the transmitter. Capacitor C_2 is a low-pass filter. It removes the carrier and sideband frequencies since they are no longer needed. Resistor R_L serves as the load for the information signal.

The diode makes a good detector because it is a nonlinear device. All nonlinear devices can be used to detect AM. Figure 12-8 is a volt-ampere characteristic curve of a solid-state diode. It shows that a diode will make a good detector and a resistor will not.

A nonlinear device will take two or more frequencies and produce sum and difference frequencies. If, for example, a 500-kHz signal and a 503-kHz signal both arrive at a nonlinear device, several new frequencies will be generated. For our purposes, the most important of these new frequencies is the difference frequency. That difference frequency will be 3 kHz. Now, refer to Fig. 12-5.

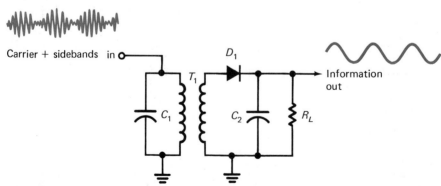

Carrier + sidebands in

Information out

Fig. 12-7 An AM detector.

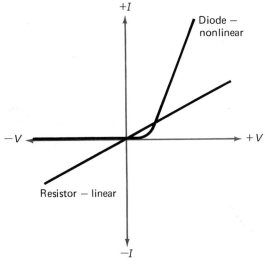

Fig. 12-8 Diodes are nonlinear devices.

pass the modulated signal (carrier plus sidebands). Capacitor C_4 is added to give a low-pass filter action since the high-frequency carrier and the sidebands are no longer needed after detection.

Transistors are also nonlinear devices. The base-emitter junction is a diode. The transistor detector has the advantage of producing gain. This means that the circuit of Fig. 12-9 will produce more information amplitude than the simple diode detector of Fig. 12-7. Both circuits are useful for detecting AM signals.

Review Questions

Choose the letter that best completes each statement.

1. A circuit used to place information on a radio signal is called
   A. An oscillator
   B. A detector
   C. An antenna
   D. A modulator

2. A CW transmitter sends information by
   A. Varying the amplitude of the audio signal
   B. Interrupting the radio signal
   C. Use of a microphone
   D. Use of a camera

3. Refer to Fig. 12-4. The voltage at the top of C_2 will
   A. Always be equal to V_{CC}
   B. Vary with the information signal
   C. Be controlled by the transistor
   D. Be 0 V with respect to ground

You will find that modulating a 500-kHz signal with a 3-kHz signal produces an upper sideband at 503 kHz. You can see that detection reverses this process and reproduces the 3-kHz signal.

What about the lower sideband shown in Fig. 12-5? This also arrives at the nonlinear device. It, too, will produce a difference frequency of 3 kHz because of the carrier. The two 3-kHz signals add in phase in the detector. Thus, both sidebands mix with the carrier and reproduce the original information frequencies.

A transistor can also serve as an AM detector (Fig. 12-9). It is actually just a common-emitter amplifier. Transformer T_1 and capacitor C_1 form a resonant circuit to

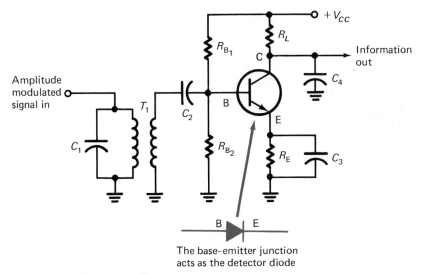

Fig. 12-9 A transistor detector.

Electronics:
Principles and
Applications
CHAPTER 12

Sensitivity

Selectivity

4. Refer to Fig. 12-4. Capacitor C_2 will reso-
nate the primary of T_2
A. At the radio frequency
B. At the audio frequency
C. at 0 Hz (dc)
D. None of the above

5. A 2-MHz radio signal is amplitude-
modulated by a 4-kHz sine wave. The fre-
quency of the lower sideband is
A. 2.004 MHz
B. 2.000 MHz
C. 1.996 MHz
D. 1.992 MHz

6. A 27-MHz radio transmitter is to be
amplitude-modulated by voice frequencies
up to 3 kHz. The bandwidth required for
the signal is
A. 0 Hz
B. 3 kHz
C. 6 kHz
D. 27 MHz

7. The electronic instrument used to show
both the carrier and the sidebands of a
modulated signal in the frequency domain
is the
A. Spectrum analyzer
B. Oscilloscope
C. Digital counter
D. Frequency meter

8. Refer to Fig. 12-7. The carrier input is
1.5000 MHz, the upper sideband (USB)
input is 1.5025 MHz, and the lower sideband (LSB) input is 1.4975 MHz. The frequency of the detected output is
A. 1.5 MHz
B. 5.0 kHz
C. 2.5 kHz
D. 0.5 kHz

9. Diodes make good AM detectors because
A. They rectify the carrier
B. They rectify the upper sideband
C. They rectify the lower sideband
D. They are nonlinear and mix the carrier
with the sidebands

12-2 SIMPLE RECEIVERS

A radio receiver can be quite simple and still
give adequate reception. An antenna is necessary to intercept the radio signal and change
it back into an electric signal. The diode detector mixes the sidebands with the carrier
and produces the audio information. The
headphones convert the audio signal into
sound. The ground completes the circuit and
allows the currents to flow. This system is
shown in Fig. 12-10.

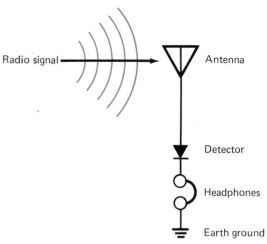

Fig. 12-10   A simple radio receiver.

Obviously, a receiver as simple as the one
shown in Fig. 12-10 must have shortcomings.
Such receivers do, indeed, work but not at the
practical level. They cannot receive weak
signals (poor *sensitivity*). They cannot separate adjacent stations well (poor *selectivity*).
They are inconvenient because they need a
long antenna, an earth ground, and headphones.

Before we leave the simple circuit of Fig.
12-10, one thing should be mentioned. You
have, no doubt, become used to the idea that
electronic circuits require some sort of power
supply. This is still the case. A radio signal is
a wave of pure energy. Thus, the signal is the
source of energy for this simple circuit.

The problem of poor sensitivity can be overcome with gain. We can add some amplifiers
to the receiver to make weak signals detectable. Of course, the amplifiers will have to be
energized. A power supply, other than the
weak signal itself, will be required. As the
gain is increased, the need for a long antenna
is decreased. A very small antenna is not efficient, but the gain will tend to overcome this
deficiency.

Gain will also do away with the need for the
headphones. Audio amplification after the
detector can make it possible to drive a loudspeaker. This makes the receiver much more
convenient to use.

What about the poor selectivity? Radio stations operate at different frequencies in any
given location. This makes it possible to use
band-pass filters to select one out of the many
on the air. The resonant point of the filter
may be adjusted to agree with the desired station frequency.

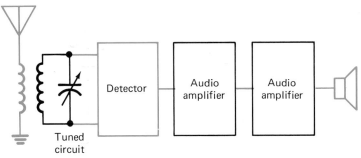

Fig. 12-11 An improved radio receiver.

Figure 12-11 shows a receiver that overcomes some of the problems of the basic receiver. A two-stage audio amplifier has been added to allow loudspeaker operation. A tuned circuit has been added to allow selection of one station at a time. This receiver will perform better.

The circuit of Fig. 12-11 is an improvement, but it is still not practical for most applications. One tuned circuit will not give enough selectivity. For example, if there is a very strong station in the area, it will not be possible to reject it. The strong station will be heard at all settings of the variable capacitor.

Selectivity can be improved by using more tuned circuits. Figure 12-12 compares the selectivity curves for one, two, and three tuned circuits. Note that more tuned circuits give a much sharper curve. This greatly improves the ability to reject unwanted signals. Figure 12-12 also shows that bandwidth is measured

3 dB down from the point of maximum gain. An AM receiver should have a bandwidth just wide enough to pass the carrier and both sidebands (about 15 kHz). Too much bandwidth means poor selectivity and possible interference. Too little bandwidth means loss of transmitted information (with high-frequency audio affected the most).

A tuned radio frequency (TRF) receiver can provide reasonably good selectivity and sensitivity (Fig. 12-13). Four amplifiers—two at radio frequencies and two at audio frequencies—give the required gain.

The TRF receiver has some disadvantages. Note in Fig. 12-13 that all three tuned circuits are gang-tuned. In practice, it is difficult to achieve perfect *tracking*. Tracking refers to how closely the resonant points will be matched for all settings of the tuning control. A second problem is in bandwidth. The tuned circuits will not have the same band-

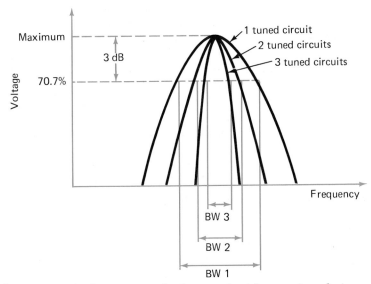

Fig. 12-12 Selectivity can be improved with more tuned circuits.

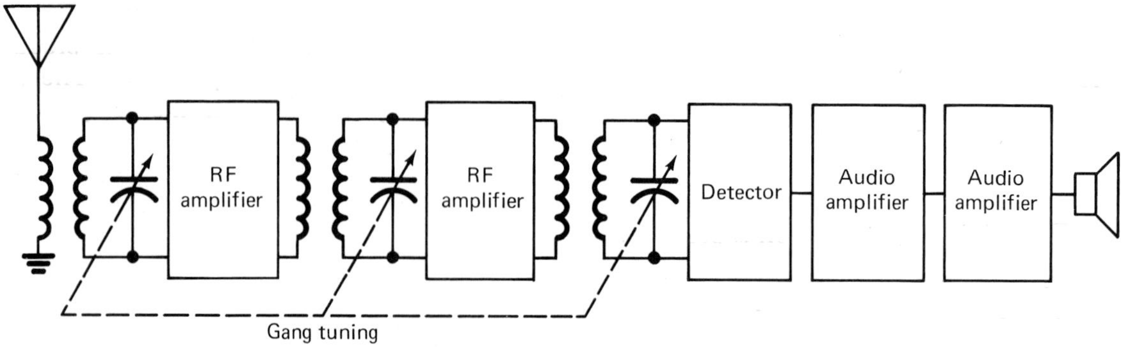

Intermediate
frequency (IF)

Heterodyne

Superheterodyne

Mixer

RF
amplifier

RF
amplifier

Detector

Audio
amplifier

Audio
amplifier

Gang tuning

Fig. 12-13   A TRF receiver.

width for all frequencies. Both of these disadvantages have been eliminated in the superheterodyne receiver design discussed in the next section.

Review Questions

Choose the letter that best answers each question.

10. Which of the following statements is *not* true concerning the radio circuit of Fig. 12-10?
    A. A very large antenna is usually required.
    B. There is no selectivity.
    C. The sensitivity is poor.
    D. No energy of any kind is required for operation.

11. To improve selectivity, the bandwidth of a receiver can be reduced by which of the following methods?
    A. Using more tuned circuits
    B. Using fewer tuned circuits
    C. Adding more gain
    D. Using a loudspeaker

12. A 250-kHz tuned circuit is supposed to have a bandwidth of 5 kHz. It is noted that the gain of the circuit drops 30 percent at 252.5 kHz and at 247.5 kHz. What can be concluded about the tuned circuit?
    A. It is not as selective as it should be.
    B. It is more selective than it should be.
    C. It is not working properly.
    D. None of the above.

13. Refer to Fig. 12-13. How is selectivity achieved in this receiver?
    A. In the detector stage
    B. In the gang-tuned circuits
    C. In the audio amplifier
    D. All the above

12-3 SUPERHETERODYNE RECEIVERS

The major difficulties with the TRF receiver design can be eliminated by fixing the tuned circuits to a single frequency. This will eliminate the tracking problem and the changing-bandwidth problem. We can call the fixed frequency the *intermediate frequency*, or simply IF. It must lie outside the band to be received. Now, any signal that is to be received can be converted to the intermediate frequency. "Heterodyne" is the word for the frequency conversion process. The *superheterodyne* receiver converts the received frequency to the intermediate frequency.

Figure 12-14 shows the basic operation of a heterodyne converter. When signals at two different frequencies are applied, new frequencies are produced. The output of the converter contains the sum and the difference signals in addition to the original signals. Any nonlinear device such as a diode can be used as a converter or mixer (different frequencies are mixed). Most superheterodyne circuits use a transistor rather than a diode. This is because the transistor can provide gain. In some cases it can also supply one of the two signals needed for mixing.

Mixing to produce new frequencies is probably the most fundamental concept in electronic communications. Unfortunately,

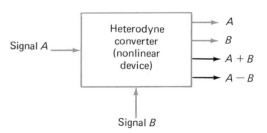

Fig. 12-14   Operation of a heterodyne converter.

many terms are used to mean the same thing. The following list will help you to reduce the confusion. Remember, the idea is the same in each case: a nonlinear device produces sum and difference frequencies.

1. Modulator (sidebands are produced)
2. Demodulator (information is produced)
3. Detector (information is produced)
4. Mixer (intermediate frequency is produced)
5. Converter (intermediate frequency is produced)

Figure 12-15 is a block diagram of a superheterodyne receiver. An oscillator provides a signal to mix with signals from the antenna. The mixer output contains sum and difference frequencies. If any of the frequencies present at the mixer output is passed by the IF stages, then that frequency will reach the detector. All other frequencies will be rejected because of the selectivity of the IF stages.

The standard intermediate frequencies are

1. AM broadcast band: 455 kHz (or 262 kHz for some automotive receivers)
2. FM broadcast band: 10.7 MHz
3. TV broadcast band: 44 MHz

Shortwave and communication receivers may use various intermediate frequencies such as 455 kHz, 1.6 MHz, 3.35 MHz, 9 MHz, 10.7 MHz, 40 MHz, and others.

The oscillator in the receiver is usually set to run above the received frequency by an amount equal to the IF. For example, to receive a station at 1020 kHz on a standard AM broadcast receiver,

Oscillator frequency = 1020 + 455
                     = 1475 kHz

The oscillator signal at 1475 kHz and the station signal at 1020 kHz will *mix* to produce sum and difference frequencies. The difference signal will be in the IF passband (those frequencies that the IF will allow to go through) and will reach the detector. Another station operating at 970 kHz can be rejected by this process. Its difference frequency will be

$$1475 - 970 = 505 \text{ kHz}$$

Since 505 kHz is out of the passband of the 455-kHz IF stages, the station is rejected.

It is clear that *adjacent* channels are rejected by the selectivity in the IF stages. However, there is a possibility of interference from a signal not even in the broadcast band. To receive 1020 kHz, the oscillator in the receiver must be adjusted 455 kHz higher. What will happen if a shortwave signal reaches the antenna at 1930 kHz? Remember, the oscillator is at 1475 kHz. Subtraction will show that

$$1930 - 1475 = 455 \text{ kHz}$$

This means that the shortwave signal at 1930 kHz will mix with the oscillator signal and reach the detector. This is called *image interference.*

The only way to reject image interference is to use selective circuits *before* the mixer. In any superheterodyne receiver, there are always two frequencies that can mix with the oscillator frequency and produce the intermediate frequency. One is the desired frequency, and the other is the image frequency. The image must not be allowed to reach the mixer.

Figure 12-16 shows how image rejection is achieved. The antenna signal is trans-

Sum and difference frequencies

Standard intermediate frequencies

Image interference

Image = carrier + LO

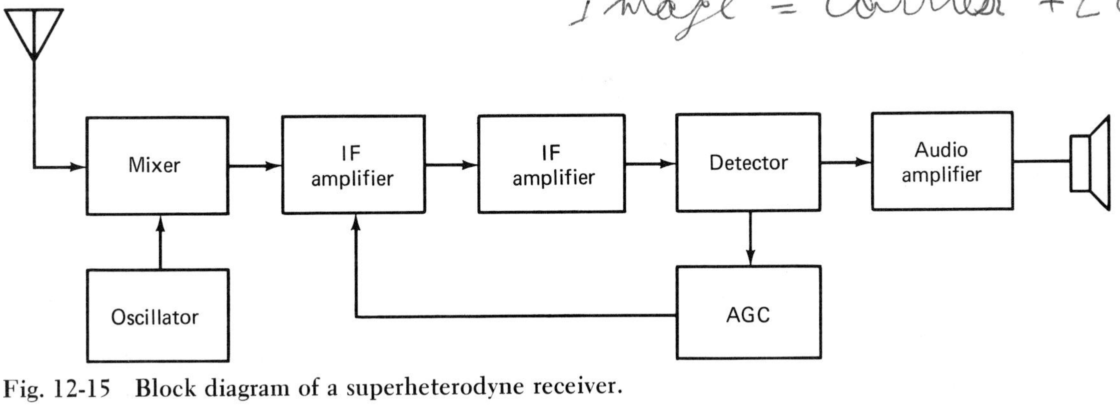

Fig. 12-15 Block diagram of a superheterodyne receiver.

Trimmer
capacitor

Automatic
volume control
(AVC)

Forward and
reverse AGC

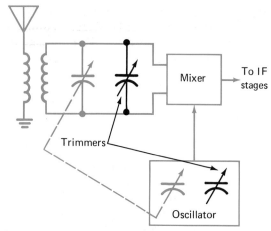

Fig. 12-16 A tuned circuit before the mixer rejects the image.

former-coupled to a tuned circuit before the mixer. This circuit is tuned to resonate at the station frequency. Its selectivity will reject the image. A dual or ganged capacitor is used to simultaneously adjust the oscillator and the mixer-tuned circuit. Trimmer capacitors are included so that the two circuits can track each other. These trimmers are adjusted only once. They are set at the factory and usually will never need to be readjusted.

The mixer-tuned circuit is not highly selective. It adds very little to the adjacent channel selectivity of the receiver. This is provided in the IF stages. The image is twice the intermediate frequency above the desired station frequency, or 910 kHz in a standard AM broadcast receiver. Such signals are easy to reject, and sharp selectivity is not required. This is good because a highly selective mixer would make the tracking critical.

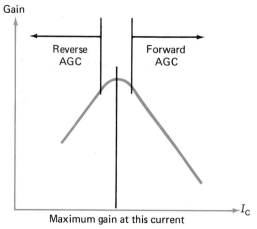

Fig. 12-17 The AGC characteristics of a transistor.

The block diagram of Fig. 12-15 shows an AGC stage. It may also be called automatic volume control (AVC). This stage develops a control voltage based on the strength of the signal reaching the detector. The control voltage, in turn, adjusts the gain of the first IF amplifier. This tends to automatically maintain a constant output from the receiver. Signal strengths can vary quite a bit as the receiver is tuned across the band. The AGC action keeps the volume from the speaker reasonably uniform.

Automatic gain control can be applied to more than one IF amplifier. It can also be applied to an RF amplifier before the mixer if the receiver has one. The control voltage is used to vary the gain of the amplifying device. If the device is a bipolar transistor, two options exist. The graph of Fig. 12-17 shows that maximum gain occurs at one value of collector current. If the bias is increased and current increases, the gain tends to drop. This is called *forward* AGC. The bias can be reduced, the current decreases, and so does the gain. This is called *reverse* AGC. Both types of AGC are used with bipolar transistors.

Different transistors vary quite a bit in their AGC characteristics. This is very important in replacing an RF or IF transistor in a receiver. If AGC is applied to that stage, an exact replacement is a must. A substitute transistor may cause loss of control and, in some cases, instability.

Dual-gate MOSFETs are often used when AGC is desired. These transistors have excellent AGC characteristics. The control voltage is usually applied to the second gate. You may wish to refer to Sec. 7-2 on field-effect transistor amplifiers.

Integrated circuits are also available which have excellent AGC characteristics. These are becoming more popular in receiver design.

Appendix C-1 contains a description of a superheterodyne receiver for the AM broadcast band. You should refer to this now and read the circuit description.

Review Questions

Choose the letter that best answers each question.

14. It is desired to receive a station at 1290 kHz on a standard AM receiver. What

must the frequency of the local oscillator be?
- A. 455 kHz
- B. 590 kHz
- C. 1745 kHz
- D. 2000 kHz

15. A standard AM receiver is tuned to 970 kHz. Interference is heard from a shortwave transmitter operating at 1880 kHz. What is the problem?
- A. Poor image rejection
- B. Poor AGC action
- C. Inadequate IF selectivity
- D. Poor sensitivity

16. Which of the following statements about the oscillator in a standard AM receiver is true?
- A. It is fixed at 455 kHz.
- B. It oscillates 455 kHz above the dial setting.
- C. It is controlled by the AGC circuit.
- D. It oscillates at the dial frequency.

17. Refer to Fig. 12-15. The receiver is properly tuned to a station at 1020 kHz that is transmitting a 1-kHz audio test signal. What frequency or frequencies are present at the input of the detector stage?
- A. 1 kHz
- B. 454, 455, and 456 kHz
- C. 1020 kHz
- D. 1020 and 1475 kHz

18. In question 17 what frequency or frequencies are present at the input of the audio amplifier?
- A. 1 kHz
- B. 454, 455, and 456 kHz
- C. 1020 kHz
- D. 1020 and 1475 kHz

19. It is noted that a receiver uses an NPN transistor in the first IF stage. Tuning a strong station causes the base voltage to become more positive. The stage
- A. Is defective
- B. Uses reverse AGC
- C. Uses forward AGC
- D. Is not AGC-controlled

12-4 FREQUENCY MODULATION AND SINGLE SIDEBAND

Frequency modulation is an alternative to amplitude modulation. Frequency modulation has some advantages that make it attractive for some commercial broadcasting and two-way radio work. One problem with AM is its sensitivity to noise. Lightning, automotive ignition, and sparking electric circuits all produce radio interference. This interference is spread over a wide frequency range. It is not easy to prevent such interference from reaching the detector in an AM receiver. An FM receiver can be made insensitive to noise interference. This noise-free performance is highly desirable.

Figure 12-18 shows how frequency modulation can be realized. Transistor Q_1 and its associated parts make up a series-tuned Colpitts oscillator. Capacitor C_3 and coil L_1 have the greatest effect in determining the frequency of oscillation. Diode D_1 is a varicap diode. It is connected in parallel with C_3. This means that as the capacitance of D_1 changes, so will the resonant frequency of the tank circuit. Resistors R_1 and R_2 form a voltage divider to bias the varicap diode. Some positive voltage

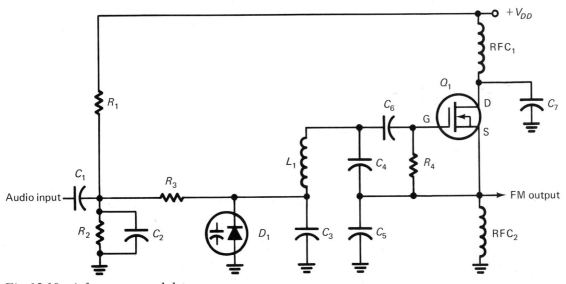

Fig. 12-18 A frequency modulator.

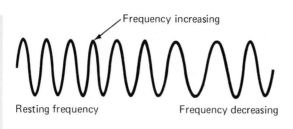

Frequency increasing

Resting frequency   Frequency decreasing

Modulating
waveform

Fig. 12-19   Frequency modulation waveforms

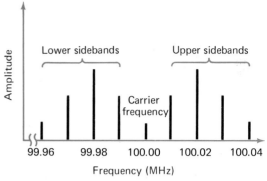

Lower sidebands   Upper sidebands

Carrier
frequency

99.96   99.98   100.00   100.02   100.04

Frequency (MHz)

Fig. 12-20   Frequency modulation produces
sidebands.

(a portion of V_{DD}) is applied to the cathode of D_1. Thus, D_1 is in reverse bias.

A varicap diode uses its depletion region as the dielectric. More reverse bias means a wider depletion region and less capacitance. Therefore, as an audio signal goes positive, D_1 should reduce its capacitance. This will shift the frequency of the oscillator up. A negative-going audio input will reduce the reverse bias across the diode. This will increase its capacitance and shift the oscillator to some lower frequency. The audio signal is modulating the frequency of the oscillator.

The relationship between the modulating waveform and the oscillator signal can be seen in Fig. 12-19. Note that the amplitude of the waveform is constant. Compare this with the AM waveform in Fig. 12-3.

Amplitude modulation produces sidebands, and so does FM (Fig. 12-20). Suppose a commercial FM station is being modulated with a steady 10-kHz (0.01-MHz) tone. This station is assigned an operating (carrier) frequency of 100 MHz. The frequency domain graph shows that several sidebands appear. These sidebands are spaced 10 kHz apart. They ap-

pear above and below the carrier frequency. This is one of the major differences between AM and FM. It should be obvious that an FM signal will require more bandwidth than an AM signal.

The block diagram for an FM superheterodyne receiver (Fig. 12-21) is quite similar to that for the AM receiver. However, you will notice that a *limiter* stage appears after the IF stage and before the detector stage. This is one way that an FM receiver can reject noise. Figure 12-22 shows what happens in a limiter stage. The input signal is very noisy. The output signal is noise-free. By limiting or by amplitude clipping all the noise spikes have been eliminated. Some FM receivers will use two stages of limiting to eliminate most noise interference.

Limiting cannot be used in an AM receiver. The amplitude variations carry the information to the detector. In an FM receiver, the frequency variations contain the information. Amplitude clipping in an FM receiver will not remove the information, just the noise.

Detection in FM is more complicated than in AM. Since FM contains several sidebands

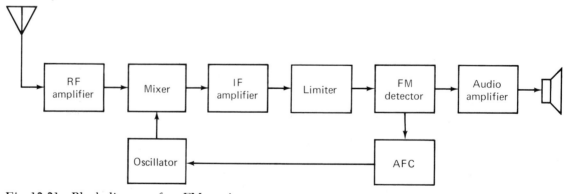

Fig. 12-21   Block diagram of an FM receiver.

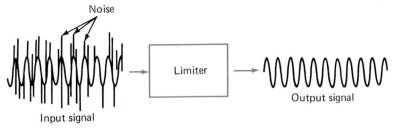

Fig. 12-22   Operation of a limiter.

above and below the carrier, a single non-linear detector will not demodulate the signal. A double-tuned *discriminator* circuit is shown in Fig. 12-23. It can serve as an FM detector. The discriminator works by having two resonant points. One is above the carrier frequency, and one is below the carrier frequency.

In the frequency response curves for the discriminator circuit (Fig. 12-24), f_o represents the correct point on the curves for the carrier. In a receiver, the station's carrier frequency will be heterodyned to f_o. This represents a frequency of 10.7 MHz for broadcast FM receivers. The heterodyning process allows one discriminator circuit to demodulate any signal over the FM band.

Refer to Figs. 12-23 and 12-24. When the carrier is unmodulated, D_1 and D_2 will conduct an equal amount. This is because the circuit is operating where the frequency response curves cross. The amplitude is equal for both tuned circuits at this point. The current through R_1 will equal the current through R_2. If R_1 and R_2 are equal in resistance, the voltage drops will also be equal.

Since the two voltages are series-opposing, the output voltage will be zero. When the carrier is at rest, the discriminator output is zero.

Suppose the carrier shifts higher in frequency because of modulation. This will increase the amplitude of the signal in L_2C_2 and decrease the amplitude in L_1C_1. Now there will be more voltage across R_2 and less across R_1. The output goes positive.

What happens when the carrier shifts below f_o? This moves the signal closer to the resonant point of L_1C_1. More voltage will drop across R_1, and less will drop across R_2. The output goes negative. The output from the discriminator circuit is zero when the carrier is at rest, positive when the carrier moves higher, and negative when the carrier moves lower in frequency. The output is a function of the carrier frequency.

The output from the discriminator can also be used to correct for any drift in the receiver oscillator. Note in Fig. 12-21 that the FM detector feeds a signal to the audio amplifier and to a stage marked *AFC*. The letters AFC stand for automatic frequency control. If the oscillator changes frequency, f_o will not equal

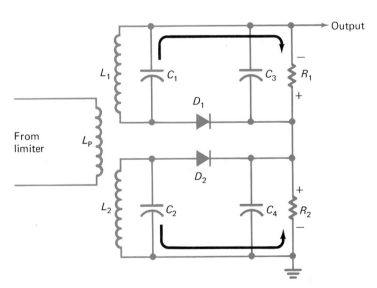

Fig. 12-23   A discriminator.

Electronics:
Principles and
Applications
CHAPTER 12

Ratio detector

Single sideband
(SSB)

Balanced
modulator

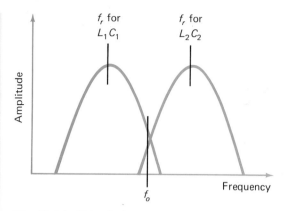

Fig. 12-24   Discriminator response curves.

exactly 10.7 MHz. There will be a steady dc output voltage from the discriminator. This dc voltage can be used as a control voltage to change the oscillator frequency automatically. Some receivers use the discriminator output to drive a tuning meter as well. A zero-center meter shows the correct tuning point. Any tuning error will cause the meter to deflect to the left or to the right of zero.

Frequency modulation discriminator circuits work well, but they are sensitive to amplitude. This is why one or two limiters are needed for noise-free reception. An improved FM detector is the *ratio detector*. It is not nearly as sensitive to the amplitude of the signal. This makes it possible to build receivers without limiters and still provide good noise rejection.

Figure 12-25 shows a typical ratio detector circuit. Its design is based on the idea of dividing a signal voltage into a ratio. This ratio is equal to the ratio of the voltages on either side of L_2. With frequency modulation, the

ratio shifts and an audio output signal is available at the center tap of L_2. Since the circuit is ratio-sensitive, the input signal amplitude may vary over a wide range without causing any change in output. This makes the detector insensitive to amplitude variations such as noise.

There are quite a few other FM detector circuits being used. Some of the more popular ones are the quadrature detector, the phase-locked-loop detector, and the pulse-width detector. These circuits are likely to be used in conjunction with integrated circuits. They usually have the advantage of requiring no alignment or only one adjustment. Alignment for discriminators and ratio detectors is more time-consuming.

Single sideband (SSB) is another alternative to amplitude modulation. Single sideband is a subclass of AM. It is based on the idea that both sidebands in an AM signal carry the same information. Therefore, one of them can be eliminated in the transmitter. This should not cause any loss of information at the receiver. It has been found that the carrier can also be eliminated at the transmitter. Therefore, an SSB transmitter sends one sideband and no carrier.

Energy is saved by not sending the carrier and the other sideband. Also, the signal will occupy only half the original bandwidth. Single sideband is much more efficient than AM. It has an effective gain of 9 dB. This is equivalent to increasing the transmitter power 8 times!

The carrier is eliminated by using a balanced modulator at the transmitter (Fig. 12-26). The diodes are connected so that no

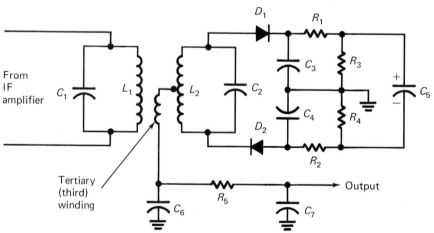

Fig. 12-25   A ratio detector.

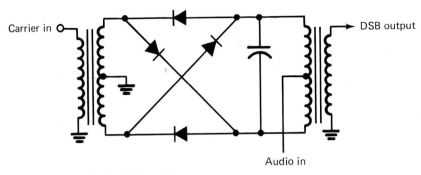

Fig. 12-26   A balanced modulator.

carrier can reach the output. However, when audio is applied, the circuit balance is upset and sidebands appear at the output. This is called a double sideband (DSB) signal.

A band-pass filter can be used to eliminate the unwanted sideband. Figure 12-27 shows that only the upper sideband reaches the output of the transmitter. The carrier is shown as a broken line since it has already been eliminated by the balanced modulator circuit.

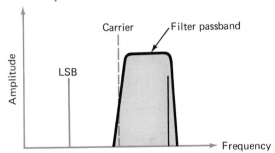

Fig. 12-27   Eliminating one of the sidebands.

A receiver designed to receive SSB signals is only a little different from an ordinary AM receiver. However, the cost can be quite a bit more. There are two important differences in the SSB receiver: (1) the bandwidth in the IF amplifier will be narrower, and (2) the car-

rier must be replaced by a second (local) oscillator so detection can occur. You will recall that the carrier is needed to mix with the sidebands to produce the difference (audio) frequencies.

Single sideband receivers usually achieve the narrow IF bandwidth with crystal or mechanical filters. These are more costly than inductor-capacitor filters. An SSB receiver must be very stable. Even a small drift in any of the receiver oscillators will change the quality of the received audio. A small drift, say 50 Hz, will not even be noticeable in an ordinary AM receiver. Stable oscillators are more expensive. This, along with filter costs, makes an SSB receiver more expensive.

Notice in the block diagram for the SSB receiver (Fig. 12-28) that the detector is called a *product detector*. This name is used since the audio output from the detector is the difference product between the IF signal and the beat frequency oscillator (BFO) signal. Actually, all AM detectors are *product* detectors. They all use the difference frequency product as their useful output. An ordinary diode detector will detect the SSB so long as it is supplied a BFO signal in addition to the IF signal.

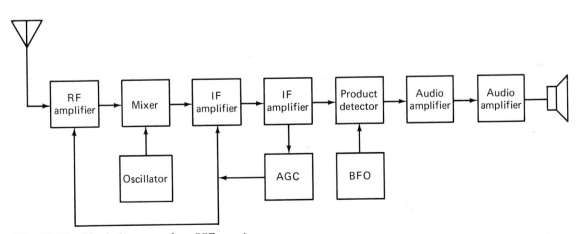

Fig. 12-28   Block diagram of an SSB receiver.

The BFO in an SSB receiver can be fixed at one frequency. In fact, it is often crystal-controlled for the best stability. Any error between the BFO frequency and the carrier frequency of the transmitted signal can be corrected by adjusting the main tuning control. The main difference between tuning an AM receiver and an SSB receiver is the need for *critical* tuning in an SSB receiver. Even a slight tuning error of 50 Hz will make the received audio sound very unnatural.

The critical tuning of the SSB makes it undesirable for most radio work. It is useful when maximum communication effectiveness is needed. Since it is so efficient, in terms of both power and bandwidth, it is popular in citizens' band radio, amateur radio, and military radio.

Review Questions

Choose the letter that best completes each statement.

20. Refer to Fig. 12-18. Resistors R_1 and R_2
    A. Form a voltage divider for the audio input
    B. Set the gate voltage for Q_1
    C. Divide V_{DD} to forward-bias D_1
    D. Divide V_{DD} to reverse-bias D_1

21. Refer to Fig. 12-18. A negative-going signal at the audio input will
    A. Increase the capacitance of D_1 and raise the frequency
    B. Increase the capacitance of D_1 and lower the frequency
    C. Decrease the capacitance of D_1 and raise the frequency
    D. Decrease the capacitance of D_1 and lower the frequency

22. Frequency modulation as compared to amplitude modulation
    A. Can provide better noise rejection
    B. Requires more bandwidth
    C. Requires complex detector circuits
    D. All the above

23. The function of the limiter stage in Fig. 12-21 is to
    A. Reduce amplitude noise
    B. Prevent overdeviation of the signal
    C. Limit the frequency response
    D. Compensate for tuning error

24. The purpose of the AFC stage in Fig. 12-21 is to
    A. Reduce noise

B. Maintain a constant audio output (volume)
    C. Compensate for tuning error and drift
    D. Provide stereo reception

25. Refer to Figs. 12-23 and 12-24. Assume that f_o is 10.7 MHz. If the signal from the limiter is at 10.75 MHz,
    A. The output voltage will be 0 V
    B. The output voltage will be negative
    C. The output voltage will be positive
    D. Resistor R_1 will conduct more current than R_2

26. Refer to question 25. If the signal from the limiter is at 10.7 MHz, then
    A. Diode D_1 will conduct the most current
    B. Diode D_2 will conduct the most current
    C. The output voltage will be positive
    D. None of the above

27. Refer to Fig. 12-25. The advantage of this FM detector as compared to a discriminator is that it
    A. Is less expensive (uses fewer parts)
    B. Can drive a tuning meter
    C. Can provide AFC
    D. Rejects amplitude variations

28. Refer to Fig. 12-26. The carrier input is 455 kHz, and the audio input is a 2-kHz sine wave. The output frequency or frequencies are
    A. 2 kHz
    B. 455 kHz
    C. 453, 455, and 457 kHz
    D. 453 and 457 kHz

29. Single sideband as compared to amplitude modulation is
    A. More efficient in terms of bandwidth
    B. More efficient in terms of power
    C. More critical to tune
    D. All the above

30. Refer to Fig. 12-28. The purpose of the BFO circuit is
    A. To correct for tuning error
    B. To replace the missing carrier so detection can occur
    C. To provide noise rejection
    D. All the above

31. Refer to Fig. 12-28. The IF bandwidth of this receiver, compared with an ordinary AM receiver, is
    A. Narrower
    B. The same
    C. Wider
    D. Indeterminate

12-5 RECEIVER TROUBLESHOOTING

Radio receiver troubleshooting is very similar to amplifier troubleshooting. Most of the circuits in a receiver are amplifiers. All the material covered in Chap. 10 on amplifier troubleshooting is relevant to receiver troubleshooting. For example, Sec. 10-1 on preliminary checks should be followed in exactly the same way.

You should view a receiver as a signal chain. If the receiver is dead, the problem is to find the broken link in the chain. Signal injection should begin at the output (speaker) end of the chain. However, a receiver involves gain at different frequencies. Several signal generator frequencies will be involved. You must use both an audio generator and an RF generator. Figure 12-29 shows the general scheme of signal injection in a superheterodyne receiver.

It is also possible to make a click test in most receivers. Use the same procedure discussed in Chap. 10 on amplifier troubleshooting. This will work in the audio and IF stages. The noise generated by the sudden shift in transistor bias should reach the speaker. It is also possible to test the mixer with the click test. The oscillator may respond to the click test, but the results would not be conclusive. It is possible that the oscillator is not oscillating, or it may be oscillating at the wrong frequency.

If we assume that the signal chain is intact from the first IF to the speaker, then the problem must be in the mixer or the oscillator. Checking the oscillator is not too difficult. An oscilloscope or frequency counter could be used. A voltmeter with an RF probe is an-

other possibility, but there would be no way to tell whether the frequency were correct. Many technicians prefer to tune for the oscillator signal by using a second receiver. Bring a second receiver very close to the receiver being tested. Set the dial on the second receiver above the dial frequency on the receiver under test. The difference should equal the IF of the receiver under test. This is based on the fact that the oscillator is supposed to run above the dial setting by an amount equal to the IF. Now, rock one of dials back and forth a little. You should hear a carrier (no modulation). This tells you the oscillator is working, and it also indicates whether the frequency is nearly correct.

If the receiver distorts badly on some stations, the problem could be in the AGC circuit. This is easy to check with a sensitive voltmeter. Monitor the control voltage as the receiver is tuned across the band. You should be able to see changes in the control voltage from no station (clear frequency) to a strong station. The service notes for the receiver usually will indicate the normal AGC range.

If the receiver has poor sensitivity, again it is possible that the AGC circuit is defective. Since AGC can produce several symptoms, it is recommended that it be checked early in the troubleshooting process.

Poor sensitivity can be difficult to troubleshoot. A weak stage is more difficult to find than a dead stage. Signal injection may work. It is normal to expect less injection for a given speaker volume as the injection point moves toward the antenna. Some technicians disable the AGC circuit when making this test. This can be done by clamping the AGC control line with a fixed voltage from a power supply. Do not use this technique until you

Click test

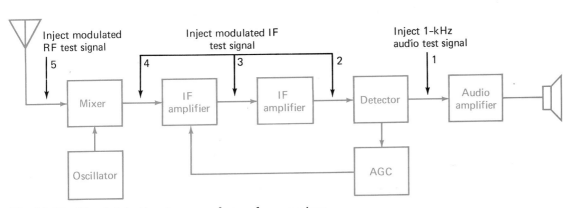

Fig. 12-29 Signal injection in a superheterodyne receiver.

Improper
alignment

have checked the circuit to make sure damage will not result. A current-limited supply is the safest.

Poor sensitivity can be caused by a leaky detector diode. Lift one end of the diode from the circuit, and check the resistance with an ohmmeter. Use the $R \times 10$ or $R \times 100$ range to check the forward resistance. It should be around 1000 Ω or so. The reverse resistance is also important. A germanium diode should show 100 kΩ or more. A silicon diode should measure infinity.

Improper alignment is another possible cause of low gain and poor sensitivity. All the IF stages must be adjusted for the correct frequency. Also, the oscillator and mixer-tuned circuits must track for good performance across the band. If the receiver has an RF stage, then three tuned circuits must track across the band.

Alignment is usually good for the life of the receiver. However, someone may have misadjusted the tuned circuits, or a part may have been replaced that upsets the alignment. Do not attempt alignment unless the service notes and the proper equipment are available.

Intermittent receivers and noisy receivers should be approached by using the same techniques described in Chap. 10 on amplifier troubleshooting. In addition, you should realize that receiver noise may be due to some problem outside the receiver itself. Some locations are very noisy, and poor receiver performance is typical. Most technicians carry a standard receiver with them to check for this problem. If the receiver is brought to them, they have the opportunity to check its performance in a standard location.

It should also be mentioned that receiver performance can vary considerably from one model to another. Many complaints of poor performance cannot be resolved with simple repairs. Some receivers simply do not work as well as others.

A superheterodyne receiver may have a total gain in excess of 100 dB. Unwanted feedback paths or coupling of circuits may cause oscillations. If the receiver squeals only when a station is tuned in, the problem is likely to be in the IF amplifier. If the receiver squeals or motorboats constantly, a bypass capacitor or AGC filter capacitor may be open. Always check to be sure that all grounds are good and that all shields are in place. In some cases, poor alignment can also cause oscillation.

Interference from nearby transmitters is becoming an increasingly complex problem.

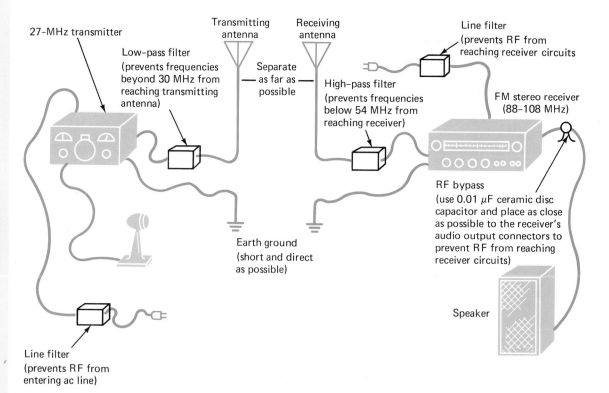

Fig. 12-30 Steps to prevent radio interference.

Many people are on the air with transmitters today. When their transmitting antenna is located close to other receiving equipment, problems are likely to occur. These interference problems can be very difficult to solve. Figure 12-30 shows some of the possible solutions that may be successful. Solving these problems is a process of trying various things until progress is noted. Try the easiest and least expensive cures first.

Review Questions

Choose the letter that best answers each question.

32. A 1-kHz test signal can be used for testing which stage of a superheterodyne receiver?
   A. Mixer
   B. IF
   C. Detector
   D. Audio

33. It is desired to check the oscillator of a superheterodyne receiver by using a second receiver. If the dial is set at 700 Hz, where should the oscillator be heard on the second receiver?
   A. 455 kHz
   B. 700 kHz
   C. 1155 kHz
   D. 1610 kHz

34. A receiver sounds distorted only on the strongest signals. Where would the fault be likely to be found?
   A. In the AGC system
   B. In the loudspeaker
   C. In the audio amplifier
   D. In the volume control

35. What would cause poor sensitivity in a receiver?
   A. A defective mixer
   B. A defective IF amplifier
   C. A weak detector
   D. Any of the above

36. What results from improper alignment?
   A. Poor sensitivity
   B. Dial error
   C. Oscillation
   D. Any of the above

Radio Receivers
CHAPTER 12

Summary

1. A high-frequency oscillator signal becomes a radio wave at the antenna.
2. Modulation is the process of putting information on the radio signal.
3. Turning the signal on and off with a key is called CW modulation.
4. When AM is used, the signal has three components: a carrier, a lower sideband, and an upper sideband.
5. The total bandwidth of an AM signal is twice the highest modulating frequency.
6. Demodulation is usually called detection.
7. A diode makes a good AM detector.
8. Other nonlinear devices, such as transistors, can also be used as AM detectors.
9. An AM receiver can consist of an antenna, a detector, headphones, and a ground.
10. Sensitivity is the ability to receive weak signals.
11. Selectivity is the ability to separate stations operating at different frequencies.
12. Gain provides sensitivity.
13. Tuned circuits provide selectivity.
14. Optimum bandwidth for an ordinary AM receiver is about 15 kHz.
15. A superheterodyne receiver converts the received frequency to an intermediate frequency.
16. The intermediate frequency permits most of the tuned circuits to be fixed.
17. The mixer output will contain several frequencies. Only those in the IF passband will reach the detector.
18. The standard IF for the AM broadcast band is 455 kHz.
19. The receiver oscillator will usually run above the received frequency by an amount equal to the intermediate frequency.
20. Two frequencies will always mix with the oscillator frequency and produce the IF: the desired frequency and the image frequency.
21. Adjacent-channel interference is rejected by the selectivity of the IF stages.
22. The AGC circuit compensates for different signal strengths.
23. In an FM transmitter, the audio information modulates the frequency of the oscillator.
24. Frequency modulation produces several sidebands above the carrier and several sidebands below the carrier.
25. Frequency modulation detection can be achieved by a discriminator circuit.

211

26. Discriminators are sensitive to amplitude; thus limiting must be used before the detector.
27. A ratio detector has the advantage of not requiring a limiter circuit for noise rejection.
28. Single sideband (SSB) is a subclass of AM.
29. Receiver troubleshooting is similar to amplifier troubleshooting.

30. The signal chain can be checked stage by stage by using signal injection.
31. A leaky detector can cause poor sensitivity.
32. Good alignment is necessary for proper receiver performance.

Chapter Review Questions

Choose the letter that best answers each question.

12-1. Which portion of a transmitting station converts the high-frequency signal into a radio wave?
(A) The modulator (B) The oscillator (C) The antenna (D) The power supply

12-2. What is the modulation used in radio telegraphy called?
(A) CW (B) AM (C) FM (D) SSB

12-3. An AM transmitter is fed audio as high as 4.5 kHz. What is the bandwidth required for its signal?
(A) It is dependent on the carrier frequency (B) 4.5 kHz (C) 9.0 kHz (D) 455 kHz

12-4. An AM demodulator uses the difference frequency between what two frequencies?
(A) USB and the LSB (B) Sidebands and the carrier (C) IF and the detector (D) All the above

12-5. Which of the following components is useful for AM detection?
(A) A tank circuit (B) A resistor (C) A capacitor (D) A diode

12-6. Refer to Fig. 12-4. Assume the audio input is zero. The carrier output to the antenna will
(A) Fluctuate in frequency (B) Fluctuate in amplitude (C) Be zero (D) None of the above

12-7. Refer to Fig. 12-10. What serves as the energy source for this receiver?
(A) The radio signal (B) The detector (C) The headphones (D) There is none

12-8. Refer to Fig. 12-11. How may the selectivity of this receiver be improved?
(A) Add more audio gain (B) Add more tuned circuits (C) Use a bigger antenna (D) All the above

12-9. A tuned circuit has a center frequency of 455 kHz and a bandwidth of 20 Hz. At what frequency or frequencies will the response of the circuit drop to 70 percent?
(A) 475 kHz (B) 435 kHz (C) 435 and 475 kHz (D) 445 and 465 kHz

12-10. An AM receiver has an IF amplifier with a bandwidth that is too narrow. What will the symptom be?
(A) Loss of high-frequency audio (B) Poor selectivity (C) Poor sensitivity (D) All the above

12-11. The major advantage of the superheterodyne design over the TRF receiver design is that

(A) It eliminates the image problem (B) It eliminates tuned circuits (C) It eliminates the need for an oscillator (D) The fixed IF eliminates tracking problems and bandwidth changes

12-12. A superheterodyne receiver is tuned to 940 kHz. Where is the image?
(A) 455 kHz (B) 485 kHz (C) 940 kHz (D) 1850 kHz

12-13. Refer to Fig. 12-16. The dial of the receiver is set at 1190 kHz. Which statement is true?
(A) The mixer-tuned circuit should resonate at 1190 kHz (B) The oscillator circuit should resonate at 1645 kHz (C) The difference mixer output should be at 455 kHz (D) All the above

12-14. An FM receiver is set at 97 MHz. Interference is received from a station transmitting at 118.4 MHz. What is the problem due to?
(A) Poor selectivity in the RF and mixer-tuned circuits (B) Poor selectivity in the IF stages (C) Poor limiter performance (D) Poor ratio detector performance

12-15. What FM receiver circuit is used to correct for frequency drift in the oscillator?
(A) AGC (B) AVC (C) AFC (D) All the above

12-16. A transistor in an FM receiver is controlled by decreasing its current as the received signal becomes stronger. What is this an example of?
(A) Forward AGC (B) Reverse AGC (C) Stereo reception (D) None of the above

12-17. How does frequency modulation compare to amplitude modulation with regard to the number of sidebands produced?
(A) FM produces the same number of sidebands (B) FM produces fewer sidebands (C) FM produces more sidebands (D) FM produces no sidebands

12-18. What is the function of a limiter stage in an FM receiver?
(A) It rejects adjacent-channel interference (B) It rejects image interference (C) It rejects noise (D) It rejects drift

12-19. The output of Fig. 12-23 is connected to a zero-center tuning meter. How will the meter respond when a station is correctly tuned?
(A) It will indicate in the center of its scale (B) It will deflect maximum to the right (C) It will deflect to the left (D) It depends on the station

12-20. Which of the following circuits is not used for FM demodulation?
(A) Diode detector (B) Discriminator (C) Ratio detector (D) Quadrature detector

12-21. The output of a balanced modulator is called
(A) SSB (B) DSB (C) FM (D) None of the above

12-22. An SSB transmitter runs 5 W. What power will be required in an AM transmitter to achieve the same range?
(A) 1 W (B) 5 W (C) 20 W (D) 40 W

12-23. What is the bandwidth of an SSB signal as compared to that of an AM signal?
(A) About 2 times as great (B) About the same (C) About half as great (D) About 10 percent

12-24. What must be done to demodulate an SSB signal?
(A) Replace the missing carrier (B) Use two diodes (C) Use a phase-locked-loop detector (D) Convert it to an FM signal

213

12-25. Which of the following test signals would be the *least* useful for troubleshooting an AM broadcast receiver?
(A) 1-kHz audio (B) 455-kHz modulated RF (C) 1-MHz modulated RF (D) 10.7-MHz frequency-modulated RF

12-26. An FM receiver works very well, but the dial accuracy is very poor. The problem is most likely in the
(A) Detector (B) Oscillator (C) IF amplifiers (D) Limiter

Answers to Review Questions

1. *D*	13. *B*	25. *C*
2. *B*	14. *C*	26. *D*
3. *B*	15. *A*	27. *D*
4. *A*	16. *B*	28. *D*
5. *C*	17. *B*	29. *D*
6. *C*	18. *A*	30. *B*
7. *A*	19. *C*	31. *A*
8. *C*	20. *D*	32. *D*
9. *D*	21. *B*	33. *C*
10. *D*	22. *D*	34. *A*
11. *A*	23. *A*	35. *D*
12. *D*	24. *C*	36. *D*

Linear Integrated Circuits

■ An integrated circuit (IC) can be the equal of dozens of separate electronic parts. Some very complex ICs have thousands of transistors on a small chip of silicon. It is easy to see why these amazing devices are taking over electronics.

   Integrated circuits can be used to perform all the circuit functions that you have studied. This chapter will help you apply your knowledge to circuits based on integrated technology.

13-1 INTRODUCTION

The integrated circuit was introduced in 1958. It has been called the most significant technological development of this century. It has allowed electronics to expand at an amazing rate. Most of the growth has been in the area of digital electronics. Developments in linear integrated circuits lagged behind those of digital ICs for the first ten years or so. Now, linear ICs are receiving increasing attention, and quite a variety of linear ICs are available.

Electronics is growing rapidly for several reasons. One major reason is that electronics continues to advance in performance while the cost remains stable and may even decrease from time to time. Another reason for the growth in electronics is that circuits and systems have become increasingly reliable over the years. Integrated circuits have much to do with the steady progress of electronics.

Discrete circuits use individual resistors, capacitors, diodes, transistors, and other devices to achieve the circuit function. These individual or discrete parts must be interconnected. The usual approach is to use a circuit board. This method, however, increases the cost of the circuit. The board, assembly, soldering, and testing all make up a part of the cost.

Integrated circuits do not eliminate the need for circuit boards, assembly, soldering, and testing. However, with ICs the number of discrete parts can be reduced. This means that the circuit boards can be smaller and that they will cost less to produce. It may also be possible to reduce the overall size of the equipment with integrated circuits, which can reduce costs in the chassis and cabinet.

Integrated circuits may lead to the creation of circuits requiring fewer alignment steps at the factory. Alignment is expensive, and therefore fewer steps would mean lower costs. Also, variable components are more expensive than fixed components. Again, savings are realized.

Integrated circuits may also increase performance. Certain ICs work better than equivalent discrete circuits. A good example is a modern integrated voltage regulator. A typical unit may offer 0.03 percent regulation, excellent ripple and noise suppression, automatic current limiting, and thermal shutdown. An equivalent discrete regulator may contain dozens of parts, cost 6 times as much, and still not work as well!

Reliability is related *indirectly* to the number of parts in the equipment. As the number of parts goes up, the reliability comes down. Integrated circuits make it possible to reduce the number of discrete parts in a piece of equipment. Thus, electronic equipment can be made more reliable by the use of more ICs and fewer discrete components.

Integrated circuits are available in a variety of package styles (Fig. 13-1). Although only the popular packages used for linear ICs are shown, digital ICs also use these packages

From page 215:
Discrete circuit

Integrated circuit

Package style

On this page:
Dual-in-line
package

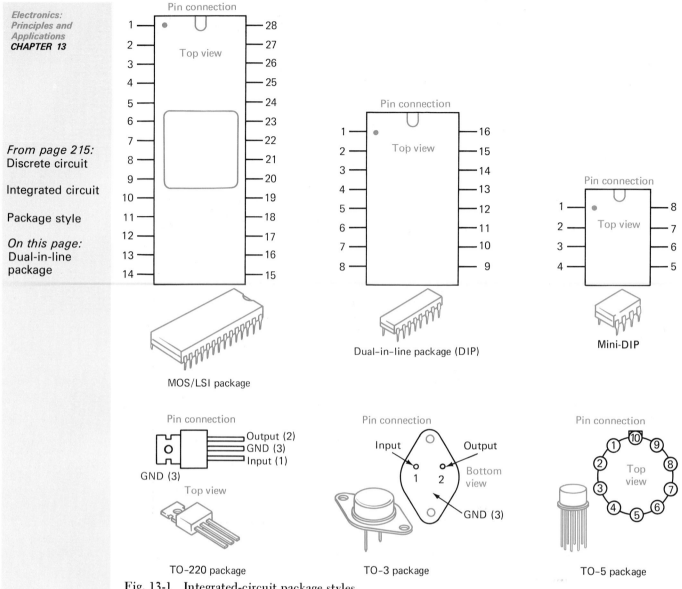

Fig. 13-1 Integrated-circuit package styles.

(with the exception of the TO-220 and TO-3 styles). The MOS/LSI package is used for complex circuits. The letters stand for metallic oxide semiconductor, and the letters LSI stand for large-scale integration. The MOS/LSI package can have as many as 40 pins.

The dual-in-line package shown in Fig. 13-1 is very popular. It may have 14 or 16 pins. The mini-DIP is a shorter version of the dual-in-line package. It has 8 pins. The TO-5 package is available with 8, 10, or 12 pins.

The TO-3 and TO-220 packages are used mainly for voltage regulator ICs. Their appearance is identical to packages used for power transistors. This is a good example of how valuable service literature is when you are

troubleshooting equipment for the first time. Positive component identification cannot be based on a visual check alone.

Schematics seldom show any of the internal features for integrated circuits. The technician usually does not need to know exactly how the IC does its job. It is more important to know what it is supposed to do and how it functions as a part of the overall circuit. Figure 13-2 shows the internal schematic for a μA7812 IC voltage regulator. Note the complexity of the circuit. Most diagrams will show the IC as in Fig. 13-3. Note the simplicity. The voltage regulator function is simple and direct. Figure 13-3 plus a few voltage specifications is all that a technician would need to check for proper operation of the IC.

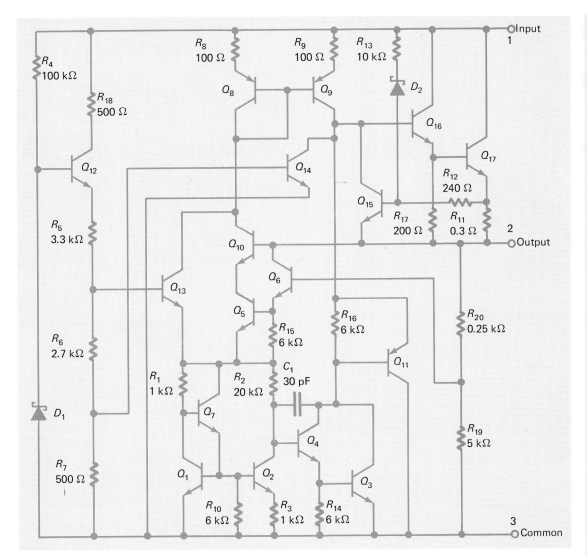

Fig. 13-2   Schematic for an IC voltage regulator.

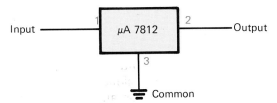

Fig. 13-3   Normal way of showing an IC.

Review Questions

Choose the letter that best answers each question.

1. When was the integrated circuit developed?
   A. 1920
   B. 1944
   C. 1958
   D. 1977

2. What is an electronic circuit that is made up of individual resistors, capacitors, transistors, and other parts called?
   A. An integrated circuit
   B. A chassis
   C. A circuit board
   D. A discrete circuit

3. The use of ICs in a design can
   A. Decrease the number and size of parts
   B. Lower cost
   C. Increase reliability
   D. All the above

4. What is the only sure way to identify a part as an integrated circuit?
   A. Look at the package style.
   B. See how it is connected to other parts.
   C. Check the schematic.
   D. Count the pins.

5. When will a technician need the internal schematic for an integrated circuit?
   A. Very seldom
   B. When troubleshooting
   C. When making circuit adjustments
   D. When taking voltage and waveform readings

P-type silicon crystal

Wafers

Photolithography

Substrate

Photoresist

Batch-processed

13-2 FABRICATION

The fabrication of integrated circuits begins in a radio-frequency furnace. Silicon that has been doped with a P-type impurity is melted in a quartz crucible. A large crystal of P-type silicon is then pulled from the molten silicon (Fig. 13-4). The crystal is then sliced into wafers 10 mils thick [0.025 centimeters (cm) or 0.01 in].

The P-type silicon wafers are processed using *photolithography*, as shown in Fig. 13-5. The wafer is called the *substrate*. When it is exposed to heat and water vapor the surface is oxidized. Then the oxide surface is coated with *photoresist*, a sensitive material that hardens when exposed to ultraviolet light. The exposure is made through a photomask. After exposure, the unhardened areas wash away in the developing process. Now the wafer is etched to expose areas of the substrate. These exposed areas are then penetrated with N-type impurity atoms. Thus, PN junctions have been formed in the silicon wafer. The wafer can now be reoxidized, and the process repeated several times to make the desired P- and N-type zones in the substrate.

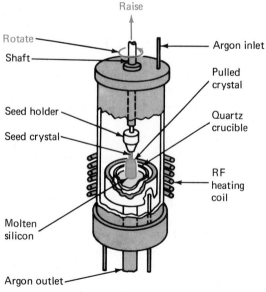

Fig. 13-4 Forming the crystal.

(a) Crystalline silicon

(b) Oxidize surface of substrate

(c) Coat oxide with positive photoresist

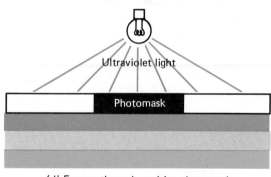

(d) Expose through positive photomask

(e) Develop, removing the unexposed photoresist

(f) Etch through oxide (silicon dioxide)

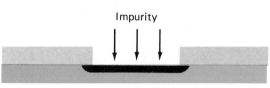

(g) Impurity penetrates substrate and a PN junction is formed

Fig. 13-5 The photolithographic process.

Integrated circuits are *batch-processed*. As shown in Fig. 13-6, one silicon wafer will produce a batch of chips. This keeps the costs down. Also, the wafer is much easier to handle and process since it is large compared to a single IC.

Not all the chips in the batch will be good. They are tested, with needle-sharp probes that

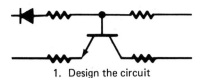

1. Design the circuit

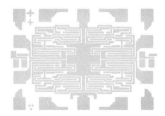

2. Design the layout

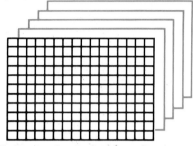

3. Prepare the photomasks — five or more will be required

4. Expose the silicon wafer using each photomask

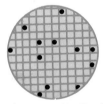

5. Run probe test and scribe the wafer

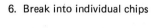

6. Break into individual chips

7. Mount chip into package — bond and seal

Fig. 13-6 The major steps in making an IC.

Ball bonding

Epitaxial

Isolation diffusion

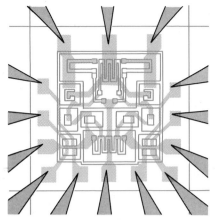

Fig. 13-7 The probe test. Needle-sharp probes touch each pad on the chip.

contact each chip on the wafer (Fig. 13-7). Bad chips are marked with a dot. Now the wafer is scribed and broken apart. The good chips are mounted on a metal header. The pads on the chip are wired to the tabs on the header. Figure 13-8 shows how this is done. The process is called *ball bonding* since a ball forms on the end of the wire when the wire is cut by the gas flame.

It is interesting to examine how some of the specific circuit functions are achieved in the IC. The steps used to form an NPN-junction transistor are seen in Fig. 13-9. It begins with the P-type substrate. An N+ diffusion layer is added (N+ indicates that it is heavily doped with an N-type impurity). This N+ layer will improve the collector characteristics of the transistor. Now, an N-type layer is grown over the substrate. This *epitaxial* process produces a uniform crystalline structure. The epitaxial layer is oxidized, and photolithography is used to expose an area surrounding the transistor. Boron, a P-type impurity, is diffused through the opening until the substrate is reached. This *isolates* a region of the N-type epitaxial layer. The various circuit functions in the IC can be separated by isolation diffusion.

Refer again to Fig. 13-9. It is time to produce the base of the transistor. Again, photolithography opens up the desired area in the oxide layer. A P-type impurity penetrates, and the base region is formed. The last diffusion will use an N-type impurity to form the emitter of the transistor. Polarity reversal by diffusion will eventually saturate the crystal. A maximum number is three diffusions. Review Fig. 13-9; you will see that the collector of the transistor will be lightly doped and the

219

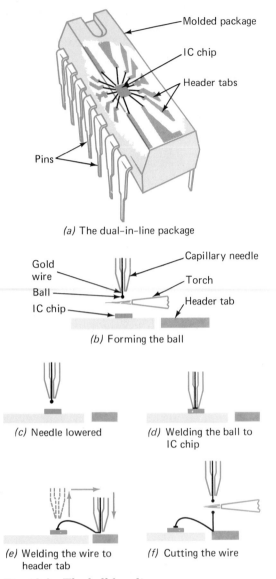

(a) The dual-in-line package

(b) Forming the ball

(c) Needle lowered

(d) Welding the ball to IC chip

(e) Welding the wire to header tab

(f) Cutting the wire

Fig. 13-8   The ball-bonding process.

emitter will be heavily doped. This is as it should be for proper junction transistor action.

The transistor has been formed. Now it must be connected, as in Fig. 13-10. Again, the wafer is oxidized, and photolithography is used to produce openings that expose parts of the wafer. This time, the openings expose the collector, base, and emitter of the transistor. Aluminum is evaporated onto the surface of wafer. It makes contact with the transistor. The unwanted aluminum is etched away. This leaves a separate metal path for the collector, base, and emitter of the transistor. The paths will connect the transistor to other parts of the integrated circuit.

At the same time as transistors are being

formed, other circuit functions are being created as well. The PN-junction diode of Fig. 13-11 looks very much like the junction transistor. The emitter diffusion has been eliminated. The collector-base junction is the diode.

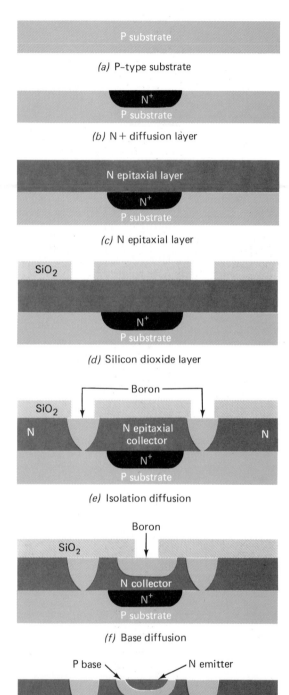

(a) P-type substrate

(b) N+ diffusion layer

(c) N epitaxial layer

(d) Silicon dioxide layer

(e) Isolation diffusion

(f) Base diffusion

(g) Emitter diffusion

Fig. 13-9   Forming an NPN-junction transistor.

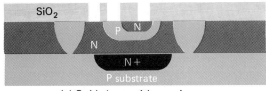

(a) Oxide layer with openings

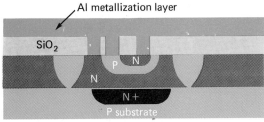

(b) Aluminum is evaporated onto the wafer

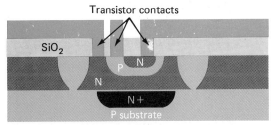

(c) The unwanted aluminum is etched away

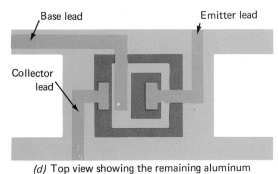

(d) Top view showing the remaining aluminum

Fig. 13-10   Connecting the transistor.

This diode (Fig. 13-11) could also work as a capacitor. You will recall that the depletion region can act as the dielectric for a capacitor. This approach is currently used for some IC capacitors. A MOS (metallic oxide semiconductor) capacitor can also be formed. Figure 13-12 shows this approach. The N-type region is one plate, the aluminum layer is another, and the silicon dioxide layer forms the dielectric.

The formation of a resistor in the integrated circuit is shown in Fig. 13-13. It is possible to control the size of the N channel and the amount of impurities in order to achieve different values of resistance.

Note the silicon dioxide layer between the gate lead and the channel in the MOS transistor of Fig. 13-14. A MOS transistor has the advantage of using less space in the IC chip as compared to a junction transistor.

IC components have certain disadvantages compared to discrete components:

1. Resistor accuracy is limited.
2. Very low and very high values of resistors are not feasible.
3. Inductors are not feasible.
4. Only small values of capacitance are practical.
5. PNP transistors have poor performance.
6. Power dissipation is limited.

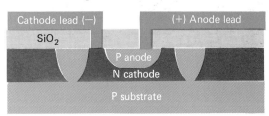

Fig. 13-11   Forming a junction diode.

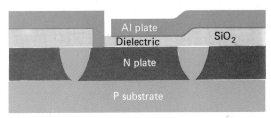

Fig. 13-12   Forming a MOS capacitor.

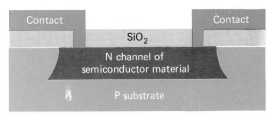

Fig. 13-13   Forming a resistor.

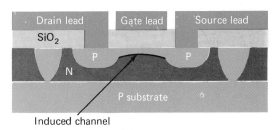

Fig. 13-14   Forming a MOS transistor.

221

Monolithic
integrated circuit

Hybrid
integrated circuit

Linear
integrated circuit

On the other hand, IC components have advantages, too:

1. Since all components are formed at the same time, matched characteristics can be easily achieved.
2. Since all components exist in the same structure, excellent thermal tracking can be achieved.

Thus far, the discussion has been limited to *monolithic* integrated circuits. A *monolith* (single-stone) type of structure finds all the components in a single chip of silicon. *Hybrid* integrated circuits are also available. These use another substrate such as ceramic to combine various types of components. For example, a thin ceramic wafer may hold several silicon ICs, some film resistors, a few chip-type capacitors, and a power transistor. The hybrid approach is more costly, but it can eliminate some the disadvantages cited for IC components. Hybrid ICs are available with power ratings of several hundred watts.

Review Questions

Choose the letter that best answers each question.

6. How are ICs made?
A. On tiny printed boards.
B. They are batch-processed on silicon wafers.
C. They are miniature assemblies of discrete parts.
D. None of the above.

7. What is the basic process in making ICs called?
A. Photolithography
B. Wave soldering
C. Electron beam fusion
D. Acid etching

8. Refer to Fig. 13-5(*e*). How was the window produced?
A. By boron diffusion.
B. With electron beam milling.
C. The unexposed area washes away.
D. By stencil cutting.

9. What is the purpose of the probe test?
A. To check the photoresist coatings
B. To verify the ball-bonding process
C. To count how many ICs have been processed
D. To eliminate bad IC chips before packaging

10. What process is used to wire the chip pads to the header tabs?
A. Ball bonding
B. Soldering
C. Epoxy
D. Aluminum evaporation

11. Refer to Fig. 13-9. Why is this called a monolithic IC?
A. A hybrid structure is used.
B. Everything is formed in a single slab of silicon.
C. The base of the transistor is in the collector.
D. All the above.

12. Refer to Fig. 13-9. Which step prevents the transistor from shorting to other components being formed at the same time?
A. The N+ diffusion layer
B. The epitaxial layer
C. The silicon dioxide layer
D. The isolation diffusion

13. Refer to Fig. 13-10. What prevents the aluminum layer from shorting out everything in the IC?
A. The N+ diffusion layer
B. The P-type substrate
C. The silicon dioxide layer
D. The Teflon spacers

14. How are capacitors formed in monolithic integrated circuits?
A. By forming PN junctions and reverse-biasing them.
B. By using the MOS approach.
C. Both of the above.
D. Capacitors cannot be formed in ICs.

15. Which of the following components cannot be formed in a monolithic IC?
A. Resistors
B. Inductors
C. Diodes
D. Transistors

13-3 APPLICATIONS

Linear integrated circuits have wide application. They can perform almost all the functions associated with linear electronics. Additionally, quite a few ICs are neither digital nor linear but both. Since digital techniques are taking over many functions once considered strictly linear, some knowledge of both circuit functions is necessary.

An integrated circuit can be used to replace a single-transistor stage. Figure 13-15 shows an IC commonly used as an IF amplifier. You may wonder why an IC would be used

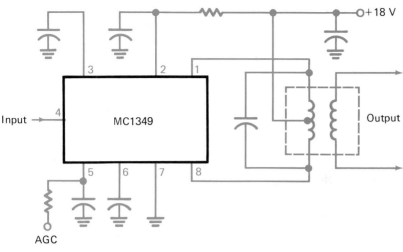

Fig. 13-15 An integrated-circuit IF amplifier.

Linear Integrated
Circuits
CHAPTER 13

Subsystem
integrated circuit

Timer IC

when a single transistor works in such a stage. The answer is performance. The MC1349 integrated circuit boasts as much as 61-dB gain at 45 MHz with an AGC range of 80 dB. No single-transistor IF stage can approach this kind of performance.

Integrated circuits can also be used to replace several stages. These are often called *subsystem integrated circuits* (Fig. 13-16). The dual-in-line package contains an IF amplifier, a limiter, an FM detector, an audio driver, a regulated power supply, and an electronic volume control. This greatly reduces the parts count in the sound section of the television receiver. Cost is reduced, performance is good, and the reliability is better than in an equivalent discrete design.

An AM radio receiver subsystem consisting of an RF amplifier, an oscillator, a mixer, an IF amplifier, and an AGC circuit can be fabricated on a single chip (Fig. 13-17). There is also an on-chip voltage regulator. The audio amplifier could be housed in one integrated circuit. This would make it possible to have a complete receiver using two ICs and a limited number of discrete components.

One of the most popular linear ICs is the operational amplifier. Many standard and special op-amps are available in integrated form. Differential amplifiers are also available. The theory and application of these circuits were covered in Chap. 9.

The timer IC was introduced in 1972 and has enjoyed wide application. It is an ex-

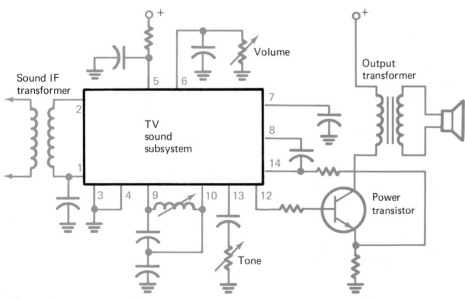

Fig. 13-16 A television sound subsystem.

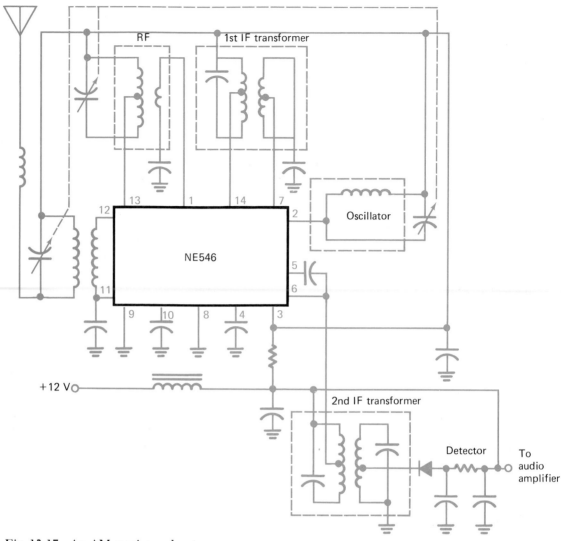

Fig. 13-17 An AM receiver subsystem.

ample of those ICs that are both digital and linear. Figure 13-18 shows the functional block diagram for the NE555 IC timer. The divider network references the threshold comparator at two-thirds of the supply and the trigger comparator at one-third of the supply. The comparator outputs go to the flip-flop. If the trigger voltage goes below one-third of the supply, its comparator sets the output of the flip-flop high. The threshold pin is normally referenced to the capacitor voltage of an external *RC* timing network. If the capacitor voltage exceeds two-thirds of the supply, then the threshold comparator resets the flip-flop. This, in turn, activates the discharge transistor which drains an external timing capacitor.

The IC timer has three operating modes:

1. Monostable, or one-shot, mode
2. Astable, or oscillator, mode
3. Time-delay mode

It is a very versatile circuit. The one-shot mode is useful for producing an accurately timed pulse when the circuit is triggered. The pulse time (width) is selected by choosing the resistor and capacitor in the external *RC* timing network. An NE555 set up for the one-shot mode is shown in Fig. 13-19. If the timing resistor is 10,000 Ω and the timing capacitor is 0.1 μF, then the output pulse width will be

$$t_{\mathrm{on}} = 1.1 \times R \times C$$
$$= (1.1) \times (10 \times 10^3) \times (0.1 \times 10^{-6})$$
$$= 1.1 \times 10^{-3} \text{ s} = 1.1 \text{ ms}$$

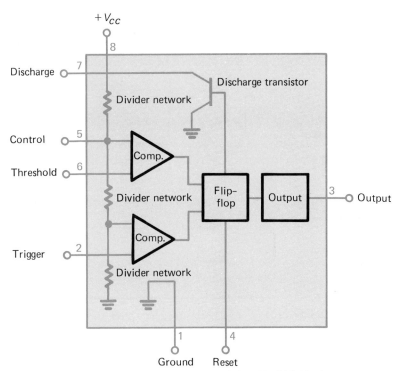

Fig. 13-18 Functional block diagram of the NE555 IC timer.

Thus, the output pulse width will be 1.1 ms *regardless* of the input pulse width.

In Fig. 13-20 the IC timer is operating in the oscillator mode. Operation is set by R_A, R_B, and C. The output is a continuous square wave (free-running). The *duty cycle* of the output refers to the width of the "on" pulse compared to the "off" pulse. Equal pulse width would be called a 50 percent duty cycle.

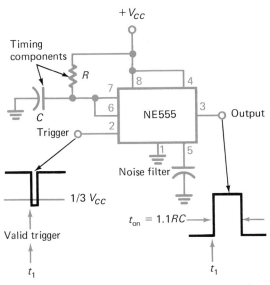

Fig. 13-19 Using the timer in the one-shot mode.

The on-pulse width is given by

$$t_1 = 0.693 \times R_A \times C$$

The width of the off pulse is given by

$$t_2 = 0.693 \times R_B \times C$$

Suppose R_A and R_B are both 10,000 Ω and $C = 0.1$ μF. The pulses would be equal (50 percent duty cycle) and could be found by

$$t = 0.693 \times 10 \times 10^3 \times 0.1 \times 10^{-6}$$
$$= 6.93 \times 10^{-4} \text{ s} = 0.693 \text{ ms}$$

One cycle would require two pulses:

$$2 \times 6.93 \times 10^{-4} = 1.39 \times 10^{-3} \text{ s} = 1.39 \text{ ms}$$

The frequency of oscillation is given by the reciprocal of the period:

$$f = \frac{1}{1.39 \times 10^{-3}} = 721.5 \text{ Hz}$$

Figure 13-21 shows the NE555 operating in the time-delay mode. This mode calls for the output to change state at some determined time after the trigger is received. Transistor Q_1 is turned off by a valid trigger signal. This

Voltage
regulator

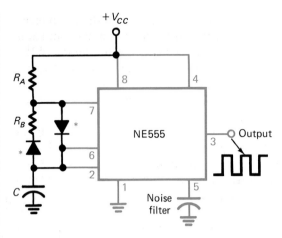

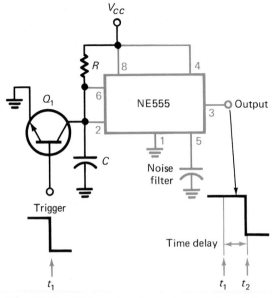

*Diodes may not be required, depending on duty
cycle desired.

Fig. 13-20 Using the timer as an oscillator.

Fig. 13-21 Using the timer in the time-delay
mode.

allows the capacitor C to begin charging
through the resistor R. When the capacitor
reaches the threshold, the output switches to a
low state. If $R = 47,000\ \Omega$ and $C = 0.5\ \mu F$,
the time delay can be found by

$$
\begin{aligned}
t_{\text{delay}} &= 1.1 \times R \times C \\
&= 1.1 \times 47 \times 10^3 \times 0.5 \times 10^{-6} \\
&= 2.59 \times 10^{-2}\ \text{s} \\
&= 25.9\ \text{ms}
\end{aligned}
$$

The IC audio power amplifier of Fig. 13-22
is capable of supplying 5 W of continuous output. Of course, heat sinking is a must at this
power level. It features low distortion, as
much as 46 dB of voltage gain, high input

impedance, low quiescent current, and
immunity to short circuits across its output.
An equivalent discrete amplifier would need
about 4 times as many parts.

Integrated-circuit voltage regulators are
popular. They are available as fixed three-
terminal devices with output voltages that
range from 5 to 50 V. They are available as
both positive and negative voltage devices.
They can supply currents from 500 mA to 3 A.
Figure 13-23 shows the ease with which a
three-terminal voltage regulator can be used.

In addition to being easy to use, the three-

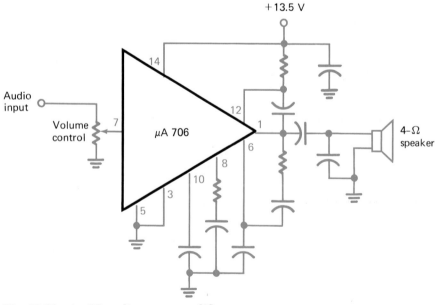

Fig. 13-22 An IC audio power amplifier.

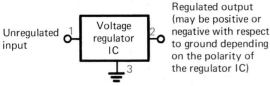

Fig. 13-23 A voltage regulator IC.

terminal regulators offer high performance. They feature internal current limiting, thermal shut-down, and excellent regulation. Of course, they will require heat sinking if their maximum current capacity is needed.

Other linear IC voltage regulators are also available in addition to the fixed three-terminal type. These make it possible to build many different types of power-supply designs, including variable-voltage/variable-current supplies, tracking supplies, and switching supplies.

Phase-locked loops are interesting circuits which are now available in integrated form. A phase-locked-loop circuit is shown in block diagram form in Fig. 13-24. The phase detector compares an input signal with the signal from a voltage-controlled oscillator. Any fre-

quency or phase difference produces an error voltage. This error voltage is filtered and amplified. It then corrects the frequency of the voltage-controlled oscillator. Eventually, the voltage-controlled oscillator will lock with the incoming signal. Once locked, it will track the input signal.

If a phase-locked-loop circuit is tracking an FM signal, the error voltage will be set by the deviation of the input signal. Thus, FM detection is realized. Figure 13-25 shows a 560B IC used as an FM detector. The variable capacitor is set so that the voltage-controlled oscillator operates at the center frequency of the FM signal. As modulation shifts the signal frequency, an error voltage is produced. This error voltage is the audio output. Phase-locked loops make very good FM detectors.

Phase-locked loops are also used as *tone decoders*. Tone decoders are useful circuits which allow remote control by selecting different tones. In Fig. 13-26 two phase-locked-loop ICs are used to build a dual-tone decoder. The output will go high only when both tones are present at the input. This type

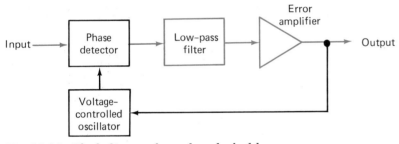

Fig. 13-24 Block diagram for a phase-locked loop.

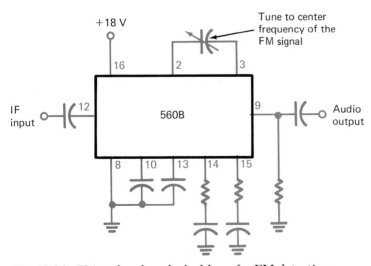

Fig. 13-25 Using the phase-locked loop for FM detection.

Block diagram

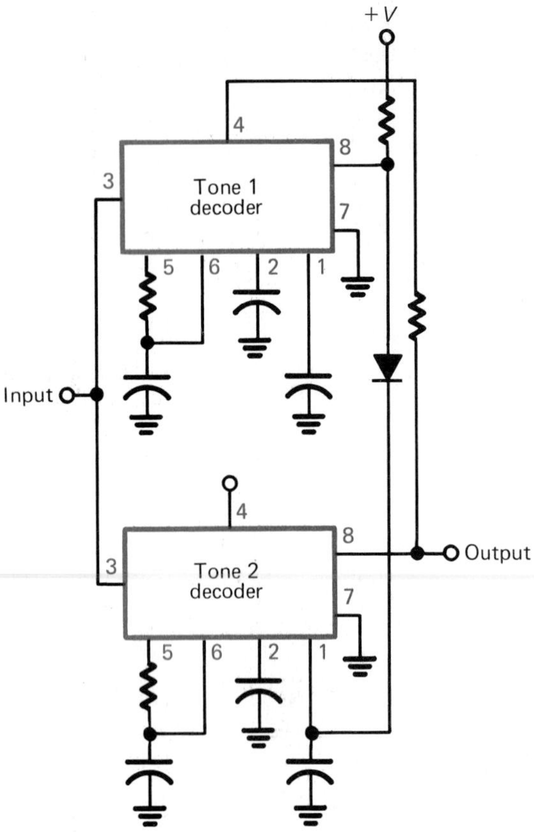

Fig. 13-26   A phase-locked-loop tone decoder.

of approach is less likely to be accidentally tripped by false signals. Touch-tone dialing systems use dual tones for this reason.

It is not possible to completely cover linear IC applications here. The few samples chosen show the vast range of circuits available. You are encouraged to refer to Appendix D-1 for more information. You should also browse through some of the many books and data sheets offered by the IC manufacturers. Their materials often include application notes to show how the ICs can be used in typical circuits. These materials are very informative and interesting.

Review Questions

Choose the letter that best answers each question.

16. Refer to Fig. 13-19. Assume that $V_{CC} = 6$ V. What will trigger the IC?
    A. A signal on pin 2 that goes below 2 V
    B. A signal on pin 7 that goes above 6 V
    C. Any negative-going signal on pin 2
    D. Any positive-going signal on pin 2

17. In Fig. 13-19, $R = 47$ kΩ and $C = 0.5$ μF. How long will pin 3 stay high after the IC is triggered?
    A. 2 ms
    B. 12.8 ms
    C. 25.9 ms
    D. 3.8 s

18. In Fig. 13-20, $R_A = R_B = 0.1$ MΩ and $C = 0.2$ μF. What is the frequency of oscillation?
    A. 9 Hz
    B. 18 Hz
    C. 36 Hz
    D. 72 Hz

19. Refer to question 18. What is the duty cycle of the square-wave oscillation?
    A. 10 percent
    B. 25 percent
    C. 35 percent
    D. 50 percent

20. Refer to Fig. 13-21. What happens when the IC is triggered?
    A. The output goes high.
    B. There is a delay before the output goes low.
    C. The output goes low for a time set by R and C.
    D. None of the above.

21. Refer to Fig. 13-24. When is an error voltage produced?
    A. At any time there is an input signal
    B. When the voltage-controlled oscillator is running
    C. Both of the above
    D. When there is a phase/frequency difference between the input and the oscillator

22. Refer to Fig. 13-26. When will the output go high?
    A. When tone 1 is present at the input
    B. When tone 2 is present at the input
    C. When tones 1 and 2 are present at the input
    D. All the above

13-4 TROUBLESHOOTING

Troubleshooting procedures for equipment using integrated circuits are about the same as those covered in Chap. 10 on amplifier troubleshooting. The preliminary checks, signal tracing, and signal injection can all be used to locate the general area of the problem.

The real key to good troubleshooting of complex equipment is a sound knowledge of the overall block diagram. This diagram gives

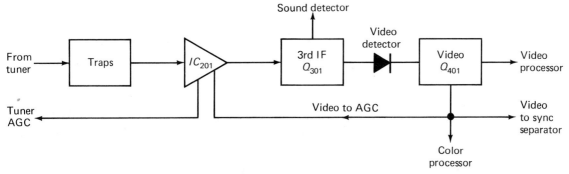

Fig. 13-27   A partial block diagram of a TV receiver.

the symptoms meaning. It is usually possible to limit the difficulty to one area quickly when the function of each stage is known. It is really not important if the stage uses ICs or discrete circuits. The function of the stage is what helps to determine if it could be causing the symptom or symptoms.

Figure 13-27 shows a portion of a block diagram for a television receiver. After the preliminary checks, this type of diagram can be used to limit the possibilities. Again, the service literature is very valuable when troubleshooting. Suppose the symptoms indicate the problem could be in IC_{201}. Now it is time to check the schematic diagram.

Figure 13-28 is the schematic diagram for IC_{201}. It shows, in block form, the major functions in the integrated circuit. It also shows the pin numbers and how the external parts are connected. Note that dc voltages are given. This is *very important* when troubleshooting linear ICs. When a particular IC is suspected, the dc voltages should be checked. The dc voltages must be correct if proper operation is to result.

Some schematics, such as in Fig. 13-28, show many of the dc voltages as two readings. These represent the acceptable voltage range at that particular point. For example, pin 3 is marked

$$\frac{3.2 \text{ V}}{3.6 \text{ V}}$$

This means that any voltage from 3.2 to 3.6 V will be acceptable. A reading outside this range indicates trouble.

The pin voltages tend to be more critical in circuits using ICs. In a discrete circuit, the voltages may vary over a ± 20 percent range. Some circuits using ICs will not function properly with an error of 5 percent. Many technicians prefer using a digital voltmeter when working on linear ICs.

Refer again to Fig. 13-28. Note that a waveform is specified for pin 7. This is another valuable servicing aid to be found on some schematics. If the waveform is missing, distorted, or low in amplitude, then a valuable clue has been found. It could indicate a faulty IC. It could also indicate that the signal arriving at the IC is faulty. The schematic usually will have enough sample waveforms that it will be possible to check this out.

Pin 15 in Fig. 13-28 is not specified for any dc voltage. This is because pin 15 is grounded. It should, therefore, be at 0 V with respect to ground. Most technicians will take a measurement at this pin in any case. The reason is simple. The ground connection could be open. Solder joints can fail. This reading will make certain that pin 15 is grounded. The oscilloscope should show a straight line at such a pin. This prevents making a false assumption that 0 V dc ensures ground.

If a dc voltage error is found, the next step is to determine whether the problem is in the IC or the surrounding circuits. It is not a good idea to immediately change the IC. Integrated circuits are not easy to unsolder, and they can be damaged. If sockets are used, it is easy to try a new IC. However, you will run the risk of damaging the new unit if certain types of faults exist in the external parts. Also, *never plug or unplug an IC with the power on*. This invites circuit damage.

You will see that $+12$ V is applied to the circuit in Fig. 13-28. If any of the pin voltages is wrong, this 12-V supply circuit should be checked *immediately*. This must be correct in order for the pin voltages to be correct.

Suppose pin 2 in Fig. 13-28 is reading only

229

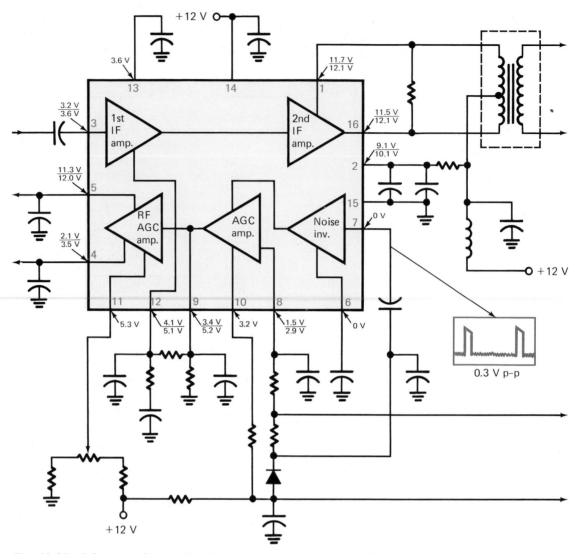

Fig. 13-28 Schematic diagram for IC$_{201}$.

3.5 V. It should not read below 9.1 V. If the 12-V supply is good, what could be wrong? Notice that there is a resistor in series with pin 2. It could be open or high in value. Also note that there are two capacitors from pin 2 to ground. One could be leaky. Since the IC has 16 pins to be unsoldered, it might be a good idea to check these parts first. It is easier and safer to unsolder one lead and lift such parts for an ohmmeter test.

If you reach the conclusion that the fault is in the IC, then it must be replaced. Sockets are the exception, not the rule. Thus, a tricky desoldering job is in store for you. It is very important to not damage the circuit board with excess heat. It is also possible to damage the board by applying the heat too long. Use the proper tools and work carefully.

Review Questions

Choose the letter that best answers each question.

23. Refer to Fig. 13-28. Pin 1 of the IC measures 11.9 V. What can you conclude?
 A. The IC is defective.
 B. There is a power-supply problem.
 C. The transformer is defective.
 D. The voltage is in the normal range.

24. Refer to Fig. 13-28. Pin 1 measures 0 V. What is the next logical step?
 A. Change the IC.
 B. Replace the transformer.
 C. Replace the choke coil in the supply line.
 D. Take a measurement at pins 16 and 2.

25. Refer to Fig. 13-28. All dc voltages check out as good. An IF signal is injected into pin 3, and no output is produced from the IC. What is the problem?
  A. The IC is defective.
  B. The capacitor at pin 14 is leaky.
  C. The capacitor at pin 13 is shorted.
  D. None of the above.

13-5 FUTURE TRENDS

The transistor was a major development in electronics, but it was not widely applied overnight. New skills and processes had to be developed to make the transistor the basic building block for electronic circuits. It took about fifteen years for the transistor to become well established.

The experience gained with transistors made the changeover to integrated circuits much easier. The IC was being widely applied just seven years after its development. It is obvious that ICs are probably the single most important type of component.

The use of ICs can decrease costs. This single fact guarantees the future growth of electronics. Many mechanical devices have been increasing in cost. Designers are discovering that ICs may be available to do the job at reduced cost. They may also be attracted by their smaller size and better reliability. It is sure that electronics will replace other types of systems and designs in the future.

Brand-new products will be available in the future. As the integrated circuits become more and more capable, new ways of using them will develop. A good example of how this happens is found in the story of the electronic calculator. In just a few short years, it became a standard product. Now many people would not think of being without one!

Integrated circuits have few limitations. For example, it was once thought that reasonable power levels were not possible with ICs. Today, standard ICs handle several watts, and some hybrid designs handle several hundred watts. More progress is expected in this area.

Electronics does, indeed, have a bright future. Some of the more obvious growth areas are as follows:

1. *Automotive electronics*. The automobile will carry a significant electronics package involving pollution control, engine efficiency, antitheft, antiskid, anticollision, malfunction indicators, safety, comfort, entertainment, and two-way communications.

2. *Industrial electronics*. There will be an emphasis on improving production through automation. Also more will be spent on energy conservation and efficiency. Integrated circuits will serve in the areas of voltage control, peak-power equalizers, temperature control, speed control, safety, protection, power conversion, and pollution monitoring and control.

3. *Aircraft electronics*. Even the small, private aircraft will become a complex electronic package. Integrated circuits will be used in radio communication, navigation, radar, altimeters, transponders, autopilot, direction-finding, ranging, beam-approach, and safety systems.

4. *Communications electronics*. The world has an enormous appetite for communicating. Personal radio and paging systems will continue to grow. Satellites will open up an era of communications not even dreamed of in the past. Integrated circuits will make all this possible. Some of the specific developments will be in the areas of improved receivers and transmitters, frequency synthesis, security (private line) transmission, signal processing, and data transmission. Telephones will continue to expand in capabilities. Picture transmission, data transmission, call routing, and more links to personal radio systems are all expected.

5. *Architectural electronics*. This is a broad area including schools, homes, institutions, and practically all buildings. It is expected that many new systems will become commonplace. The major areas are fire and smoke detection, energy management, security, temperature and humidity control, and safety systems. It is also expected that many video systems will be installed in buildings such as retail stores to discourage theft.

6. *Medical electronics*. Integrated circuits are ideally suited to this area. Their small size and high reliability are valuable assets. More emphasis will be placed on electronic implants, patient monitoring (including out-patients), diagnostic equipment, therapeutic equipment, and laboratory analysis devices. A new series of powerful instru-

Automotive electronics

Industrial electronics

Aircraft electronics

Communications electronics

Architectural electronics

Medical electronics

Entertainment
electronics

Personal
electronics

ments will allow quick and painless medical measurements. It will be possible to locate tumors and eliminate dangerous exploratory surgery in some cases.

7. *Entertainment electronics.* Here the emphasis will be on flexibility and versatility. Receivers will have various options for remote control, programming, and taping. Television screens will provide more than one picture, and other information and data will become easily accessible. Video games—even chess—will be plentiful. Access to educational libraries will make a home communications center a reality.

8. *Personal electronics.* The hobby aspects of electronics have produced a major industry. Growth will expand in this area as more people become aware of the potential and as the costs go down.

The list could go on and on. The key word is "growth." The future is bright for those skilled in electronics.

Review Question

Choose the letter that best answers the question.

26. Why will integrated circuits find more applications?
    A. They can reduce size.
    B. They can increase reliability.
    C. They can lower cost.
    D. All the above.

Summary

1. Discrete circuits use individual components to achieve a function.
2. Integrated circuits decrease the number of discrete components and reduce cost.
3. Integrated circuits can reduce the size of equipment and eliminate some factory alignment procedures.
4. Integrated circuits often outperform their discrete equivalents.
5. It is possible to increase the reliability of electronic equipment by using more ICs and fewer discrete components.
6. Linear ICs are available in a variety of package styles.
7. Integrated circuits are batch-processed into 10-mil-thick silicon wafers.
8. The key process is called photolithography.
9. Photoresist is the light-sensitive material used to coat the wafer.
10. Aluminum is evaporated onto the wafer to interconnect the various components.
11. A monolithic IC uses a single-stone type of structure.
12. A hybrid IC combines several types of components on a common substrate.
13. Some ICs are both digital and linear.
14. A linear IC may be used to replace several transistor stages.
15. Subsystem ICs may replace more than six separate discrete stages.
16. The operational amplifier is one of the most widely applied linear ICs.
17. Timer ICs use both digital and linear techniques.
18. Timers can be used in the monostable mode, the astable mode, and the time-delay mode.
19. The pulse width of a timer IC is controlled by external parts.
20. Linear ICs often are used to replace audio stages and regulator stages.
21. A phase-locked loop compares an incoming signal with a local signal and produces an error voltage according to any phase or frequency differences.
22. When troubleshooting ICs, check dc voltages at the pins.
23. Check the supply voltages.
24. Always remove and insert ICs with the power off.
25. Take care not to damage the circuit board when desoldering components.

Chapter Review Questions

Choose the letter that best completes each statement.

13-1. A monolithic integrated circuit contains all its components (A) On a ceramic substrate (B) In a single chip of silicon (C) On a miniature printed circuit board (D) On an epitaxial substrate

13-2. A discrete circuit uses
(A) Hybrid technology (B) Integrated technology (C) Individual electronic components (D) None of the above

13-3. Refer to Fig. 13-1. When troubleshooting ICs, one may find a pin by
(A) Counting counterclockwise from pin 1 (top view) (B) Counting clockwise from pin 1 (bottom view) (C) Both of the above (D) None of the above

13-4. Refer to Fig. 13-1. One may find pin 1 on an IC by
(A) Looking for the long pin (B) Looking for the short pin (C) Looking for the wide pin (D) Looking for package markings and/or using data sheets

13-5. When electronic equipment is inspected, a positive identification of ICs can be made by
(A) Using the service literature (B) Counting the package pins (C) Finding all TO-3 packages (D) All the above

13-6. Refer to Fig. 13-2. A technician needs this information
(A) Seldom (B) For choosing a replacement (C) For troubleshooting (D) To determine how to insert the replacement IC

13-7. The major process used in making monolithic ICs is
(A) Microsoldering (B) Photolithography (C) Ball bonding (D) Probe testing

13-8. When monolithic ICs are made, the following is exposed using ultraviolet light:
(A) Silicon dioxide (B) Aluminum (C) Photomask (D) Photoresist

13-9. Monolithic ICs are
(A) Batch-processed (B) Processed individually (C) Miniature discrete circuits (D) None of the above

13-10. The pads on the IC chip are wired to the header tabs
(A) Using plastic conductors (B) With photoresist (C) Using the ball-bonding process (D) In a diffusion furnace

13-11. Refer to Fig. 13-9. Assume that the last boron diffusion (step f) was not performed. The component available is
(A) An inductor (B) A diode (C) A resistor (D) A MOS transistor

13-12. The function of the isolation diffusion is
(A) To insulate the transistors from the substrate (B) To insulate the various components from one another (C) To improve the collector characteristics (D) To form PNP transistors

13-13. The various components in a monolithic IC are interconnected to form a complete circuit by
(A) The aluminum layer (B) Ball bonding (C) Printed wiring (D) Tiny gold wires

13-14. Refer to Fig. 13-11. If this structure is to be used as a capacitor, the dielectric will be
(A) The isolation diffusion (B) The silicon dioxide (C) The substrate (D) The depletion region

13-15. Refer to Fig. 13-15. Assume that signal-injection tests show that no IF signal will pass through the stage. The problem could be
(A) A defective IC (B) Improper AGC voltage (C) One of the discrete components has failed (D) Any of the above

13-16. Refer to Fig. 13-15. Assume that the schematic shows that pin 1

should measure 18 V but a reading shows 0 V. Also assume that all other pin voltages are normal. The trouble is
(A) A shorted capacitor across the transformer (B) A defective IC (C) An open in the primary of the transformer (D) Any of the above

13-17. Refer to Fig. 13-19. A check with an accurate oscilloscope shows that the output pulse is only half as long as it should be. The problem is in
(A) The timing resistor (B) The timing capacitor (C) The IC (D) Any of the above

13-18. Refer to Fig. 13-19. It is desired to make the output pulse 1 s long. A 1-μF capacitor is already in the circuit. The value of the timing resistor is
(A) 1 kΩ (B) 90 kΩ (C) 220 kΩ (D) 0.909 MΩ

13-19. Refer to Fig. 13-20. It is desired to build an oscillator with a 50 percent duty cycle and an output frequency of 38 kHz. Assume that a 0.01-μF capacitor is already in the circuit. The values for R_A and R_B are
(A) $R_A = R_B = 1899\ \Omega$ (B) $R_A = R_B = 3798\ \Omega$ (C) $R_A = 1899\ \Omega$ and $R_B = 3798\ \Omega$ (D) None of the above

13.20. In Fig. 13-21, $R = 18,000\ \Omega$ and $C = 4.7\ \mu$F. The output will switch low, after the trigger, in
(A) 18.2 ms (B) 93.1 ms (C) 188 ms (D) 0.82 s

13-21. A phase-locked-loop IC makes an excellent tone decoder or
(A) Voltage regulator (B) FM demodulator (C) Television IF amplifier (D) Power amplifier

Answers to Review Questions

1. C	10. A	19. D
2. D	11. B	20. B
3. D	12. D	21. D
4. C	13. C	22. C
5. A	14. C	23. D
6. B	15. B	24. D
7. A	16. A	25. A
8. C	17. C	26. D
9. D	18. C	

Electronic Control Devices and Circuits

- One of the major contributions of modern electronics is in the area of control. For example, it may be desired to accurately set and hold the speed of a motor. A bank of lights may need to be dimmed. It may be necessary to regulate the temperature of an electric oven. All these jobs fit under the broad category of control.

  Perhaps the most obvious approach to control in circuits is to use a rheostat. The variable resistance of this device can regulate or control the current flowing in the circuit. You will see in this chapter that several solid-state devices exist that do the job much more efficiently than rheostats.

14-1 INTRODUCTION

Figure 14-1 shows the use of a rheostat to control the brightness of an incandescent lamp. It is fairly obvious that as the rheostat is adjusted for more resistance, the lamp will dim. However, it may not be quite as obvious that this is a very wasteful way to control a lamp.

Consider what happens when some typical numbers are added so that we may analyze a rheostat control circuit. In order to dim the lamp in Fig. 14-2, the rheostat has been set for a resistance of 120 Ω. This makes the total circuit resistance

$$R_T = 120 + 120 = 240 \ \Omega$$

The circuit current can now be found by using Ohm's law:

$$I = \frac{V}{R} = \frac{120}{240} = 0.5 \ A$$

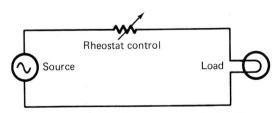

Fig. 14-1 A simple rheostat control circuit.

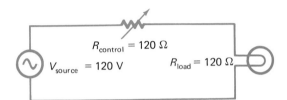

Fig. 14-2 Analyzing a rheostat control circuit.

This, of course, is much less current compared with the flow when the rheostat is set for no resistance:

$$I = \frac{V}{R} = \frac{120}{120} = 1 \ A$$

Ohm's law illustrates that setting the resistance of the rheostat equal to the load resistance halves the current flow. Now, it is time to investigate the efficiency of the circuit when the lamp is dimmed.

The current flow was found to be 0.5 A, and the load resistance is 120 Ω. So

$$P = I^2R = (0.5)^2 \times 120 = 30 \ W$$

When the rheostat is set at no resistance, the power is

$$P = 1^2 \times 120 = 120 \ W$$

From page 235:
Control

Rheostat

On this page:
Voltage control

So we can see that the rheostat is controlling the power dissipated in the load. Specifically, it is seen that the power dissipation has been dropped to one-fourth its original value by halving the current. This is to be expected since power varies as the square of the current.

Now the question is, what happens in the rheostat? At full power, the rheostat is set for no resistance. Therefore, no power will be dissipated in the rheostat:

$$P = 1^2 \times 0 = 0 \text{ W}$$

At one-fourth power, the rheostat dissipation is

$$P = (0.5)^2 \times 120 = 30 \text{ W}$$

This is not an efficient circuit. Half the power is dissipated in the control device. This makes the efficiency 50 percent. As the resistance of the control device increases, the circuit efficiency decreases. This is wasteful. In a high-power circuit, the waste will mean high cost of operation. It will also mean an expensive rheostat. Finally, it will mean that quite a bit of heat will be produced in the control device. This heat will add to the burden on air conditioning in the summer and may even do damage.

Our analysis is not totally accurate. It assumed that the resistance of the incandescent lamp remains constant. It does not. However, the conclusions are correct. Rheostat control is inefficient.

What are the alternatives? One is voltage control. Figure 14-3 shows such a circuit. As the voltage of the source is adjusted from 0 to 120 V, the power dissipated in the load will vary from 0 to 120 W. This is very efficient compared to using a rheostat. Since there is only one resistance in the circuit of Fig. 14-3, there is only one place to dissipate power. The efficiency of the circuit will always be 100 percent.

Unfortunately, voltage control is not easy to obtain. There is no simple and inexpensive

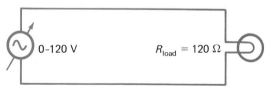

Fig. 14-3   Voltage control.

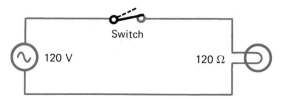

Fig. 14-4   Switch control.

way to control line voltage. A variable transformer will work, but it will be quite expensive in a high-power circuit. It will also be large. Voltage control is much more efficient than resistance control but has several limitations.

Another possibility is suggested by the rheostat circuit. Remember that when the rheostat was set for no resistance, there was no power dissipation in the rheostat. It is possible to control a circuit without dissipating a lot of power in the control device. The secret is to arrange it so that the control device shows very low resistance or very high resistance. This is the concept of a *switch*.

When the switch in Fig. 14-4 is closed, 1 A of current flows. The power dissipated in the load is 120 W. If the switch has very low resistance, very little power is dissipated in the switch. When the switch is open, its resistance is very high. It is so high that no current flows. With no current, there cannot be any dissipation in the switch. Thus, there is never any significant power dissipation in the switch.

You are probably wondering what the circuit of Fig. 14-4 has to do with dimming a lamp. It seems that only basic on-off control is available. This is usually the case with ordinary mechanical switches. However, think for a moment about a very fast switch. Suppose this fast switch can open and close 60 times per second. Suppose this fast switch is closed only half the time. What do you suppose the condition of the lamp will be? Since the lamp will be connected to the source only half the time, it should operate at reduced intensity and the control device should run cool.

Mechanical switches would not serve in this capacity. Even if they could be made to operate quickly, they would wear out in a short time. It is very important to operate quickly, and so an *electronic* switch is needed. The fast operation will allow the lamp to dim without any noticeable flicker. And the rapid transition from near zero resistance to near infinite resistance will ensure that very little power is dissipated in the switch.

Review Questions

Choose the letter that best answers each question.

1. A load has a constant resistance of 60 Ω. A rheostat is connected in series with the load and set for 0 Ω. How much power is dissipated in the load if the line voltage is 120 V?
   A. 0 W
   B. 60 W
   C. 120 W
   D. 240 W

2. What is the circuit efficiency of question 1?
   A. 0 percent
   B. 25 percent
   C. 50 percent
   D. 100 percent

3. Refer to question 1. Everything is the same except the rheostat is set for a resistance of 180 Ω. How much power is dissipated in the load?
   A. 0 W
   B. 15 W
   C. 30 W
   D. 60 W

4. What is the circuit efficiency of question 3?
   A. 0 percent
   B. 25 percent
   C. 50 percent
   D. 75 percent

5. The resistance of a certain load is constant. The current through the load is doubled. By what factor will the power dissipated in the load increase?
   A. 1.25 times
   B. 2.00 times
   C. 4.00 times
   D. Not enough information is given

6. Why is it not efficient to use a control resistor or a rheostat to vary load dissipation?
   A. Much of the total power is dissipated in the control.
   B. Power is set by voltage, not by circuit resistance.
   C. Power is set by current, not by circuit resistance.
   D. Loads do not show constant resistance.

7. Why is there no power dissipation in a perfect switch?
   A. When the switch is closed, its resistance is zero.
   B. When the switch is open, the current is zero.
   C. Both of the above.
   D. None of the above.

14-2 THE SILICON-CONTROLLED RECTIFIER

Silicon-controlled rectifier (SCR)

One of the most popular electronic switches for the control of circuit power is the *silicon-controlled rectifier*, or SCR. This device is easier to understand if we first examine the two-transistor circuit shown in Fig. 14-5. The circuit shows two directly connected transistors, one an NPN and the other a PNP. The secret to understanding this circuit is to recall that bipolar transistors do not conduct until base current is applied. It can be seen in Fig. 14-5 that each transistor must be on to supply the other with base current.

The question is, how does the circuit of Fig. 14-5 come on in the first place? This is why a gate switch has been included. When the source is first connected, no current will flow through the load. Both transistors are off. When the gate switch is closed, the positive side of the supply is applied to the base of the NPN transistor. You should recall that this will forward-bias the base-emitter junction of an NPN unit. The NPN transistor turns on. Now base current begins to flow into the base of the PNP transistor. It, too, turns on. With both transistors on, current flows through the load.

What happens when the gate switch opens? Will the transistors shut off, and will this stop the load current? No, because when the transistors are on, they will supply each other with the needed base current. Once triggered by the gate circuit, the transistors in Fig. 14-5 continue to conduct until the source is removed or the load circuit is opened. The two-transistor switch can be turned on by a gating current, but removing the gating current will not turn the switch back off. Such a

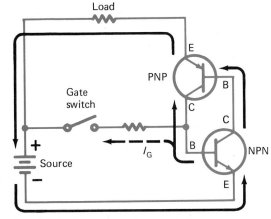

Fig. 14-5 A two-transistor switch.

Electronics:
Principles and
Applications
CHAPTER 14

Latch

Volt-ampere
characteristic
curve

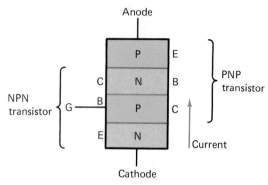

Fig. 14-6 A four-layer diode, or silicon-controlled rectifier.

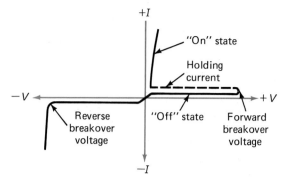

Fig. 14-8 An SCR volt-ampere characteristic curve.

circuit is often called a *latch*. Once triggered, a latch stays on. This circuit may be considered a solid-state equivalent to a latching relay.

Does the circuit of Fig. 14-5 approach the efficiency of a switch? It does show very low resistance when it is on. Both transistors operate in saturation, and the voltage drop across both will be low. It does show a very high resistance when it is off. The circuit of Fig. 14-5 is one example of how electronic switching can be accomplished.

Figure 14-6 shows a way to simplify the two-transistor latching switch. A single four-layer device will do the same job. Study Fig. 14-6 and verify that it is the equal of the two transistors shown in Fig. 14-5. The four-layer diode, as shown in Fig. 14-6, has become very popular in modern electronics. It is called a diode because it conducts in one direction and blocks in the other. It is also called a silicon-controlled rectifier (SCR).

Figure 14-7 shows the schematic symbol for a four-layer diode, or SCR. The current flow is the same as in an ordinary diode, from cathode to anode. Note the addition of the gate lead.

Figure 14-8 is a volt-ampere characteristic curve for an SCR. It shows device behavior for both forward bias ($+V$) and reverse bias ($-V$). As in ordinary diodes, very little cur-

rent flows when the device is reverse-biased until the reverse breakover voltage is reached. This would not occur in most circuits. This much reverse bias would probably destroy the SCR. The forward-bias portion of the volt-ampere curve is very different when compared to an ordinary diode. The SCR stays in the "off" state until the forward breakover voltage is reached. Then, the diode switches to the "on" state. The drop across the diode decreases rapidly, and the current increases. The holding current is the minimum flow that will keep the SCR latched on.

Figure 14-8 does not show how the gate current affects the characteristics of the SCR. In Fig. 14-9, only the forward-bias portion of the curve is shown. Gate current I_{G_1} represents the least of the three values of gate current. You can see that when gate current is low, a high forward-bias voltage is required to turn on the SCR. Gate current I_{G_2} is greater than I_{G_1}. Note that less forward voltage is now needed to turn on the SCR. Finally, I_{G_3} is the highest of the three gate currents shown. It

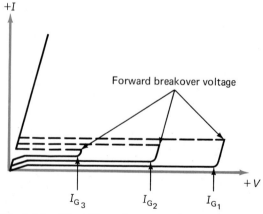

Fig. 14-9 The effect of gate current on breakover voltage.

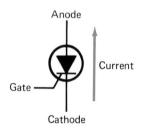

Fig. 14-7 Schematic symbol for the SCR.

requires the least forward bias to turn on the SCR. Figure 14-9 is important because it shows how gate current I_G affects the turn-on characteristic of a silicon-controlled rectifier.

In ordinary operation, SCRs are not subjected to voltages high enough to reach forward breakover. They are switched to the on state with a gate pulse large enough to guarantee turn-on independent of the forward-bias voltage. Once triggered "on" by gate current, the device remains on until the current flow is reduced to a value lower than the holding current.

Now that we know something about SCR characteristics, we can better understand some applications. Figure 14-10 shows the basic use of an SCR to control power in an ac circuit. The load could be a lamp, a heating element, or a motor. The SCR will conduct in only the direction shown. This is a half-wave circuit. The adjustable gate control allows control of the turn-on characteristic of the circuit. Turn-off is automatic when the ac source reverse-biases the SCR.

Figure 14-11 will help you understand how an SCR can control load dissipation in an ac circuit. If the SCR is gated on very late, the power dissipation will be very low. Since the SCR is off most of the time, the load is effectively disconnected from the source for most of the time. Gating the SCR on earlier increases the load dissipation. Finally, at the bottom of Fig. 14-11, full power is shown. Note that the SCR is turned on at the beginning of the cycle. However, "full power" in a simple SCR circuit is a misleading term. Since the SCR blocks the negative alternation completely, half power is really the most that can be achieved.

Figure 14-11 represents what is known as *conduction angle control*. A small conduction angle means the circuit is on for a small

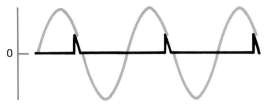

Very low power: SCR is gated on very late

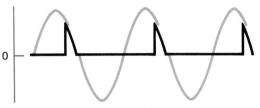

Low power: SCR is gated on late

High power: SCR is gated on early

Full power: SCR is gated on at start of cycle

Fig. 14-11 Conduction angles in an SCR power control circuit.

portion of the ac cycle. A large conduction angle means the circuit is on for a large portion of the ac cycle. The largest conduction angle shown in Fig. 14-11 is at the bottom and is equal to 180°. This can be considered full power since the circuit is half-wave. Half power would be a conduction angle of 90°, and so on. This method of controlling ac power is very efficient. Most of the power will be dissipated in the load device, and only a small fraction in the control device (the SCR).

Silicon-controlled rectifiers and related devices have become very important in power switching and power control applications. They are used at voltages from 6 to over 1000 V. They can operate at current levels from less than 0.5 to over 1000 A. The SCR

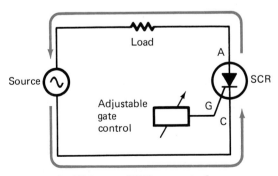

Fig. 14-10 Using an SCR to control ac power.

239

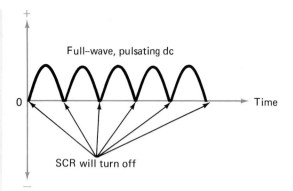

Fig. 14-12 Commutation with full-wave, pulsating dc.

can be used in the conversion of ac to dc, voltage control, current control, as a pulse modulator, and as an inverter.

Silicon-controlled rectifiers can be used in dc circuits if there is some means of turning off the device. Turn-off is automatic in ac circuits since the polarity is continually reversing. There are various methods of achieving dc turn-off. Some of them are as follows:

1. The load circuit can be interrupted by another control device.
2. The SCR current can be diverted by a parallel device.
3. Momentary reverse bias can be applied to the SCR with a capacitor, inductor, or a pulse transformer.

Turning off an SCR circuit after it has been gated on is called *commutation*. Commutation is automatic in ac or pulsating dc circuits. Special commutation circuitry may have to be added for dc control.

As mentioned, commutation is automatic in most pulsating dc circuits. A full-wave, pulsating dc waveform is shown in Fig. 14-12. At those times when the waveform drops to 0 V, the SCR will no longer be forward-biased. The current will drop to zero, which is, of course, less than the holding current, and the SCR will turn off. This makes it possible to achieve full-wave control with an SCR. Of course, it will be necessary to change the ac input to full-wave dc first. A bridge rectifier is often used for this type of application. Figure 14-13 shows how this concept is used in an SCR battery charger.

The battery charger of Fig. 14-13 uses an SCR to provide a shut-down or cut-out feature when the battery reaches full charge. The path is shown through the SCR when the battery is charging. The SCR is on because of the charge stored across capacitor C. When the battery comes up to full voltage, the PNP-NPN transistor switch latches on and provides a discharge path for capacitor C. Now, the SCR is not gated on. The only charging path is now through the transistors and the lamp. Since the lamp comes on, it indicates that the battery has reached full charge.

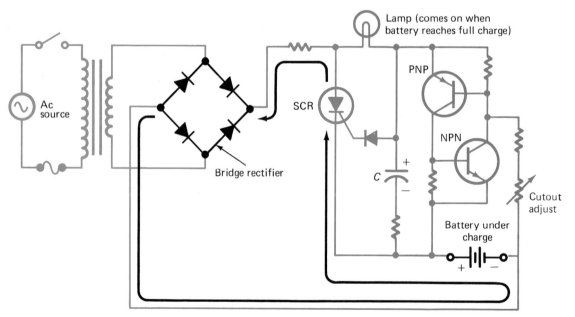

Fig. 14-13 Battery charger with SCR shut-down.

Choose the letter that best answers each question.

8. Refer to Fig. 14-5. Assume the source voltage has just been applied and the gate switch has *not* been closed. What can you conclude about the load current?
   A. The load current will equal zero.
   B. The load current will gradually increase.
   C. It will be mainly determined by V_{source} and the load resistance.
   D. The load current will flow until the gate switch is closed.

9. Refer to Fig. 14-5. Assume the gate switch has been closed and then opened again. What can you conclude about the load current?
   A. It will go off and then on.
   B. It will go on and then off.
   C. It will come on and stay on.
   D. None of the above.

10. How are SCRs normally turned on?
    A. By applying a reverse breakover voltage
    B. By applying a forward breakover voltage
    C. By a separate commutation circuit
    D. By applying gate current

11. How many PN junctions are there in an SCR?
    A. One
    B. Two
    C. Three
    D. Four

12. What happens to the value of forward breakover voltage required to turn on an SCR as more gate current is applied?
    A. It is not changed.
    B. It increases.
    C. It decreases.
    D. None of the above.

13. Refer to Fig. 14-10. Assume that the adjustable gate control is set for maximum power dissipation in the load. What should the load waveform look like?
    A. Half-wave, pulsating dc
    B. Full-wave, pulsating dc
    C. Pure dc
    D. Sinusoidal ac (same as the source)

14. How does an SCR control load dissipation in a circuit such as that shown in Fig. 14-10?
    A. The resistance of the SCR is adjustable.

B. The source voltage is adjustable.
C. The load resistance is adjustable.
D. The conduction angle is adjustable.

15. Refer to Fig. 14-13. What is the function of the SCR?
    A. It controls the current when the battery is charging.
    B. It protects the battery from polarity errors.
    C. It remains off when full charge is reached.
    D. It gates on and activates the lamp when full charge is reached.

14-3 FULL-WAVE DEVICES

The SCR is a *unidirectional* device. It conducts in one direction only. It is possible to combine the function of two SCRs in a single structure to obtain *bidirectional* conduction. The device in Fig. 14-14 is called a *triac*. The triac may be considered as two SCRs connected in parallel, but in opposite directions. When one of the SCRs is in its reverse blocking mode, the other will support the flow of load current. A triac is a full-wave device.

Figure 14-14 shows that the three triac connections are called main terminal 1, main terminal 2, and gate. The gate polarity usually is measured from gate to main terminal 1. A triac may be triggered by a gate pulse that is either positive or negative with respect to main terminal 1. Also, main terminal 2 can be either positive or negative with respect to main terminal 1 when triggering occurs. There are a total of four possible combinations

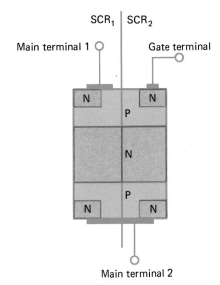

Fig. 14-14   The structure of a triac.

Table 14-1   Triac triggering mode summary

Mode	Gate to Terminal 1	Terminal 1 to Terminal 2	Gate Sensitivity
1	Positive	Positive	High
2	Negative	Positive	Moderate
3	Positive	Negative	Moderate
4	Negative	Negative	Moderate

or triggering modes for a triac. Table 14-1 summarizes the four modes for triac triggering. Note that mode 1 is the most sensitive. Mode 1 compares with ordinary SCR triggering. The other three modes require more gate current.

Figure 14-15 shows the schematic symbol for a triac. It appears as if two SCRs have been connected back to back. The arrows show that the triac is bidirectional—load current can flow in both directions.

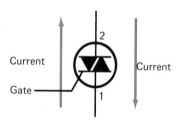

Fig. 14-15   Schematic symbol for the triac.

Triacs were developed to control ac power. Silicon-controlled rectifiers are used where ac is to be converted to dc or where dc must be switched. Both devices are included under the general heading *thyristors*. The term "thyristor" can refer to either an SCR or a triac. Thyristors may be thought of as solid-state switching devices.

Figure 14-16 shows the general use of a triac to control load dissipation in an ac circuit. The adjustable gate control allows the triac to be gated on for either alternation of the ac source. It can be seen that a triac may be gated on at different points for power control (Fig. 14-17). This is the same concept of *conduction angle* that was discussed for SCR con-

trol. However, you should compare Figs. 14-17 and 14-11. It will be obvious that the triac is capable of reaching full power at its largest conduction angle. The SCR circuit only reaches half power at its largest conduction angle.

Commutation can also be different in triac circuits. In ac power control, the triac should switch to its "off" state at each zero-power point. These points occur twice each cycle. If the triac fails to turn off, power control is lost. Commutation is automatic when the load is resistive. When the load is inductive (a motor is an example), the current lags the voltage. Thus, when the current goes to zero, a voltage is applied across the triac. This voltage can turn on the triac. Commutation and power control are lost.

Another problem area is *line transients*. Transients can cause a large voltage change in a short time. Such a transient can switch a thyristor to its on state. Recall that a PN junction when not conducting has a depletion

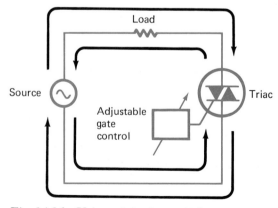

Fig. 14-16   Using a triac to control ac power.

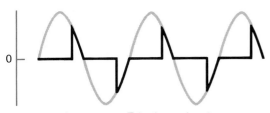

Very low power: Triac is gated on very late

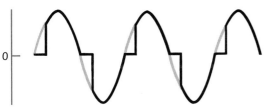

Low power: Triac is gated on late

High power: Triac is gated on early

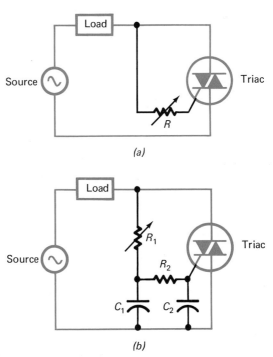

Full power: Triac is gated on at start of each alternation

Fig. 14-17 Conduction angles in a triac power control circuit.

region. Also recall that the depletion region acts as the dielectric of a capacitor. This means that a thyristor in its off state has several internal capacitances. A sudden voltage change across the thyristor terminals will cause the internal capacitances to draw charging currents. These charging currents can act as a gating current and switch on the device.

Now inductive loads and transients are both seen as problem areas in triac control. Such problems can be reduced by special networks that limit the rate of voltage change across the triac. An *RC snubber network* has been added in Fig. 14-18. Snubber networks divert the charging current from the thyristor and can prevent unwanted turn-on.

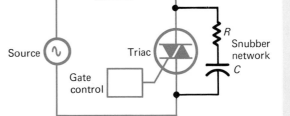

Fig. 14-18 A snubber network.

Triac gating circuits vary according to the specific application. In some applications, the triac is switched full on or full off in a manner similar to a relay. In other applications, the triac is adjusted for various conduction angles. There are many gating circuits in use, and they range from very simple to complex.

Figure 14-19 shows two of the more simple triac gating circuits. Figure 14-19(a) is a simple variable-resistor control. As R is set for less resistance, the triac will gate on sooner and the conduction angle increases. This will result in an increase in load power. Such a simple circuit does not provide control over the entire 360° range. It is also subject to poor symmetry and temperature effects.

Figure 14-19(b) is an *RC* gate control circuit. This circuit has the advantage of pro-

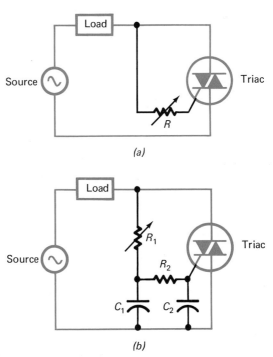

(a)

(b)

Fig. 14-19 Simple triac gating circuits.

243

Electronics:
Principles and
Applications
CHAPTER 14

Diac

Radio-frequency
interference
(RFI)

Harmonics

viding a broader range of control. The setting of R_1 will control how rapidly C_1 and C_2 charge. When the voltage across C_2 reaches the breakover point of the triac, it discharges and the triac is gated on.

The better trigger circuits use a *negative-resistance* device to gate on the triac. Such devices show a rapid decrease in resistance after some critical turn-on voltage is reached. Triggering devices with this negative-resistance quality include neon lamps, unijunction transistors, two-transistor switches, and *diacs*.

Diac

Fig. 14-20 Schematic symbol for the diac.

The schematic symbol for a diac is shown in Fig. 14-20. The symbol suggests that the diac is a bidirectional device. This makes it perfect for controlling triacs which are also bidirectional. The characteristic curve for a diac is shown in Fig. 14-21. The device shows two breakover points V_P+ and V_P-. If either a positive or negative voltage reaches the breakover value, the diac rapidly switches from a high-resistance state to a low-resistance state.

Figure 14-22 shows a popular circuit that combines a diac and a triac to give smooth power control. Resistors R_1 and R_2 determine how rapidly C_3 will charge. When the voltage

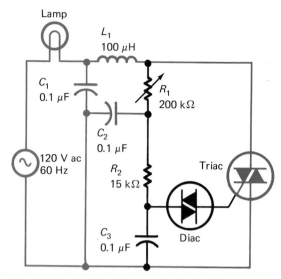

Fig. 14-22 A diac-triac control circuit.

across C_3 reaches the diac breakover point, the diac fires. This provides a complete path for C_3 to discharge into the gate circuit of the triac. The discharge of C_3 gates on the triac.

Figure 14-22 also includes two components to suppress radio-frequency interference (RFI). Triacs switch from the off state to the on state in 1 or 2 μs. This produces an extremely rapid increase in load current. Such a current step contains many *harmonics*. A harmonic is a higher multiple of some frequency. For example, the third harmonic of 1 kHz is 3 kHz. The harmonic energy in triac control circuits can extend to several megahertz. This will produce severe interference to AM radio reception. Capacitor C_1 and inductor L_1 in Fig. 14-22 form a low-pass filter to prevent the harmonic energy from radiating. This will reduce the interference to a nearby AM radio receiver.

It is often necessary to isolate one electronic circuit from another. For example, it may be a sensitive control circuit that could be damaged by a transient. Transients can be induced by loads such as motors. A collapsing field in a motor can generate a voltage transient thousands of volts over the normal line potential! Obviously, such a transient could severely damage a sensitive control circuit.

Isolation from one circuit to another may also be required to protect a human being who must come in contact with sensors or controls. The isolation is needed in this case to protect against electric shock. It is a safety feature.

A control circuit can be isolated by a relay.

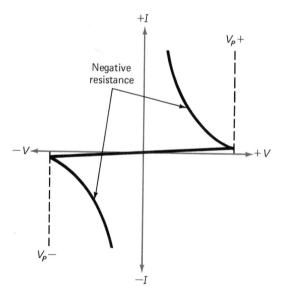

Fig. 14-21 Diac volt-ampere characteristic curve.

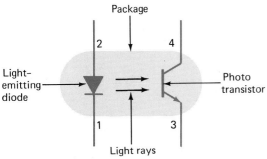

Fig. 14-23 Symbols used to represent an opto-isolator.

There are several types of opto-isolators. Some, for example, use a light-emitting diode along with a phototransistor. The base lead of the transistor in Fig. 14-23 is not shown. The transistor is controlled by light rather than by base current. A control current flowing into pin 1 of the opto-isolator will produce light output from the LED. This light will turn on the phototransistor, and load current will flow into pin 3. Since only light "connects" the two circuits, the isolation is very good.

Another opto-isolator is the photo-SCR. It uses a silicon-controlled rectifier in place of the phototransistor. Look at the device labeled PSCR in the left part of Fig. 14-24. It has an LED and a light-sensitive SCR in a single package. A current applied at the control terminal determines the output of the LED. This, in turn, determines the turn-on characteristics of the SCR.

Figure 14-24 is a circuit that provides isolated ac power control. The load could be a motor, heater, and so on. Both alternations have been traced to help show you how the triac is controlled by the photo-SCR. Note that the bridge rectifier allows the SCR to gate the triac for either alternation of the source. Resistor R_1 limits the gate current in the triac. Resistor R_2 controls the sensitivity of the photo-SCR. You should trace both the load current and the triac gating current for both alternations.

Opto-isolator

However, relays have limitations such as operating speed and reliability. Newer, solid-state devices are replacing relays and provide high-speed operation and good reliability. They also give a large amount of isolation from one circuit to another.

One way to achieve solid-state isolation is to utilize the sensitivity to light that certain materials possess. Silicon, for example, is a light-sensitive material. It is possible to control certain silicon semiconductor characteristics with light.

One of the newer devices in electronic control is the opto-isolator. An opto-isolator uses a light source and a sensor that are sealed into one package (often epoxy plastic). The only connection from one device to the other is an optical connection—the light rays. The isolation of such devices is very good and is usually rated in many millions of ohms.

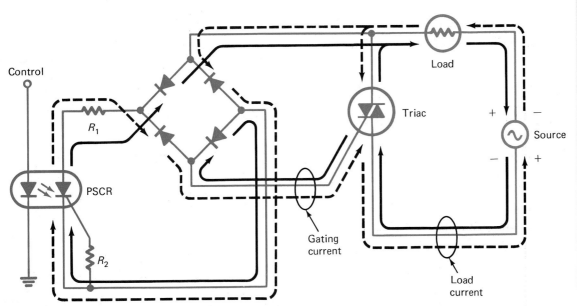

Fig. 14-24 An isolation circuit using a photo-SCR.

Review Questions

Choose the letter that best answers each question.

16. Which of the following devices was specifically developed to control ac power by varying the circuit conduction angle?
    A. The SCR
    B. The triac
    C. The diac
    D. The two-transistor, negative-resistance switch

17. In a triac circuit, how may the load dissipation be maximized?
    A. Hold the conduction angle to 0°.
    B. Hold the conduction angle to 180°.
    C. Hold the conduction angle to 270°.
    D. None of the above.

18. Suppose a triac is used to control the speed of a motor. Also assume that the motor is highly inductive and causes loss of commutation. What is the likely result?
    A. The triac will short and be ruined.
    B. The triac will open and be ruined.
    C. Power control will be lost.
    D. The motor will stop.

19. Which of the following events will turn on a triac?
    A. A rapid increase in voltage across the main terminals
    B. A positive gate pulse (with respect to terminal 1)
    C. A negative gate pulse (with respect to terminal 1)
    D. Any of the above

20. Why may a voltage transient cause unwanted turn-on in thyristor circuitry?
    A. Because of internal capacitances in the thryistor
    B. Because of arc-over
    C. Because of a surge current in the snubber network
    D. All the above

21. Which of the following is a solid-state, bidirectional, negative-resistance device?
    A. Neon lamp
    B. Diac
    C. UJT
    D. Two-transistor switche

22. How many breakover points does the volt-ampere characteristic curve of a diac show?
    A. One   C. Three
    B. Two   D. Four

23. Where will the RFI produced by a triac control circuit be the most pronounced?
    A. In a CB receiver
    B. In a TV receiver
    C. In a stereo FM receiver
    D. In a broadcast-band AM receiver

14-4 FEEDBACK IN CONTROL CIRCUITRY

Electronic control circuits can be made more effective by using feedback to automatically adjust operation should some change be sensed. For example, suppose a thyristor is used to control the speed of a motor. After the motor has been set for speed, assume the load on the motor increases. This will tend to slow down the motor. It is possible, by using feedback, to make the speed of the motor constant even though the mechanical load is changing.

Study Fig. 14-25. We see that a thyristor (SCR) is being used to control a motor. Components R_1 through R_3, C_1, and D_1 form an adjustable dc power supply. The dc is produced by rectifying the ac line. When the top of the source goes positive, the SCR is forward-biased. The SCR gate timing is a function of the setting of R_2. As the wiper arm of R_2 moves down, more of the negative dc voltage stored across C_1 will series-subtract from the positive side (top) of the source. This means the SCR will fire later and the motor will slow down. Moving the wiper arm of R_2 up will use less of the stored energy across C_1, and the motor will speed up since the SCR will fire sooner.

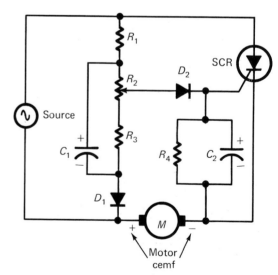

Fig. 14-25 A motor speed controller with feedback.

Let us imagine in Fig. 14-25 that the motor speed has been set. What will happen if the load on the motor is increased? The motor will tend to slow down. However, the motor generates a *counter electromotive force* (cemf). Any change in motor speed will reduce or increase the cemf. So, if the motor slows down, less cemf will be generated by the motor. Since the polarity of the cemf is negative as applied to the gate through R_4 and C_2, it acts to delay the firing of the SCR. When the motor speed drops, the SCR tends to fire sooner, and this will speed up the motor. One effect tries to cancel the other, and the speed of the motor remains reasonably constant.

Feedback is very valuable in circuits such as the one shown in Fig. 14-25. However, a motor may not develop an adequate cemf voltage that can be used as a feedback signal. In many cases, it will be necessary to use a separate circuit or device to produce the feedback signal. It is also possible to achieve more accurate control by using a separate speed sensor.

The electronic speed control in Fig. 14-26 is more elaborate but gives better speed accuracy. The motor drives a tachometer which acts as a speed sensor. The tachometer converts shaft speed into electric pulses. These pulses form a feedback signal that is compared with a reference signal in the error detector. Any change in motor speed will cause an error between the two signals. This resulting error signal is amplified and used to correct the speed of the motor.

The reference signal in Fig. 14-26 is produced in a variable-frequency oscillator (VFO). Different motor speeds can be selected by adjusting the frequency of the oscil-

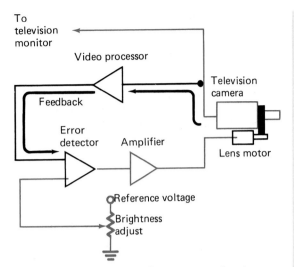

Fig. 14-27 A television lens servomechanism.

lator. When the oscillator is first adjusted, there will be an error between the feedback signal and the reference signal. This will produce a control signal which will be amplified, and the motor will change speed until the two signals agree once again. If variable speed control is not required, a fixed-frequency oscillator may be substituted for the VFO, or the reference signal may come from another source—perhaps even a computer.

Circuits such as the one in Fig. 14-26 fit into a category known as *servomechanisms*. A servomechanism is any feedback control system where one or more of the signals represent mechanical action. Mechanical action is not limited to the speed of a motor. It is possible to use a servomechanism for positioning, opening, closing, controlling, or moving any number of devices or objects. Suppose we wish to use a remote television camera for a security system. The camera will view a parking lot behind the building where we work. Because of the large change in light level, it will be necessary to adjust the camera lens from time to time. Is this a job for a servomechanism?

The television camera problem could be solved in a number of ways. It might be possible, for example, to mount a small motor on the camera to drive the lens mechanism. Then, an operator could look at a television monitor and remotely adjust the lens for the proper brightness level. This would *not* be a servomechanism. It lacks one very important element—feedback to provide automatic control.

How could a servomechanism be applied to the TV camera problem? Refer to Fig. 14-27.

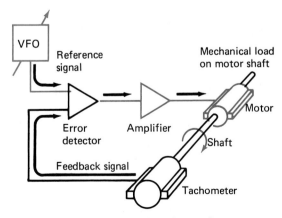

Fig. 14-26 Sensing speed with a tachometer.

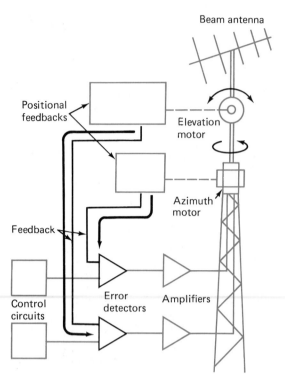

Fig. 14-28 A beam antenna servomechanism.

Part of the signal from the camera is sent to a video processor. This is a circuit that examines the signal and produces an output relative to the brightness of the image. This brightness signal is fed back into an error detector

along with a reference voltage. Any difference between the reference and the brightness signal will be amplified and used to automatically adjust the lens. Notice that the amplifier drives the lens motor. This is a servomechanism. It is automatic, it uses feedback, and mechanical action is involved.

A servomechanism may involve more than one feedback loop (Fig. 14-28). A beam antenna must be pointed in the proper direction (azimuth) and tilted at the proper angle (elevation) to receive earth satellite signals. Two motors are used to control the proper positioning of the beam antenna. Some servomechanisms are very complex.

Integrated circuits are beginning to reduce the cost and complexity of servomechanisms. Figure 14-29 shows almost a complete motor-speed servomechanism on one chip! The performance and characteristics for such an IC are very impressive:

1. 1 percent speed accuracy
2. −30 to +85°C temperature range
3. 10- to 16-V dc supply range
4. Supplies motor currents up to 2 A
5. Supplies starting currents up to 3 A
6. Very few external parts required
7. Protection for the IC built in
8. Protection for the motor built in

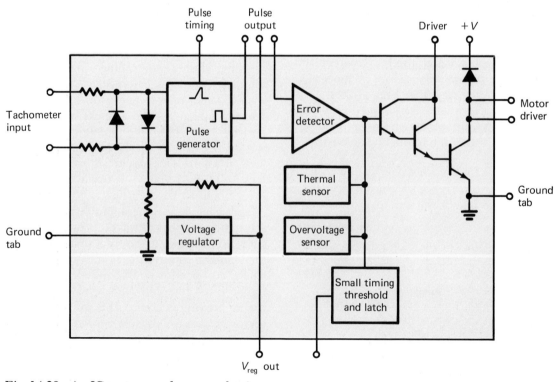

Fig. 14-29 An IC motor speed servomechanism.

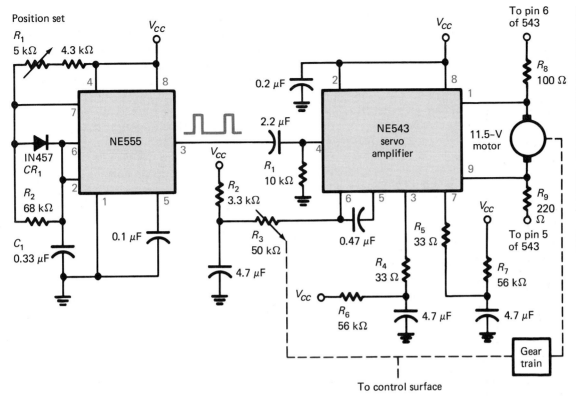

Fig. 14-30 A positioning servomechanism.

9. Protects against overvoltage, excessive temperature, and motor stalls
10. Suited for use in tape players, recorders, and various industrial controls
11. All the above in one dual-in-line IC

A positioning servomechanism based on integrated circuits is shown in Fig. 14-30. A 555 timer IC forms the reference oscillator. A 543 servo IC contains the error circuits and motor amplifiers required to complete the system.

Review Questions

Choose the letter that best answers each question.

24. Refer to Fig. 14-25. Assume the charge across C_2 is increasing. What could this be caused by?
   A. A shorted SCR
   B. The motor slowing down
   C. The motor speeding up
   D. A decrease in source voltage

25. Refer to Fig. 14-25. Assume the charge across C_2 is increasing. What should this cause?

A. The SCR to gate on earlier in the alternation
B. The SCR to gate on later in the alternation
C. No change in the conduction angle
D. D_2 to be forward-biased all the time

26. Refer to Fig. 14-25. What produces the feedback signal?
   A. D_2
   B. C_1
   C. R_4
   D. The motor

27. Suppose a light dimmer circuit uses a photosensor to provide a feedback signal for better regulation of light intensity. Why would this *not* qualify as a servomechanism?
   A. The system is not automatic.
   B. No mechanical action is involved.
   C. The circuit has no load.
   D. All the above.

28. Suppose a chemical plant operator uses a remote pressure sensor to monitor gas flow. When the pressure goes too high, the operator closes a circuit which runs a motor and controls a valve. Why would this *not* qualify as a servomechanism?
   A. The system is not automatic.
   B. No mechanical action is involved.

Electronics:
Principles and
Applications
CHAPTER 14

Thyristor control
circuit

C. Gas pressure has nothing to do with servomechanisms.
D. All the above.

29. Refer to Fig. 14-26. Assume that you have measured the output of the error detector for 1 minute and noted no change. If the servomechanism is working properly, what can you conclude?
A. The error detector has only one input signal.
B. The motor is gradually slowing down.
C. The motor speed is stable.
D. The motor is gradually speeding up.

30. Refer to Fig. 14-26. Assume that you are monitoring the output of the error detector. You note that as the mechanical load increases, thje output goes positive. As the mechanical load decreases, the output goes negative. What can you conclude about the servomechanism?
A. It is working properly.
B. It is missing its reference signal.
C. It is not working at all.
D. It is connected to a defective power supply.

14-5 TROUBLESHOOTING ELECTRONIC CONTROL CIRCUITS

Before we give specific troubleshooting hints, it is necessary to review some rules for the proper use of electronic control circuits. It is also necessary to establish some general safety precautions.

The thyristor control circuits in this chapter can be used with only certain kinds of loads. Severe damage to the load and the control circuit may result if improper connections are attempted. The general rule for SCR and triac control circuits is this: never attempt to use them with ac-only equipment. Ac-only equipment would include

1. Fluorescent lamps (unless specially designed for thyristor control)
2. Radios
3. Television receivers
4. Induction motors (including those on fans, record players, tape players, washing machines, large equipment such as air compressors, and so on)
5. Transformer-operated devices (such as soldering guns, model-train power supplies, battery chargers, and so on)

In general, it is safe to use thyristor control circuits with resistive loads. This would include incandescent lamps, soldering irons, heating pads, and so on. It is also safe to use thyristor control circuits with universal (ac/dc) motors. These motors usually are found in portable power tools such as drills, saber saws, and sanders. When in doubt, check the manufacturer's specifications. Also be sure that the wattage rating of the load does not exceed the wattage rating of the control circuit.

The general safety rules for analyzing and troubleshooting electronic control circuits are the same as those for any line-operated circuit. It is very dangerous to use oscilloscopes and other similar test equipment in these circuits. A ground loop is very likely to cause damage and perhaps severe electric shock. Even if the test equipment is battery-operated, danger exists. A battery-operated oscilloscope may seem safe, but remember that the cabinet and the probe grounds may reach a dangerous potential when directly connected into power circuits.

If a control circuit is light-duty, it may be possible to use an isolation transformer. Then it is safe to use test instruments for analyzing the circuit. Be sure the wattage rating on the isolation transformer is adequate before attempting this approach.

Always remember that troubleshooting is a logical and orderly process. Think of "GOAL." Good troubleshooting involves Observation, Analysis, and finally Limiting the possible causes. Suppose you are troubleshooting a thyristor motor speed control. You notice that the motor always runs at top speed. The speed control has no effect. Assume that you have already run the usual preliminary checks and have found nothing obvious. What is the next step?

Once the preliminary checks and observations have been made, it is time for analysis. Ask what kinds of problems could cause the motor to always run at top speed. Could the thyristor be open? No, because that should stop the motor. Could the thyristor be shorted? Yes, that is a definite possibility. Is it time to change the thyristor? No, the analysis is not over yet. Are there any other causes for the observed symptoms? What if the gating circuit is defective? Could this make the motor run at top speed? Yes, it could.

The last part of the troubleshooting process

is to limit the possibilities. How can this be done? One way is to shut off the power. Then disconnect the gate lead of the thyristor. Turn the power back on. Does the motor run at top speed again? If it does, the thyristor is, no doubt, shorted. It should be replaced. What if the motor will not run at all? This means the thyristor is good. With its gate lead open, it will not come on. The problem is in the gate circuit.

There are many types of gate circuits. It will be necessary to study the circuit and determine its principle of operation. If the circuit uses a unijunction transistor, it will be necessary to determine whether the UJT pulse generator is working as it should. If the circuit uses a diac, it will be necessary to determine whether the diac is operating properly. It may be possible to use resistance analysis (with the power off) to find a defect. A resistor may be open. A capacitor may be shorted. Some solid-state device may have failed.

At this point, you should begin to realize that the answers are usually not in the manuals or the textbooks. A good troubleshooter understands the basic principles of electronic devices and circuits. This knowledge will allow a logical and analytic process to flow. It is not always easy. Highly skilled technicians "get stuck" from time to time. However, usually they do not keep retracing the same steps over and over. Once a particular fact is confirmed, it is noted on paper or mentally. Using paper is best because another job or quitting time can interfere. It is too easy to forget what has and has not been checked.

Different technicians use somewhat different approaches to troubleshooting. All good technicians have these things in common, however:

1. They work safely.
2. They follow the manufacturer's recommendations.
3. They find and use the proper service literature.
4. They use a logical and orderly process.
5. They observe, analyze, and limit the possibilities.
6. They keep abreast of technology.
7. They understand how devices and circuits work in general terms. They understand what each major stage is supposed to do.
8. They are skilled in the use of test equipment and tools.
9. They are neat, use the proper replacement parts, and put *all* the shields, covers, and fasteners back where they belong.
10. They check their work carefully to make sure nothing was overlooked.
11. They never consider modifying a piece of equipment or defeating a safety feature just because it is convenient at the time.

Technicians who develop these skills and habits are in great demand. They always will be.

Review Questions

Choose the letter that best answers each question.

31. Which of the following loads should never be connected to a thyristor control circuit?
    A. Incandescent lamps
    B. Soldering irons
    C. Soldering guns
    D. Soldering pencils

32. Why is it not safe to connect a line-operated oscilloscope across a triac in a light dimmer?
    A. A ground loop could cause damage.
    B. The cabinet and controls of the scope could assume line potential.
    C. Both of the above.
    D. None of the above.

33. Refer to Fig. 14-25. Assume the SCR is open. What is the most likely symptom?
    A. The motor will run at top speed (no control).
    B. The motor will speed up and slow down.
    C. Diode D_1 will be burned out by the overload.
    D. The motor will not run.

34. In Fig. 14-25, assume D_2 is open. What is the most likely symptom?
    A. The motor will run at top speed (no control).
    B. The motor will be damaged.
    C. The motor will not run at all.
    D. Resistor R_4 will burn up.

35. Refer to Fig. 14-25. The SCR is shorted from cathode to anode. What is the most likely symptom?

A. The motor will run at top speed (no control).
B. The motor will stop.
C. The motor will run slow, but with no speed regulation.
D. The potentiometer R_2 will overload and smoke.

36. Refer to Fig. 14-27. When the power is turned on, the motor runs the lens wide open and then stops. Light level and the brightness-adjust control seem to have no effect. What is wrong?
A. Defective error detector.

B. Defective amplifier.
C. Defective reference signal.
D. Any of the above could be defective.

37. Refer to Fig. 14-27. Suppose the reference voltage is unstable because of a faulty regulator in the power supply. What is the most likely symptom?
A. The brightness will change.
B. The motor will run in only one direction.
C. The motor will not run at all.
D. The feedback voltage from the video processor will refuse to change.

Summary

1. A rheostat can be used to control circuit current.
2. Rheostat control is not efficient since much of the total circuit power is dissipated in the rheostat.
3. Voltage control is much more efficient than resistance control.
4. Switches dissipate little power when open or when closed.
5. A fast switch can control power in a circuit without producing undesired effects such as flicker.
6. Switch control is much more efficient than resistance control.
7. A latch circuit can be formed from two transistors: one an NPN and the other a PNP.
8. A latch circuit is normally off. It can be turned on with a gating current.
9. Once the latch is on, it cannot be turned off by removing the gate current.
10. A latch can be turned off by interrupting the load circuit or by applying reverse bias.
11. A four-layer diode or silicon-controlled rectifier (SCR) is equivalent to the NPN-PNP latch.
12. An SCR, like an ordinary diode, conducts from cathode to anode.
13. An SCR, unlike an ordinary diode, does not conduct until turned on by a breakover voltage or a gate current.
14. In ordinary operation, SCRs are gated on and not operated by breakover voltage.
15. The SCR is basically a half-wave device.
16. Commutation refers to turning off the SCR.
17. The SCR is a unidirectional device since it conducts in only one direction.

18. The triac is a bidirectional device since it conducts in both directions.
19. Triacs are capable of full-wave ac power control.
20. Triacs have four triggering modes.
21. The term "thyristor" is general and can be used in referring to SCRs or triacs.
22. A snubber network may be needed when triacs are used with inductive loads or when line transients are expected.
23. Negative-resistance devices are often used to trigger thyristors.
24. A diac is a bidirectional, negative-resistance device.
25. Opto-isolators use light-sensitive devices and LED emitters sealed into one package.
26. Feedback can be used in control circuits to provide automatic correction for any error.
27. A load such as a motor may provide its own feedback signal.
28. A separate sensor such as a tachometer may be required to provide the necessary feedback signal.
29. A servomechanism is any control system using feedback that represents mechanical action.
30. Servomechanisms provide automatic control.
31. Thyristor control circuits may be safely used with universal (ac/dc) motors.
32. The wattage rating of thyristor control circuit must be greater than its connected load.
33. A problem in a thyristor control circuit may be isolated by opening the gate lead.

Chapter Review Questions

Choose the letter that best answers each question.

14-1. Refer to Fig. 14-1. Suppose the load resistance is constant at 80 Ω and the source voltage is 240 V. What will the load dissipation be if the rheostat is set for 60 Ω?
(A) 85 W (B) 168 W (C) 235 W (D) 411 W

14-2. What is the dissipation in the rheostat in question 14-1?
(A) 62 W (B) 176 W (C) 345 W (D) 590 W

14-3. What is the efficiency of the circuit in question 14-1?
(A) 57 percent (B) 68 percent (C) 72 percent (D) 83 percent

14-4. Suppose the resistance of a load is constant. What will happen to the power dissipation in the load if the current is increased 3 times its original value?
(A) The power will drop to one-third its original value (B) The power will remain constant (C) The power will increase 3 times (D) The power will increase 9 times

14-5. Why is resistance control so inefficient?
(A) Resistors are very expensive (B) The control range is too restricted (C) Much of the circuit power dissipates in the control device (D) None of the above

14-6. Refer to Fig. 14-5. What is the purpose of the gate switch?
(A) To turn the transistor switch on and off (B) To commutate the NPN transistor (C) It provides an emergency shut-down feature (safety) (D) It turns on the transistor switch

14-7. Refer to Fig. 14-5. The transistors are on. How may they be shut off?
(A) Open the gate switch (B) Close the gate switch (C) Open the load circuit (D) Increase the source voltage

14-8. Refer to Fig. 14-5. Which of the following terms best describes the way the circuit works?
(A) Latch (B) Resistance controller (C) Rheostat controller (D) Linear amplifier

14-9. How is a silicon-controlled rectifier similar to a diode rectifier?
(A) Both can be classed as thyristors (B) Both support only one direction of current flow (C) Both are used to change ac to pulsating dc (rectify) (D) They both have one PN junction

14-10. What is the effect of increasing the gate current in an SCR?
(A) The reverse breakover voltage is improved (B) The forward breakover voltage is increased (C) The forward breakover voltage is decreased (D) The internal resistance of the SCR increases

14-11. Refer to Fig. 14-10. What is the maximum conduction angle of this circuit?
(A) 45° (B) 90° (C) 180° (D) 360°

14-12. Refer to Fig. 14-10. If the load is a motor, what should the motor do if the conduction angle is increased?
(A) Slow down (B) Stop (C) Gradually slow down (D) Speed up

14-13. Why is thyristor control more efficient than resistance control?
(A) Thyristors are less expensive (B) Thyristors are easier to mount on a heat sink (C) Thyristors vary their resistance automatically (D) Thyristors are switches

14-14. Refer to Fig. 14-13. What turns on the lamp when the battery reaches full charge?
(A) The two-transistor switch (B) The bridge rectifier (C) The SCR (D) The capacitor

14-15. Refer to Fig. 14-13. What happens to the SCR when the two-transistor switch turns on?
(A) It turns on for every positive alternation of the source (B) It remains off (C) It turns off and goes on again when its anode is positive (D) It charges the battery at maximum current

14-16. Turning off a thyristor is known as
(A) Gating (B) Commutating (C) Forward biasing (D) Interrupting

14-17. Which of the following devices was developed specifically for the control of ac power?
(A) The SCR (B) The UJT (C) The snubber (D) The triac

14-18. Refer to Fig. 14-16. Suppose the load is an incandescent lamp and the conduction angle of the circuit is decreased. What will happen to the lamp?
(A) Nothing (B) It will dim (C) It will output more light (D) It will flicker violently

14-19. Refer to Fig. 14-16. The load is operating at full power. What is the conduction angle of the circuit?
(A) 45° (B) 90° (C) 180° (D) None of the above

14-20. What is the chief advantage of a triac as compared with a silicon-controlled rectifier?
(A) It costs less to buy (B) It runs much cooler (C) The triac is bi-directional (D) All the above

14-21. Refer to Fig. 14-18. What is the function of the snubber network?
(A) It prevents false commutation (B) It reduces television interference (C) It helps reduce unwanted gate-on (D) It helps the gate control circuit work sooner

14-22. Refer to Fig. 14-22. Which component turns on and then gates the triac?
(A) Capacitor C_3 (B) Capacitor C_1 (C) Resistor R_2 (D) The diac

14-23. Refer to Fig. 14-22. Which component or components have been added to reduce radio interference?
(A) Inductor L_1 and capacitor C_1 (B) Capacitor C_3 (C) The diac (D) Capacitor C_2 and resistor R_2

14-24. Some devices exhibit a rapid decrease in resistance after some turn-on voltage is reached. What are they called?
(A) Negative-resistance devices (B) FETs (C) Linear resistive elements (D) Voltage-dependent resistors (VDRs)

14-25. Refer to Fig. 14-24. Assume the current flowing through the LED toward the point marked control is increased. What should the effect be?
(A) The load should shut off (B) The SCR should gate sooner and increase the triac conduction angle (C) The triac should be commutated sooner (D) None of the above

14-26. Refer to Fig. 14-24. Where is the electrical path from the control point to the load circuit?
(A) Through the PSCR (B) Through R_1 and the bridge rectifier (C) Through R_2 and the triac (D) There is none

14-27. Refer to Fig. 14-26.   The shaft that drives the tachometer is slipping badly.   What do you suppose the symptom will be?
(A) None because the speed is regulated (B) The motor will slow down and stop (C) The motor will run fast (D) The reference signal will become unstable

14-28. Which of the following devices should never be operated from a thyristor power control device?
(A) A washing machine motor (B) A heater (C) Christmas tree lights (D) A soldering iron

Answers to Review Questions

1. *D*	14. *D*	26. *D*
2. *D*	15. *C*	27. *B*
3. *B*	16. *B*	28. *A*
4. *B*	17. *D*	29. *C*
5. *C*	18. *C*	30. *A*
6. *A*	19. *D*	31. *C*
7. *C*	20. *A*	32. *C*
8. *A*	21. *B*	33. *D*
9. *C*	22. *B*	34. *C*
10. *D*	23. *D*	35. *A*
11. *C*	24. *C*	36. *D*
12. *C*	25. *B*	37. *A*
13. *A*		

Appendix

257

Appendix A

ELECTRONIC WIRING AND SOLDERING

A1 Chassis Wiring

To Install a Part:

1. Cut the leads to the proper length.

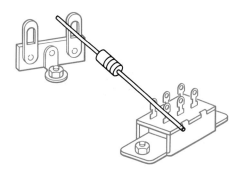

2. Fasten the lead ends.

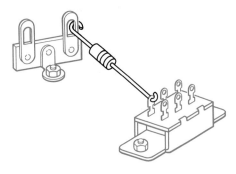

NOTE: Use sleeving when it is called for to provide insulation.

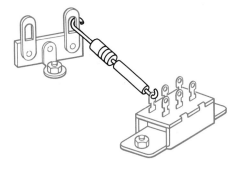

This material excerpted from the Heathkit Builder's Guide by permission of the Heath Company.

259

To Solder a Connection:

1. Position the circuit board with the plain side (not the foil side) up.

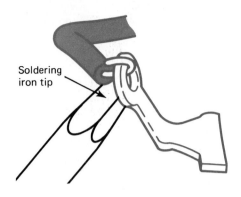

Soldering iron tip

2. Apply only enough solder to thoroughly wet both the tip and the connection.

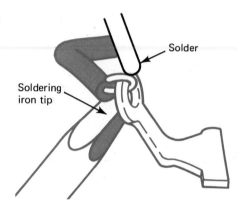

Solder

Soldering iron tip

3. Let the connection harden before moving the wire. The connection should be smooth and bright.

4. Check the connection. Poor connections look crystalline and grainy, or the solder tends to blob. Reheat the connection if it does not look smooth and bright.

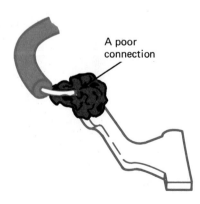

A poor connection

Remember:

Keep the soldering iron tip clean. Wipe it often on a wet sponge or cloth; then apply solder to it to give the entire tip a wet look. This "tinning" process will protect the tip and enable you to make good connections. When the solder tends to "ball" or not stick to the tip, the tip needs to be cleaned and retinned. Use rosin core, radio-type solder (60:40 or 50:50 tin-lead content) for all soldering.

A2 Printed Circuits

To Install a Part:

The following example uses a resistor, since resistors are usually installed first.

1. Position the circuit board with the plain side (not the foil side) up.

2. Hold the resistor by the body as shown and bend the leads straight down.

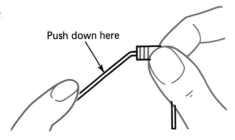

3. Push the leads through the holes at the proper location on the circuit board. The end with color bands may be positioned either way.

4. Press the resistor against the circuit board. Then bend the leads outward slightly to hold the resistor in place.

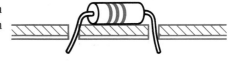

To Solder a Connection:

1. Place the soldering iron tip against both the lead and the circuit board foil. Heat both for 2 or 3 seconds.

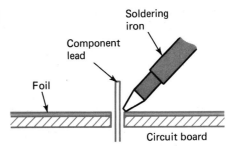

2. Then apply solder to the other side of the connection. IMPORTANT: Let the heated lead and the circuit board foil melt the solder.

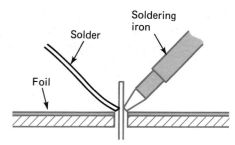

3. As the solder begins to melt, allow it to flow around the connection. Then remove the solder and the iron and let the connection cool.

4. Hold the lead with one hand while you cut off the excess lead length close to the connection. This will keep you from being hit in the eye by the flying lead.

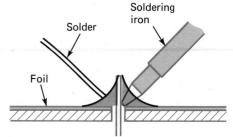

To Check a Connection:

Be sure the solder made a good electric connection. When both the lead *and* the circuit board foil are heated at the same time, the solder will flow onto the lead and the foil evenly. The solder will then make a good electrical connection between the lead and the foil.

Solder flows outward and gradually blends with the foil and the lead.

Foil

Soldering iron positioned correctly

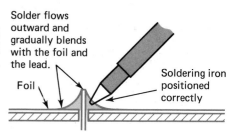

When the *lead* is not heated sufficiently, the solder will not flow onto the lead as shown. Reheat the connection and, if necessary, apply a small amount of additional solder to obtain a good connection.

Solder does not flow onto lead. A dark rosin bead surrounds and insulates the lead from the connection.

Burned rosin

Foil

Solding iron positioned incorrectly

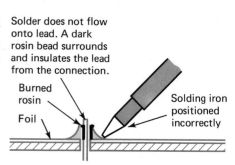

When the *foil* is not heated sufficiently, the solder will blob on the circuit board as shown. Reheat the connection and, if necessary, apply a small amount of additional solder to obtain a good connection.

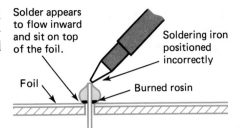

Solder appears to flow inward and sit on top of the foil.

Foil

Soldering iron positioned incorrectly

Burned rosin

Be sure you did not make any solder bridges. Due to the small foil area around the circuit board holes and the small areas between foils, you must use the utmost care to prevent solder bridges between adjacent foil areas.

A solder bridge may occur if you accidentally touch an adjacent connection, if you use too much solder, or if you "drag" the soldering iron across other foils as you remove it from the connection. Always take a good look at the foil area around each lead before you solder it. Then, when you solder the connection, make sure the solder remains in this area and does not bridge to another foil. This is especially important when the foils are small and close together.

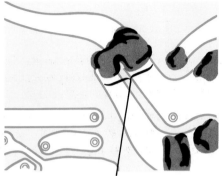

A solder bridge between two adjacent foils

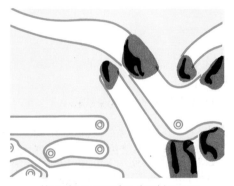

How the connection should appear

A3 Printed Circuit Desoldering

It is sometimes necessary to remove a defective part from a printed circuit board. This can be difficult to do when the part has several leads. Several tools and aids have been developed to make the job easier. There are two popular vacuum type tools for this job. The vacuum desoldering pencil melts the joint and then the bulb is released to draw the solder off of the board (Fig. A3-1).

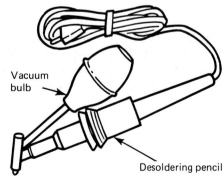

Vacuum bulb

Desoldering pencil

Fig. A3-1

After all the leads have been desoldered, the part can be removed. A separate vacuum desoldering bulb (Fig. A3-2) can be used with a separate soldering pencil to accomplish the same job.

Fig. A3-2

Some technicians prefer to heat all connections of a component at the same time. This will also allow removal of the part. Special desoldering tips are available to accomplish this. Different tip styles are needed for the various transistor and integrated circuit parts (Fig. A3-3).

Yet another technique is to use finely braided wire. The wire and the tip are both applied to the connection. Capillary action causes the solder to flow off of the board and into the braided wire. Special wire made just for this purpose is available.

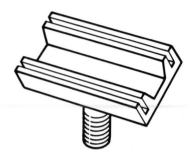

Fig. A3-3

Appendix B

MAJOR SEMICONDUCTOR COMPONENTS

Name of Device	Circuit Symbol	Electrical Characteristics		Max. Ratings Available	Major Applications
Diode or Rectifier	Anode / Cathode		Conducts easily in one direction, blocks in the other.	1500 A 3000 V	Rectification Blocking Detecting Steering
Avalanche (Zener) Diode	Anode / Cathode		Constant voltage characteristic in negative quadrant.	22 V 1 W	Regulation Reference Clipping
Integrated Voltage Regulator (IVR)			Programmed to desired V_{21} by two resistors.	40 V 100 mA 0.4 W	Shunt voltage regulator Reference element Error modifier Level sensing Level shifting
Tunnel Diode	Positive electrode / Negative electrode		Displays negative resistance when current exceeds peak point current I_P.	Peak point current = 100 mA; Resistive cutoff frequency = 40 GHz	UHF converter Logic circuits Microwave circuits Level sensing
Back Diode	Anode / Cathode		Similar characteristics to conventional diode except very low forward voltage drop.	5 mA 400 mV	Microwave mixers and low power oscillators
Thyrector			Rapidly increasing current above rated voltage in either direction.	70 A peak pulse (2-in^2 cell)	Transient voltage suppression and arc suppression

Adapted by permission of Radio Shack, a division of Tandy Corporation.

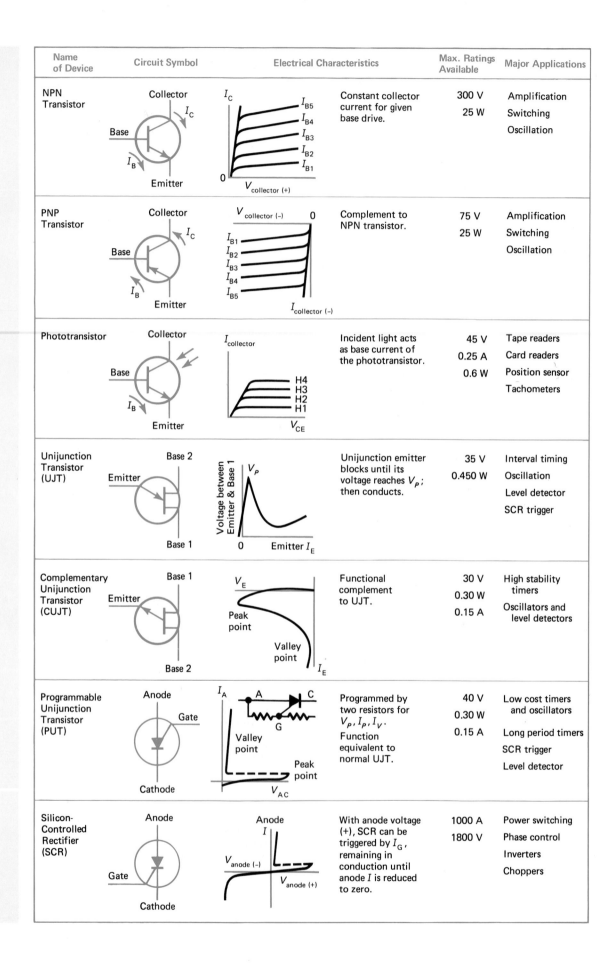

Name of Device	Circuit Symbol	Electrical Characteristics	Max. Ratings Available	Major Applications
NPN Transistor	Collector, I_C, Base, I_B, Emitter	I_C vs $V_{collector (+)}$; $I_{B5}, I_{B4}, I_{B3}, I_{B2}, I_{B1}$	300 V, 25 W	Amplification, Switching, Oscillation
Constant collector current for given base drive.				
PNP Transistor	Collector, I_C, Base, I_B, Emitter	$V_{collector (-)}$, $I_{B1}, I_{B2}, I_{B3}, I_{B4}, I_{B5}$ vs $I_{collector (-)}$	75 V, 25 W	Amplification, Switching, Oscillation
Complement to NPN transistor.				
Phototransistor	Collector, Base, I_B, Emitter	$I_{collector}$ vs V_{CE}; H4, H3, H2, H1	45 V, 0.25 A, 0.6 W	Tape readers, Card readers, Position sensor, Tachometers
Incident light acts as base current of the phototransistor.				
Unijunction Transistor (UJT)	Base 2, Emitter, Base 1	Voltage between Emitter & Base 1 vs Emitter I_E; V_P	35 V, 0.450 W	Interval timing, Oscillation, Level detector, SCR trigger
Unijunction emitter blocks until its voltage reaches V_P; then conducts.				
Complementary Unijunction Transistor (CUJT)	Base 1, Emitter, Base 2	V_E, Peak point, Valley point vs I_E	30 V, 0.30 W, 0.15 A	High stability timers, Oscillators and level detectors
Functional complement to UJT.				
Programmable Unijunction Transistor (PUT)	Anode, Gate, Cathode	I_A, A, C, G, Valley point, Peak point vs V_{AC}	40 V, 0.30 W, 0.15 A	Low cost timers and oscillators, Long period timers, SCR trigger, Level detector
Programmed by two resistors for V_P, I_P, I_V. Function equivalent to normal UJT.				
Silicon-Controlled Rectifier (SCR)	Anode, Gate, Cathode	Anode, I, $V_{anode (-)}$, $V_{anode (+)}$	1000 A, 1800 V	Power switching, Phase control, Inverters, Choppers
With anode voltage (+), SCR can be triggered by I_G, remaining in conduction until anode I is reduced to zero.				

Name of Device	Circuit Symbol	Electrical Characteristics	Max. Ratings Available	Major Applications
Complementary Silicon-Controlled Rectifier (CSCR)	Anode / Gate / Cathode	Anode I / $V_{AC(-)}$ / $V_{AC(+)}$	Polarity complement to SCR.	50 V / 0.25 A / 0.45 W	Ring counters / Low speed logic / Lamp driver
Light-Activated SCR (LASCR)	Anode / Gate / Cathode	Anode I / $V_{anode(-)}$ / $V_{anode(+)}$	Operates similar to SCR, except can also be triggered into conduction by light falling on junctions.	1.6 A / 200 V	Relay replacement / Position controls / Photoelectric applications / Slave flashes
Silicon-Controlled Switch (SCS)	Anode / Anode gate / Cathode gate / Cathode	Anode I / $V_{anode(-)}$ / $V_{anode(+)}$	Operates similar to SCR except can also be triggered on by a negative signal on anode gate. Also several other specialized modes of operation.	100 V / 200 mA	Logic applications / Counters / Nixie drivers / Lamp drivers
Silicon Unilateral Switch (SUS)	Anode / Gate / Cathode	$I_{anode(+)}$ / $V_{anode(+)}$	Similar to SCS but Zener added to anode gate to trigger device into conduction at ~ 8 V. Can also be triggered by negative pulse at gate lead.	0.350 W / 0.200 A / 10 V	Switching circuits / Counters / SCR trigger / Oscillator
Silicon Bilateral Switch (SBS)	Anode 2 / Gate / Anode 1	$I_{anode\,2}$ / $V_{anode\,2(-)}$ / $V_{anode\,2(+)}$	Symetrical bilateral version of the SUS. Breaks down in both directions as SUS does in forward.	0.350 W / 0.200 A / 10 V	Switching circuits / Counters / TRIAC phase control
Triac	Anode 2 / Gate / Anode 1	I / $V_{anode\,2(-)}$ / $V_{anode\,2(+)}$	Operates similar to SCR except can be triggered into conduction in either direction by (+) or (−) gate signal.	25 A / 500 V	AC switching / Phase control / Relay replacement
Diac Trigger		I / V	When voltage reaches trigger level (about 35 V), abruptly switches down about 10 V.	40 V / 2 A peak	Triac and SCR trigger / Oscillator

Appendix C

HEATHKIT GR-1008 AM PORTABLE RADIO

Circuit Description

Figure C-1 shows the schematic diagram and Fig. C-2 shows the block diagram of the Heathkit GR-1008 AM portable radio. The Radio consists of RF amplifier Q_1, oscillator-mixer Q_2, IF amplifiers Q_3 and Q_4, detector D_1 and D_2, audio amplifier Q_5, audio driver Q_6, and push-pull audio output amplifier Q and Q_8.

Tuner—RF Amplifier

The radio-frequency (RF) signals transmitted by broadcast stations are picked up by antenna L_1. The desired signal is selected by a tuned circuit consisting of the primary winding of the antenna and capacitors C_1 and C_{2A}. This selected signal is coupled through the secondary winding of L_1 to the base of RF amplifier transistor Q_1. Bias voltage is supplied to transistor Q_1 through resistor R_1 and the AGC (automatic gain control) circuit. Transistor Q_1 amplifies the selected RF signal before it is coupled through capacitor C_3 to the base of oscillator-mixer transistor Q_2.

Oscillator—Mixer

The selected RF signal is applied to oscillator-mixer transistor Q_2 where it is mixed with the oscillator signal. Coil L_2 and capacitors C_{2B}, C_7, and C_8 determine the frequency of the oscillator signal. Transistor Q_2, because of the mixing of the two signals, produces several different frequencies at its collector output. Among these several different frequencies is an IF (intermediate frequency) of 455 kHz. Transformer T_1 contains a tuned circuit that passes only this IF frequency on to filters FL_1 and FL_2.

IF Amplifiers

Filters FL_1 and FL_2 serve to get rid of any frequencies other than the IF. Therefore, only a pure IF of 455 kHz is applied to the base of IF amplifier transistor Q_3. Bias voltage for transistor Q_3 is developed across the voltage divider network of resistors R_{11} and R_{12}. Transistor Q_3 amplifies the IF signal. From transistor Q_3 the IF signal is applied to the base of Q_4 where it is once again amplified before being coupled through capacitor C_{13} to the detector circuit of D_1 and D_2.

Detector

Detector diode D_2 converts the IF signal into a dc voltage that varies in amplitude at an audio frequency rate. During this conversion, D_1 gets rid of the unwanted portion of the IF signal. Capacitor C_{16} filters any remaining unwanted signals from the audio line. Resistor R_{16} and capacitor C_{11} filter out the audio and leave a dc voltage that is proportional to the strength of the original RF signal voltage picked up by the antenna. This dc voltage potential is used for AGC as explained in the following paragraph.

AGC (Automatic Gain Control)

The AGC voltage is applied through resistor R_{16} and the secondary of antenna L_1 to the base of transistor Q_1. Here, the AGC voltage helps control the bias of Q_1. The overall effect of the AGC system is to control the gain of Q_1 to prevent excessively large signals picked up by the antenna from being amplified to the point of distortion.

Volume Control-Audio Amplifier-Audio Driver

The audio signal from detector diode D_2 is applied to one end of volume control resistor R_{17}. Here, some portion of the audio signal (as determined by the setting of the volume control) is picked from R_{17} by the variable

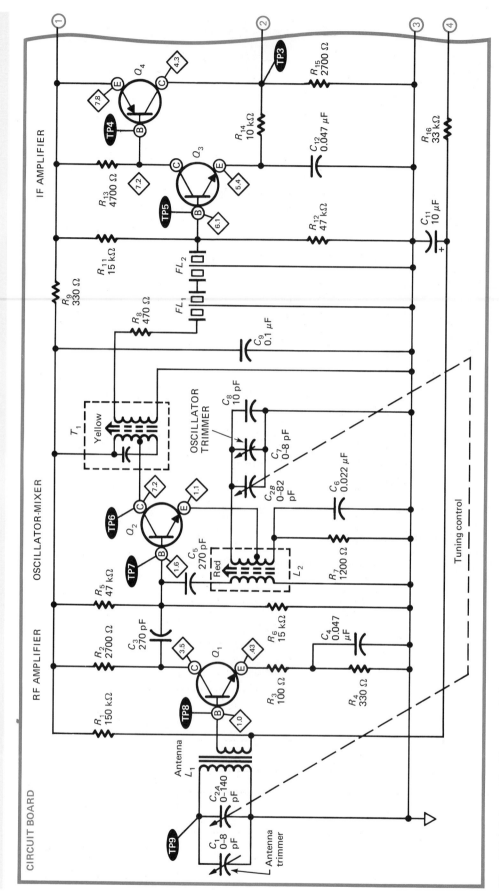

Fig. C-1 Schematic diagram of the Heathkit GR-1008 AM portable radio. (*Courtesy of Heath Company.*)

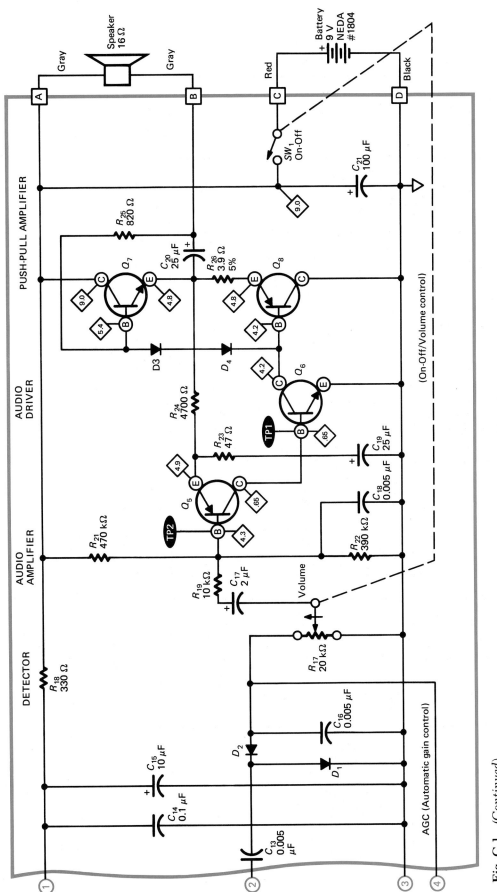

Fig. C-1 (Continued)

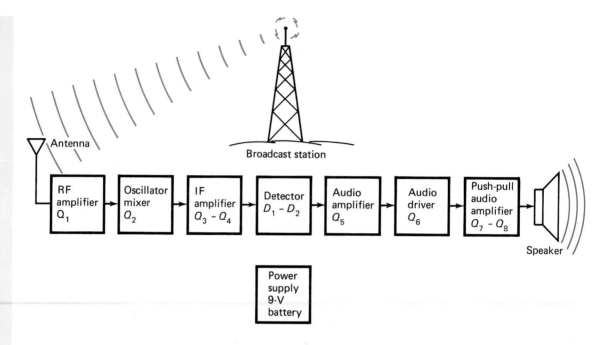

Fig. C-2   Block diagram of the Heathkit GR-1008 AM portable radio. (*Courtesy of Heath Company.*)

contact of the control. This audio signal is then coupled through capacitor C_{17} and applied through resistor R_{19} to the base of audio amplifier transistor Q_5. The audio signal is amplified by transistor Q_5 and then applied directly to the base of audio driver transistor Q_6. Transistor Q_6 again amplifies the audio signal before it is coupled directly to the base of transistor Q_8 and through diodes D_4 and D_3 to the base of transistor Q_7.

Push-Pull Amplifier

Transistors Q_7 and Q_8 make up a push-pull amplifier that provides the audio signal with the necessary current amplification to drive the speaker. A negative-going audio signal applied to the base of transistor Q_8 causes the transistor to conduct, charging capacitor C_{20} through the speaker voice coil. When the audio signal goes positive, Q_8 is cut off and Q_7 conducts. The conduction of transistor Q_7 causes capacitor C_{20} to discharge back through the speaker coil. This charging and discharging of capacitor C_{20} through the speaker voice coil produces the sound from the speaker.

Appendix D

LINEAR INTEGRATED CIRCUIT SPECIFICATIONS

DUAL LOW NOISE PREAMPLIFIER

General Description

A dual low-noise preamplifier consisting of two identically matched 68-dB gain amplifiers fed from an internal Zener regulated power supply. Operation requires only a single external power supply and a minimum number of external frequency-shaping components.

Features

- Low audio noise
- Wide power supply range
- Built-in power supply filter
- Low distortion
- High channel separation

Applications

- Stereo tape player/recorder
- Stereo radio receiver
- Movie projector
- Phonograph
- TV remote control receiver
- Microphone amplifier

Absolute Maximum Ratings

Supply Voltage . . . . 16 V dc
Temperature
    Storage . . . . − 55°C to plus 150°C
    Operating . . . . − 30°C to plus 85°C

QUAD OP AMP

General Description

The 324 series consists of four independent, high gain, internally frequency compensated operational amplifiers which were designed specifically to operate from a single power supply over a wide range of voltages. Operation

DUAL LOW NOISE PREAMPLIFIER

Pin Connection

Top view

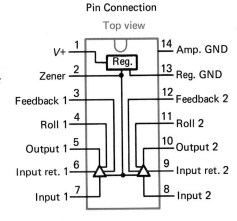

Adapted by permission of Radio Shack, a division of Tandy Corporation.

from split power supplies is also possible and the low power-supply current drain is independent of the magnitude of the power-supply voltage.

Application areas include transducer amplifiers, dc gain blocks and all the conventional op-amp circuits which now can be more easily implemented in single power supply systems. For example, the 324 series can be directly operated off of the standard +5 V dc power-supply voltage which is used in digital systems and will easily provide the required interface electronics without requring the additional ±15 V dc power supplies.

Features

- Internally frequency compensated for unity gain
- Large dc voltage gain: 100 dB
- Wide bandwidth (unity gain): 1 MHz (temperature compensated)
- Wide power-supply range:
  Single supply 3 V to 30 V dc
  or dual supplies ±1.5 V to ±15 V dc
- Very low supply current drain (800 μA)—essentially independent of supply voltage (1 mW/op-amp at +5 V dc)
- Low input biasing current 45 nA dc (temperature compensated)
- Low input offset voltage 2 mV dc and offset current 5 nA dc
- Input common-mode voltage range includes ground
- Differential input voltage range equal to the power-supply voltage
- Large output voltage swing 0 V to V+ — 1.3 V dc

QUAD OP AMP

Pin Connection
Top view

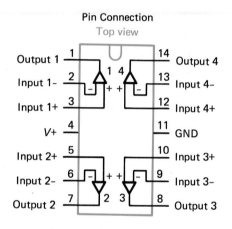

Dual-in-line and flat package

Absolute Maximum Ratings

Supply Voltage, V+ .... 32 V dc or ± 16 V dc
Differential Input Voltage .... 32 V dc
Input Voltage .... − 0.3 V dc to + 32 V dc
Power Dissipation
  Molded DIP .... 570 mW
  Cavity DIP .... 900 mW
Output Short-Circuit to GND
  (One Amplifier) .... Continuous
  V+ ≤ 15 V dc and T_A = 24°C
Input Current (V_{in} < − 0.3 V_{OL}) .... 50 mA
Operating Temperature Range .... 0°C to + 70°C
Storage Temperature Range .... − 65°C to + 150°C
Lead Temperature (Soldering, 10 seconds) ...... 300°C

QUAD COMPARATOR

General Description

The 339 series consists of four independent voltage comparators which were designed specifically to operate from a single power supply over a wide range of voltages. Operation from split power supplies is also possible and the low power-supply current drain is independent of the magnitude of the power-supply voltage. These comparators also have a unique characteristic in that the input common-mode voltage range includes ground, even though operated from a single power supply voltage.

Features

- Wide single supply:
  Voltage range 2 V to 36 V dc or dual supplies ±1V to ±18 V dc
- Very low supply current drain (0.8 mA)—independent of supply voltage (1 mW/comparator at +5 V dc)
- Input common-mode voltage range includes ground
- Differential input voltage range equal to the power-supply voltage
- Low output 1 mV at 5 μA; saturation voltage 70 mV at 1 mA
- Output voltage compatible with TTL (fanout of 2), DTL, ECL, MOS and CMOS logic systems

Absolute Maximum Ratings

Supply Voltage, V+ . . . . 36 V dc or ±17 V dc
Differential Input Voltage . . . . 36 V dc
Input Voltage . . . . −0.3 V dc to +36 V dc
Power Dissipation
    Molded DIP . . . . 570 mW
    Cavity DIP . . . . 900 mW
Output Short-Circuit to GND . . . . Continuous
Input Current ($V_{IN} < −0.3$ V dc) . . . . 50 mA
Operating Temperature Range . . . . 0°C to +70°C
Storage Temperature Range . . . . −65°C to +150°C
Lead Temperature (Soldering, 10 seconds) . . . . . . 300°C

Pin Connection

Top view

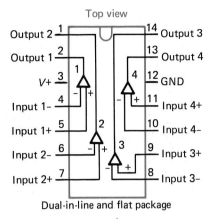

Dual-in-line and flat package

Schematic Diagram

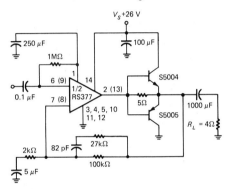

DUAL TWO-WATT AUDIO AMPLIFIER

General Description

The 377 is a monolithic dual power amplifier which offers high quality performance for stereo phonographs, tape players, recorders, and AM-FM stereo receivers, etc.

    The 377 will deliver 2 W/channel into 8- or 16-Ω loads. The amplifier is designed to operate with a minimum of external components and contains an internal bias regulator to bias each amplifier. Device overload protection consists of both internal current limit and thermal shutdown.

Features

- A_{vo} typical 90 dB
- 2 W per channel
- 70-dB ripple rejection
- 75-dB channel separation
- Internal stabilization
- Self-centered biasing
- 3-MΩ input impedance
- 10- to 26-V operation
- Internal current limiting
- Internal thermal protection

Applications

- Multi-channel audio systems
- Tape recorders and players
- Movie projectors
- Automotive systems
- Stereo phonographs
- Bridge output stages
- AM-FM radio receivers
- Intercoms
- Servo amplifiers
- Instrument systems

Absolute Maximum Ratings

Supply Voltage	26 V dc
Input Voltage	0 V–V supply
Operating Temperature	0°C to +70°C
Storage Temperature	−65°C to +150°C
Junction Temperature	150°C
Lead Temperature (Soldering, 10 seconds)	300°C

TIMER

General Description

The 555 is a highly stable device for generating accurate time delays or oscillation. Additional terminals are provided for triggering or resetting if desired. In the time delay mode of operation, the time is precisely controlled by one external resistor and capacitor. For astable operation as an oscillator, the free-running frequency and duty cycle are accurately controlled with two external resistors and one capacitor. The circuit may be triggered and reset on falling waveforms, and the output circuit can source or sink up to 200 mA or derive TTL circuits.

DUAL TWO-WATT AUDIO AMPLIFIER

Pin Connection

Top View

Bias	1	14	V+
Output 1	2	13	Output 2
GND	3	12	GND
GND	4	11	GND
GND	5	10	GND
Input 1	6	9	Input 2
Feedback 1	7	8	Feedback 2

Dual-in-line package

Schematic Diagram

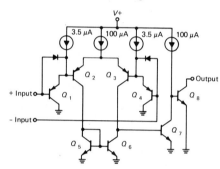

TIMER

Pin Connection

Top view

GND	1	8	+V_{CC}
Trigger	2	7	Discharge
Output	3	6	Threshold
Reset	4	5	Control voltage

Dual-in-line package

TONE DECODER

General Description

The 567 is a general purpose tone decoder designed to provide a saturated transistor switch to ground when an input signal is present within the passband. The circuit consists of an *I* and *Q* detector driven by a voltage-controlled oscillator which determines the center frequency of the decoder. External components are used to independently set center frequency, bandwidth, and output delay.

Pin Connection

Top view

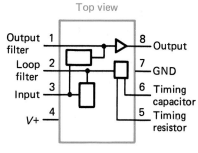

Dual-in-line package

Features

- 20 to 1 frequency range with an external resistor
- Logic compatible output with 100-mA current sinking capability
- Bandwidth adjustable from 0 to 14%
- High rejection of out-of-band signals and noise
- Immunity to false signals
- Highly stable center frequency
- Center frequency adjustable from 0.01 Hz to 500 kHz

Applications

- Touch tone decoding
- Precision oscillator
- Frequency monitoring and control
- Wide bank FSK demodulation
- Ultrasonic controls
- Carrier current remote controls
- Communications paging decoders

Absolute Maximum Ratings

Supply Voltage . . . . 10 V dc
Power Dissipation . . . . 300 mW
V_8 . . . . 15 V
V_3 . . . . -10 V
V_3 . . . . $V_8 + 0.5$ V
Storage Temperature Range . . . . $-65°C$ to $+150°C$

VOLTAGE REGULATOR

General Description

The 723 is a voltage regulator designed primarily for series regulator applications. By itself, it will supply output currents up to 150 mA; but external transistors can be added to provide any desired load current. The circuit features extremely low standby current drain, and provision is made for either linear or foldback current limiting.

Features

- 150-mA output current without external pass transistor
- Output currents in excess of 10 A possible by adding external transistors
- Input voltage 40 V max.
- Output voltage adjustable from 2 V to 37 V
- Can be used as either a linear or a switching regulator

The 723 is also useful in a wide range of other applications such as a shunt regulator, a current regulator, or a temperature controller.

Absolute Maximum Ratings

Pulse Voltage from V+ to V− (50 ms) .... 50 V
Continous Voltage from V+ to V− .... 40 V
Input-Output Voltage Differential .... 40 V
Maximum Amplifier Input Voltage (Either Input) 7.5 V
Maximum Amplifier Input Voltage (Differential) 5 V
Current from V_z .... 25 mA
Current from V_{ref} .... 15 mA
Internal Power Dissipation Metal Can .... 800 mW
  Cavity DIP .... 900 mW
  Molded DIP .... 660 mW
Operating Temperature Range .... 0°C to +70°C
Storage Temperature Range
  Metal Can .... −65°C to +150°C
  DIP .... −55°C to +125°C
Lead Temperature (Soldering, 10 sec) .... 300°C

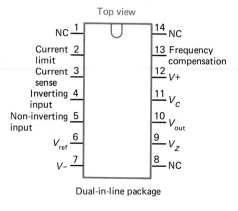

Pin Connection

Top view

Current limit

Metal can package
Note: Pin 5 connected to case.

Top view

Dual-in-line package

OPERATIONAL AMPLIFIER

General Description

The 741 series are general purpose operational amplifiers which feature improved performance over industry standards.

The amplifiers offer many features which make their application nearly foolproof: overload protection on the input and output, no latch-up when the common-mode range is exceeded, as well as freedom from oscillations.

Pin Connection

Top view

NC

Offset null   V+

Inverting input   Output

Non-inverting input   Offset null

V−

Metal can package
Note: Pin 4 connected to case.

Absolute Maximum Ratings

Supply Voltage . . . . ± 18 V dc
Power Dissipation . . . . 500 mW
Differential Input Voltage . . . . ± 30 V
Input Voltage . . . . ± 15 V
Output Short Circuit Duration . . . . Indefinite
Operating Temperature Range . . . . 0°C to +70°C
Storage Temperature Range . . . . −65°C to +150°C
Lead Temperature (Soldering, 10 seconds) . . . . . . 300°C

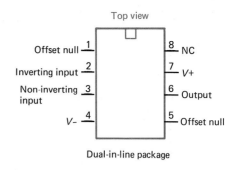

Dual-in-line package

LED FLASHER / OSCILLATOR

General Description

The 3909 is a monolithic oscillator specifically designed to flash light-emitting diodes. By using the timing capacitor for voltage boost, it delivers pulses of 2 or more volts to the LED while operating on a supply of 1.5 V or less. The circuit is inherently self-starting, and requires addition of only a battery and capacitor to function as an LED flasher.

It has been optimized for low power drain and operation from weak batteries so that continuous operation life exceeds that expected from battery rating.

Application is made simple by inclusion of internal timing resistors and an internal LED current-limiting resistor.

Timing capacitors will generally be of the electrolytic type, and a small 3-V rated part will be suitable for any LED flasher using a supply up to 6 V. However, when picking flash rates, it should be remembered that some electrolytics have very broad capacitance tolerances, for example −20% to +100%.

LED FLASHER/OSCILLATOR

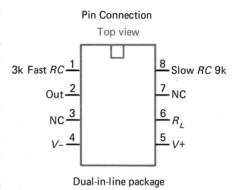

Dual-in-line package

Features

- Operation over one year from one G-size flashlight cell
- Bright, high-current LED pulse
- Minimum external parts
- Low voltage operation, from just over 1 V to 5 V
- Low current drain, averages under 0.5 mA during battery life
- Powerful; as an oscillator directly drives an 8-Ω speaker

Absolute Maximum Ratings

Power Dissipation . . . . 500 mW
V+ Voltage . . . . 6.4 V
Operating Temperature Range . . . . . . −25°C to +70°C

TEMPERATURE CONTROLLER

General Description

The 3911 is a highly accurate temperature measurement and/or control system for use over a $-25°C$ to $+85°C$ temperature range. Fabricated on a single monolithic chip, it includes a temperature sensor, a stable voltage reference and an operational amplifier.

The output voltage of the 3911 is directly proportional to temperature in Kelvins at 10 mV/K. Using the internal op-amp with external resistors, any temperature scale factor is easily obtained. By connecting the op-amp as a comparator, the output will switch as the temperature transverses the set-point making the device useful as an on-off temperature controller.

An active shunt regulator is connected across the power leads of the 3911 to provide a stable 6.8-V voltage reference for the sensing system. This allows the use of any power-supply voltage with suitable external resistors.

Features

- Uncalibrated accuracy $\pm 10°C$
- Internal op-amp with frequency compensation
- Linear output of 10 mV/K (10 mV/°C)
- Can be calibrated in Kelvins, Celsius or Fahrenheit degrees
- Output can drive loads up to 35 V

Absolute Maximum Ratings

Supply Current (Externally Set) . . . . 10 mA
Output Collector Voltage V+ . . . . 36 V
Feedback Input Voltage Range . . . . 0 V to +7.0 V
Output Short Circuit Duration . . . . Indefinite
Operating Temperature Range . . . . . . $-25°C$ to $+85°C$
Storage Temperature Range . . . . $-65°C$ to $+150°C$
Lead Temperature (Soldering, 10 seconds) . . . . . . 300°C

Pin Connection
Top view

V–	1	8	NC
Out	2	7	NC
In	3	6	NC
V+	4	5	NC

Dual-in-line package

5-V VOLTAGE REGULATOR
12-V VOLTAGE REGULATOR
15-V VOLTAGE REGULATOR

General Description

The 78XX series of three-terminal regulators is available with several fixed output voltages making them useful in a wide range of applications. One of these is local on-card regulation, eliminating the distribution problems associated with single-point regulation. The voltages available allow these regulators to be used in logic systems, instrumentation, hifi, and other solid-state electronic equipment. Although designed primarily as fixed voltage regulators these devices can be used with external components to obtain adjustable voltages and currents.

The 78XX series is available in two different packages which will allow over 1.5-A load current if adequate heat sinking is provided. Current limiting is included to limit the peak output current to a safe value. Safe area protection for the output transistor is provided to limit internal power dissipation. If internal power dissipation becomes too high for the heat sinking provided, the thermal shutdown circuit takes over preventing the IC from overheating.

Features

- Internal thermal overload protection
- No external components required
- Output transistor safe area protection
- Internal short circuit current limit

Voltage Range

7805 . . . . 5 V
7812 . . . . 12 V
7815 . . . . 15 V

Absolute Maximum Ratings

Input Voltage
   (Output Voltage Options 5V through 18V) .. 35 V dc
   (Output Voltage Option 24V) . . . . 40 V dc
Internal Power Dissipation . . . . Internally Limited
Operating Temperature Range . . . . 0°C to +70°C
Maximum Junction Temperature . . . . 150°C
Storage Temperature Range . . . . −65°C to +150°C
Lead Temperature (Soldering, 10 seconds) . . . . . . 300°C

MONOLITHIC JFET INPUT OPERATIONAL AMPLIFIER

General Description

The 13741 is a 741 with BI-FET input followers on the same die. Familiar operating characteristics—those of a 741—with the added advantage of low input bias current make the 13741 easy to use. Monolithic fabrication makes this "drop-in-replacement" operational amplifier economical as well as easy to use.

Features

- Low input bias current 50 pA
- Low input noise current 0.01 pA/$\sqrt{\text{Hz}}$
- High input impedance $5 \times 10^{11}\Omega$
- Familiar operating characteristics

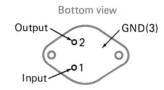

Pin Connection

Bottom view

Metal can package
Aluminum TO-3 (KC)

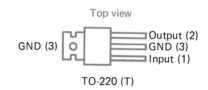

Top view

TO-220 (T)

MONOLITHIC JFET INPUT
OPERATIONAL AMPLIFIER

Pin Connection

Top view

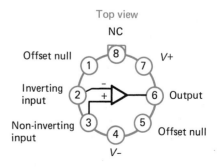

Metal can package

Note: Pin 4 connected to case.

Absolute Maximum Ratings

Supply Voltage . . . . ± 18 V dc
Power Dissipation . . . . 500 mW
TO-99 (H Package)
Operating Temperature Range . . . . 0°C to +70°C
$T_{j,\,(max)}$ . . . . 100°C
Differential Input Voltage . . . . ± 30 V
Input Voltage Range . . . . ± 16 V
Output Short Circuit Duration . . . . Continuous
Storage Temperature Range . . . . −65°C to +150°C
Lead Temperature (Soldering, 10 seconds) . . . . . . 300°C

Appendix E

THERMIONIC DEVICES

Thermionic devices (vacuum tubes) dominated electronics until the early 1950s. Since that time, solid-state devices have all but completely taken over. Today, vacuum tubes are used only in special applications such as high-power RF amplifiers, cathode-ray tubes (including television picture tubes), and some microwave devices.

Thermionic emission involves the use of heat to liberate electrons from an element called a *cathode*. The heat is produced by energizing a filament or heater circuit within the tube. A second element, called the *anode*, can be used to attract the liberated electrons. Since unlike charges attract, the anode is made positive with respect to the cathode.

A third electrode can be placed between the cathode and the anode. This third electrode can exert control over the movement of electrons from cathode to anode. Thus, it is called the *control grid*. The control grid is often negative with respect to the cathode. This negative state repels the cathode electrons and prevents them all from reaching the anode. In fact, the tube can be cut off by a high negative grid potential. Figure E-1 shows the schematic symbol for a three-electrode vacuum tube and the polarities involved.

The vacuum tube shown in Fig. E-1 is an amplifier. The signal to be amplified can be applied to the control grid. As the signal goes in a positive direction, more plate current will flow. As the signal goes in a negative direction, less plate current will flow. Thus, the plate current is a function of the signal applied to the grid. The signal power in the grid circuit is much less than the signal power in the plate circuit. The vacuum tube is capable of good power gain.

Vacuum tubes may use extra grids located between the control grid and the plate to provide better operation. The extra grids improve gain and high-frequency performance. The tube in Fig. E-1 is called a *triode* vacuum tube (the heater is not counted as an element). If a screen grid is added, it becomes a *tetrode* (four electrodes). If a screen grid and a suppressor grid are added, it becomes a *pentode* (five electrodes).

Vacuum tubes make excellent high-power amplifiers. It is possible to run some vacuum tubes with plate potentials measured in thousands of volts and plate currents measured in amperes. These tubes offer output powers of several thousand watts. It is even possible to develop 2,000,000-W amplitude-modulated RF output by using four special tetrodes. This is an example of the outstanding power capacity of vacuum tubes.

The cathode-ray tube is a vacuum tube used for the display of graphs, pictures, or data.

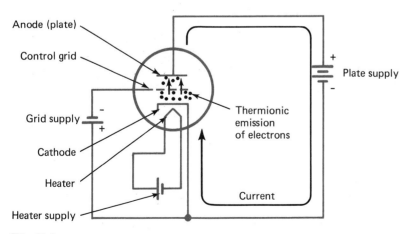

Fig. E-1

285

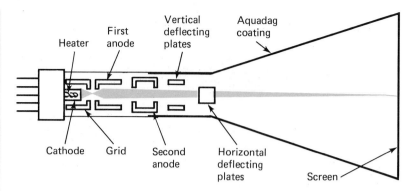

Fig. E-2   A cathode-ray tube using electrostatic deflection.

Figure E-2 shows the basic structure. The cathode is heated and produces thermionic emission. A positive potential is applied to the first anode, the second anode, and the aquadag coating. This positive field accelerates the electrons toward the screen. The inside of the screen is coated with a chemical phosphor that emits light when hit by a stream of electrons.

As shown in Fig. E-2, the electrons are focused into a narrow beam. This makes it possible to produce a small dot of light on the screen. The deflecting plates can move the beam vertically and horizontally. For example, a positive voltage applied to the top vertical deflecting plate will attract the beam and move it up. The dot of light can be positioned anywhere on the screen.

The grid shown in Fig. E-2 makes it possible to control the intensity of the beam. A negative voltage applied to the grid will repel the cathode electrons and prevent them all from reaching the screen. A high negative voltage will completely stop the electrons, and the dot of light will go out.

By controlling the position and the intensity of the dot, any type of picture information can be presented on the screen. Because the phosphor will retain its brightness momentarily and because the eye will retain the image for a brief period, the effect of the moving dot is to produce what seems to be a complete picture on the screen. If this is done repeatedly, a movie effect is produced. This is how a television picture tube works. Colors can be shown by using several different chemical phosphors.

The deflection system may be different from that shown in Fig. E-2. *Magnetic deflection* uses coils around the neck of the cathode-ray tube. When a current flows through the coil, the resulting magnetic field will deflect the electron beam. Television picture tubes generally use magnetic deflection. Oscilloscopes generally use electrostatic deflection.

Index